Cellular Trafficking of Cell Stress Proteins in Health and Disease

HEAT SHOCK PROTEINS

Volume 6

Series Editors:

A. A. A. Asea

Department of Microbiology, Biochemistry and Immunology (MBI),
Morehouse School of Medicine, 720 Westview Ave., SW, Atlanta, GA 30310, USA

S. K. Calderwood

Division of Molecular and Cellular Radiation Oncology,
Beth Israel Deaconness Medical Center and Harvard Medical School

For further volumes:
http://www.springer.com/series/7515

Brian Henderson • A. Graham Pockley
Editors

Cellular Trafficking of Cell Stress Proteins in Health and Disease

Springer

Editors
Prof. Brian Henderson
Department of Microbial Diseases
UCL-Eastman Dental Institute
University College
London
United Kingdom

Prof. A. Graham Pockley
John van Geest Cancer Research Centre
Nottingham Trent University
Nottingham
United Kingdom

ISSN 1877-1246 ISSN 1877-1254 (Electronic)
ISBN 978-94-007-4739-5 ISBN 978-94-007-4740-1 (eBook)
DOI 10.1007/978-94-007-4740-1
Springer Dordrecht Heidelberg London NewYork

Library of Congress Control Number: 2012951299

Springer is part of Springer Science+Business Media (www.springer.com)

Preface

The establishment that cells responded to stress by the induction of specific gene expression led to the identification, in the 1980s, of heat shock or cell stress proteins and the realisation that these proteins are involved in the folding, re-folding, and prevention of aggregation, of client proteins. Such protein-folding proteins are now known as molecular chaperones, with the term protein-folding catalysts (PFCs) being applied to proteins such as thioredoxin and peptidyl prolyl isomerases that involve an enzymic step in the protein folding mechanism. Initially, these proteins were thought to be exclusively intracellular. However, in 1988 evidence was presented for the secretion and uptake of Hsp70 proteins by cultured cells suggesting that at least one molecular chaperone underwent aberrant cellular trafficking. Some years later the aberrant cytoplasmic and cell surface location of the mitochondrial Hsp60 protein began to be defined. These finding suggested a potentially wider remit for the function of molecular chaperones within the cell than had previously been considered.

A growing number of reports in the 1990s established that a number of molecular chaperones and PFCs had cell signalling actions when applied externally to cultured cells, and this realisation has prompted a rapid expansion of work in this field. Despite being dogged by suggestions that the biological and immunological properties of extracellular stress proteins results from contaminants in the preparations used, it is now becoming accepted that cell stress proteins can indeed be released from cells and that, in the extracellular environment, they elicit a number of regulatory functions. This line of work has culminated in the finding that a number of molecular chaperones and PFCs are present in the circulation and that levels of these proteins may reflect tissue and organismal pathology.

Since the beginning of the twenty-first century there has been a rapid increase in our understanding of the cellular trafficking mechanisms of molecular chaperones both in eukaryotes and in prokaryotes. In the former, molecular chaperone trafficking can occur between the various cellular compartments, with concomitant movement of other proteins and in some instances at least, the release of molecular chaperones from cells. In bacteria, molecular chaperones are involved in the trafficking of other proteins and are themselves released into the external milieu. There is an increasing appreciation of the role of molecular chaperones and PFCs in the interplay between bacteria and the cells of their hosts and this is now an important area of research for understanding the mechanisms of infectious diseases.

This volume brings together experts in the biochemistry, cellular biology, immunology and molecular biology of molecular chaperones and PFCs with a focus on the mechanisms of cellular trafficking of these proteins and the role of these variegated trafficking mechanisms in both human and animal health and disease. To guide readers who may be unfamiliar with this, now voluminous, field of research, this book starts with a number of introductory chapters which provide a historical background to the key aspects of molecular chaperone biology. The second section focuses on intracellular trafficking of molecular chaperones and their interactions with different cellular compartments and cellular components and the roles that such trafficking plays in the maintenance of cell health and in controlling the death of the cell. The third section deals with the roles played by molecular chaperones in the control of selected receptors that can play roles in immunological homeostasis. Section 4 deals with the unexpected finding that molecular chaperones can actually exist in the extracellular milieu and the consequences of such release for health and disease.

This book should be of interest to a wide range of biomedical scientists.

Brian Henderson
A. Graham Pockley

Contents

Contributors

Alexzander Asea Department of Microbiology, Biochemistry and Immunology (MBI), Morehouse School of Medicine, 720 Westview Ave., SW, Atlanta, GA 30310, USA
e-mail: aasea@msm.edu

Stuart K. Calderwood Molecular and Cellular Radiation Oncology, Beth Israel Deaconess Medical Center, Harvard Medical School, 99 Brookline Avenue, Boston, MA 02215, USA
e-mail: scalderw@bidmc.harvard.edu

Daniel C. Campbell Centre for Endocrinology, William Harvey Research Institute, Barts and the London School of Medicine and Dentistry, Queen Mary University of London, London, UK

J. Paul Chapple Centre for Endocrinology, William Harvey Research Institute, Barts and the London School of Medicine and Dentistry, Queen Mary University of London, London, UK
e-mail: j.p.chapple@qmul.ac.uk

Mei Chen The Department of Dermatology and the USC/Norris Comprehensive Cancer Center, The University of Southern California Keck School of Medicine, 1441 Eastlake Avenue, Room-6320, CA 90033, Los Angeles, USA

Jianlin Gong Boston University Medical School, 650 Albany Street, Boston, MA 02118, USA

Imre Gombos Institute of Biochemistry, Biological Research Centre, Hungarian Academy of Sciences, Temesvári krt 62, Szeged, H6726, Hungary

Mario Gonzalez-Gronow Department of Pathology, Duke University Medical Center, Box 3712, Durham, NC 27710, USA
e-mail: gonza002@mc.duke.edu

Brian Henderson Department of Microbial Diseases, UCL-Eastman Dental Institute, University College London, 256 Gray's Inn Road, London WC1X 8LD, UK
e-mail: brian.henderson@ucl.ac.uk

Lawrence E. Hightower Department of Molecular and Cell Biology, The University of Connecticut, Storrs, CT 06269, USA
e-mail: lawrence.hightower@uconn.edu

Enikö Hocsak Department of Biochemistry and Medical Chemistry, University of Pécs, Pécs, Hungary

Ibolya Horváth Institute of Biochemistry, Biological Research Centre, Hungarian Academy of Sciences, Temesvári krt 62, Szeged, H6726, Hungary

Punit Kaur Department of Microbiology, Biochemistry and Immunology (MBI), Morehouse School of Medicine, 720 Westview Ave., SW, Atlanta, GA 30310, USA

Sergiy Kostenko Institute of Medical Biology, Faculty of Health Sciences, University of Tromsø, NO-9037 Tromsø, Norway
e-mail: sergiyk@uit.no

Wei Li The Department of Dermatology and the USC/Norris Comprehensive Cancer Center, The University of Southern California Keck School of Medicine, 1441 Eastlake Avenue, Room-6320, CA 90033, Los Angeles, USA
e-mail: wli@usc.edu

Andrew P. Lieberman Department of Pathology, University of Michigan Medical School, 1301 MSRB III, Ann Arbor, MI 48109-0630, USA

Uma K. Misra Department of Pathology, Duke University Medical Center, Box 3712, Durham, NC 27710, USA

Ugo Moens Institute of Medical Biology, Faculty of Health Sciences, University of Tromsø, NO-9037 Tromsø, Norway
e-mail: ugo.moens@uit.no

Yoshihiro Morishima Department of Pharmacology, University of Michigan Medical School, 1301 MSRB III, Ann Arbor, MI 48109-0630, USA

Gabriele Multhoff Klinikum rechts der Isar, TU-Munchen, Department of Radiation Oncology and Helmholtz Center Munich, National Research Center for Environmental Health GmbH, 85764 Munich, Germany
e-mail: Gabriele.multhoff@lrz.tu-muenchen.de

Ayesha Murshid Molecular and Cellular Radiation Oncology, Beth Israel Deaconess Medical Center, Harvard Medical School, 99 Brookline Avenue, Boston, MA 02215, USA

Emily J. Noonan National Cancer Institute, Cancer Prevention Fellowship Program, Center for Cancer Training, Bethesda, MD, USA

Kathryn O'Brien The Department of Dermatology and the USC/Norris Comprehensive Cancer Center, The University of Southern California Keck School of Medicine, 1441 Eastlake Avenue, Room-6320, CA 90033, Los Angeles, USA

Yoichi Osawa Department of Pharmacology, University of Michigan Medical School, 1301 MSRB III, Ann Arbor, MI 48109-0630, USA
e-mail: osawa@umich.edu

David A. Parfitt Centre for Endocrinology,William Harvey Research Institute, Barts and the London School of Medicine and Dentistry, Queen Mary University of London, London, UK

Ana-Maria Pilbat Institute of Biochemistry, Biological Research Centre, Hungarian Academy of Sciences, Temesvári krt 62, Szeged, H6726, Hungary

Salvatore V. Pizzo Department of Pathology, Duke University Medical Center, Box 3712, Durham, NC 27710, USA

A. Graham Pockley John van Geest Cancer Research Centre, Nottingham Trent University, Clifton Lane, Nottingham, NG11 8NS, UK
e-mail: graham.pockley@ntu.ac.uk

Appukuttan R. Pradeep Department of Microbiology, Biochemistry and Immunology (MBI), Morehouse School of Medicine, 720 Westview Ave., SW, Atlanta, GA 30310, USA

William B. Pratt Department of Pharmacology, University of Michigan Medical School, 1301 MSRB III, Ann Arbor, MI 48109-0630, USA

Zoltán Prohászka Research Laboratory, 3rd Department of Medicine, Semmelweis University, Budapest, Kútvölgyi st. 4, 1125 Budapest, Hungary
e-mail: prohoz@kut.sote.hu

Richard W. Stokes Department of Microbiology & Immunology, University of British Columbia, Life Sciences Centre, 2504–2350 Health Sciences Mall, Vancouver, British Columbia, V6T 1Z3, Canada
e-mail: rstokes@mail.ubc.ca

Balázs Sümegi Department of Biochemistry and Medical Chemistry, University of Pécs, Pécs, Hungary

Zsolt Török Institute of Biochemistry, Biological Research Centre, Hungarian Academy of Sciences, Temesvári krt 62, Szeged, H6726, Hungary

Kathy Triantafilou Cardiff University, School of Medicine, University Hospital of Wales, Heath Park, Cardiff CF14 4XN, UK
e-mail: TriantafilouK@cardiff.ac.uk

Martha Triantafilou Cardiff University, School of Medicine, University Hospital of Wales, Heath Park, Cardiff CF14 4XN, UK

Peter Vaupel Department of Radiotherapy and Radiooncology, Klinikum rechts der Isar, Technical University of Munich, Ismaninger Strasse 22, 81675 Munich, Germany
e-mail: vaupel@uni-mainz.de

László Vígh Institute of Biochemistry, Biological Research Centre, Hungarian Academy of Sciences, Temesvári krt 62, Szeged, H6726, Hungary
e-mail: vigh@brc.hu

David T. Woodley The Department of Dermatology and the USC/Norris Comprehensive Cancer Center, The University of Southern California Keck School of Medicine, 1441 Eastlake Avenue, Room-6320, CA 90033, Los Angeles, USA

Note on Nomenclature

Understanding the biology of molecular chaperones and protein folding catalysts is complicated by the fact that there are several nomenclatures for these proteins, often with different names being given to prokaryotic and eukaryotic homologues. In addition, a new nomenclature has been introduced for some of the major human molecular chaperones (Kampinga et al. 2009). In this nomenclature the following names are given for particular molecular chaperones:

Old Name	New Nomenclature
Hsp10	HSPE1
Hsp27	HSPB3 (plus 10 other designations for human small HSPs)
Hsp40	DNAJB1 (plus 49 other designations for the human proteins)
Hsp60	HSPD1
Hsp70	HSPA1A (plus 12 other designations for the remaining human Hsp70 proteins)
Hsp90	HSPC1 (plus 4 other designations for the Hsp90 proteins)

Kampinga HH, Hageman J, Vos MJ et al (2009) Guidelines for the nomenclature of the human heat shock proteins. Cell Stress Chaperones 14:105–111

Chapter 1
Discovery of the Cellular Secretion of Cell Stress Proteins

Lawrence E. Hightower and Emily J. Noonan

Abstract This chapter describes the discovery of extracellular Hsps or molecular chaperones. The first part, written by LEH, provides the historical context and a personal account of this discovery. The second part, written by EJN, brings these findings up to date by describing new discoveries that support and extend the original observations.

1.1 Introduction

A number of the major discoveries in the field known historically as the heat shock response, and more broadly as the cellular stress response, initially were met with quizzical expressions, if not outright disbelief. And no wonder, since several of these discoveries caused paradigm shifts in our view of how cells, tissues and, ultimately, animals function. This was one of the themes of a recent review article by Antonio De Maio (De Maio 2011) that he dedicated to the pioneer of the heat shock response Ferruccio Ritossa with the statement "It is never known how far a controversial finding will go!". In this article, Dr. De Maio pointed out that Ritossa's manuscript was rejected by the editor of a prestigious journal for insufficient relevance to the scientific community. Eventually it was accepted and published in the Swiss journal *Experientia* (Ritossa 1962). He also noted that my observation that heat shock proteins (Hsp) are released from mammalian cells "was initially deemed irrelevant and impossible". Our objectives in the present chapter are to provide the background on how this discovery came about and to link it to more recent studies.

L. E. Hightower (✉)
Department of Molecular and Cell Biology, The University of Connecticut,
Storrs, CT 06269, USA
e-mail: Lawrence.hightower@uconn.edu

E. J. Noonan
National Cancer Institute, Cancer Prevention Fellowship Program,
Center for Cancer Training, Bethesda, MD, USA

B. Henderson, A. G. Pockley (eds.), *Cellular Trafficking of Cell Stress Proteins in Health and Disease*, Heat Shock Proteins 6, DOI 10.1007/978-94-007-4740-1_1,

1.2 The Path to a New Discovery

In the early 1980s researchers in my laboratory and others were searching for functions of the Hsps (particularly Hsp70 now termed HSPA1A—see nomenclature section) inside of cultured animal cells and tissues as well as carrying out biochemical studies on purified Hsp70 in an effort to obtain clues to its functions based on its molecular properties. Two events led us to look for extracellular Hsps. The first was our collaboration with Fredric P. White. Fred had found evidence for a protein translocation system in incubated rat brain slices. This system was inhibited by vinblastine, colchicine and low concentrations of calcium ions (White 1979). Hsp70, which he called SP71, was among the proteins that rode this translocation pathway away from blood capillaries (White 1980). The mechanism was not known but he suggested two possibilities: synthesis of Hsp70 in the end feet of astrocytes that abut microvessels with intracellular transport away from the capillary bed, and synthesis of Hsp70 in vascular endothelial cells followed by extracellular transfer to astrocyte glial cells. This latter proposal was clearly heretical at the time, when Hsp70 was considered by almost all investigators to spend its entire useful lifetime inside of cells. The second stimulus was a localization study from the Lindquist lab. Susan and I were graduate students in the same program at Harvard University and even though we did not meet each other then, I knew her mentor Matthew Meselson. In fact, we had even managed to lunch together at the Hotel Continental in Saigon during the Vietnam War, when I was a soldier drafted out of graduate school into the US Army and Professor Meselson was on a USAID mission to evaluate the effects of Agent Orange on the civilian population. We discussed his plans and a possible collaboration. By the time I returned to Cambridge, MA, he was already presenting a dramatic talk on the effects of Agent Orange on the rural families and their crops in South Vietnam. I had a rough idea what was going on in his laboratory, which was supplemented by my friend and fellow graduate student Thomas Cline who worked just down the hall in Woody Hastings' laboratory at the Biological Laboratories. When I moved to the University of Connecticut to begin my faculty career, it was Tom who suggested to me over Thanksgiving turkey that I may have stumbled into the heat shock field, which was known only in *Drosophila* at that time. My description of the proteins induced by the amino acid canavanine in culture chicken embryo cells as a kind of stress protein, triggered Tom's brain to make the association with *Drosophila* Hsps. Interestingly, Susan and Tom, the two fellow graduate students who helped me the most in trying to understand what I was studying, both went on to win the US National Medal of Science!

I promptly ordered Susan's Ph.D. thesis on interlibrary loan. My summer student and I read it cover to cover. After that, I sought out every paper that subsequently flowed from the Lindquist lab. In a 1980 paper, they showed that Hsps were not only concentrated in the nuclei of *Drosophila* salivary gland tissues, but also at the cell boundaries, especially over the lumen of the gland (Velazquez 1980). This latter point had been virtually ignored and certainly neglected by investigators in the field. It suggested membrane associations and possibly secretion into an extracellular compartment. Peter Guidon, Jr., one of my Ph.D. students, had already found fatty acids noncovalently associated with Hsp70 and its cognate Hsc70 purified from tissues of heat shocked rats (Guidon and Hightower 1986). We had begun to think about these Hsps as lipoproteins. The ideas of membrane associations and release into the extracellular space fitted this hypothesis and we started to look. In 1985 we published our first two-dimensional gel patterns of proteins released into the medium over cultured rat embryo cells using a radioisotopic pulse-chase protocol (Hightower et al. 1985). The newly made radioactive proteins that chased into the culture medium were mostly the same from control and heat shocked cells, based on molecular masses and isoelectric points. For example, the rat heat shock cognate protein Hsc73, which we had identified previously, was present in both patterns. However, medium from heat shocked cells contained two additional inducible proteins, Hsp71 and Hsp110, not detected in medium from control cells. Rat Hsp71 was the major inducible protein generally known as Hsp70 and the other protein was later studied in great detail by John Subjeck and his co-workers and shown to be a phylogenetic relative of Hsp70 with some of the same activities that made it even better at molecular chaperoning (Leeyoon et al. 1995). It was still early for us in this line of investigation so we could not yet rule out the possibility of release of these proteins by necrotic cells in the cultures. But already it was clear to us that the overall gel patterns of released proteins were different than those from cytoplasmic extracts prepared by detergent lysis and that both the sodium ionophore, monensin and the antimicrotubule drug, colchicine did not block this release. We began thinking about the possibility of a new release mechanism different from the ER-Golgi pathway.

Later that same year, Fred White called me to discuss a poster from the laboratory of Michael Tytell that he had seen at a neuroscience meeting. On the poster were gel patterns of proteins isolated from squid giant axons known as glia-axon transfer proteins. Within these patterns, Fred had noticed polypeptides that looked a lot like those he had seen in patterns of stressed rat tissues in his work. This was intriguing and hinted at the possibility of tissue level functions and cell to cell transfer, if indeed they were Hsps. Approximately a year later, Tytell and co-workers published their work done in the squid giant axon model and titled "Heat shock-like protein is transferred from glia to axon" (Tytell et al. 1986). They noted the similarity in molecular mass and isoelectric point between traversin, the most abundant protein transferred from glia to axons, and Hsp70. In addition, they had found the publications from the Lasek laboratory showing that a traversin-like protein was produced in the heat shocked nerves of *Aplysia californica.* The excitement about the Tytell paper for us was the fact that here was a new tissue level property of Hsp70 that suggested a general function which we began calling 'altruistic cytoprotection'. Two heat-inducible proteins were transferred into the axon, traversin (Hsp70) and Hsp95. They carried out careful controls to show that these Hsps were not produced by the axoplasm and must have been transferred in. The transfer mechanism was not known but Tytell and coworkers speculated that it might involve the engulfment by the axon of cell surface membrane protrusions of the glial cells. This mechanism would still find favor today. But transfer for what purpose? They suggested that in the squid the transfer of these proteins might help the axon cope with the acute effects of injury or stress. They also used Fred White's favorite word for the response trigger in mammalian tissue, 'trauma'. They then did a very clever thing. They cited the work by White and Currie describing these proteins in the mammalian central nervous system and generalized their conclusions. Such cell-to-cell transfer could be a mechanism for distribution of Hsps throughout a tissue or entire organism. All cells need not make Hsps, particularly cells with limited protein synthesis capacity or unusual processes like nerve cells that required a long time to get newly made proteins from cell body to deep into the axon. At the time, one could practically count on the fingers of one hand the number of published papers analyzing the heat shock response as a tissue level response.

1.3 Hsps are Released from Mammalian Cells by an Active Process Other than the ER-Golgi Pathway: The Odyssey of Peer Review

In the same year that our book chapter appeared, I attended a heat shock meeting in Santa Fe, NM. I met at the poster session one of the first cellular immunologists whom I had seen in the field, and of course he was interested in the possibility of extracellular Hsps as potential cytokines. I told him that I had found several Hsps in the culture medium over mammalian cells but I did not know the mechanism of release. He said that as an immunologist, he did not care how the Hsps were released, he was interested in what they did afterward. This point of view was in stark contrast to that of a prominent biochemist and cell biologist who told me at the same meeting that I would never convince colleagues of the biological relevance of extracellular Hsps until I had determined the molecular mechanism of their release from cells. This should have been a big clue to what would lie ahead when we tried to publish our observations, and indeed, we worked to include controls to rule out release from dead and dying cells.

1.3.1 Hightower and Guidon 1989

We then proceeded to develop our preliminary study described above into a full length paper. My approach to studying the release of Hsps from late stage rat embryo cells was based on years of work with chicken embryo fibroblasts infected by the avian paramyxovirus, Newcastle Disease Virus. We had investigated the synthesis and processing of viral glycoproteins in the ER using radioisotopic pulse-chase protocols and inhibitors of glycoprotein biosynthesis (Schwalbe and Hightower 1982). This protocol allowed us to follow the migration of viral glycoproteins to the cell surface where they were incorporated into budding virus particles at the plasma membrane. Essentially the paramyxoviral particles can be viewed as specialized vesicles so we were keyed into thinking about these pathways as well. Several modifications of the viral protocols were necessary for the Hsp release protocol. Confluent cultures of rat embryo cells were heat shocked at 45 °C for ten minutes followed by a 2 h recovery at 37 °C to allow protein synthesis to recover. It became important that these initial manipulations were done without medium changes. Following recovery, the cultures were washed and placed in label medium containing ^{35}S-methionine at high specific activity for ten minutes at 37 °C. The cultures were washed again and placed in chase medium containing 10-fold excess unlabeled methionine for further incubation at 37 °C. A key modification of the chase medium was the omission of serum. Albumin with a molecular of mass of about 68,000 had to be eliminated because it overloaded the two dimensional gels and distorted the region containing Hsp71 and Hsc73. In addition, ovalbumin at 43,000 Da was used as a carrier in the medium precipitation step instead of albumin. The pH was carefully controlled throughout.

Quite unexpectedly, we found that the medium washes stimulated the release of Hsp71 (inducible Hsp70), Hsc73 (constitutive Hsc70), Hsp110 and nonmuscle actin from heat shocked cells and Hsc73 along with nonmuscle actin from control cultures. This confirmed a previous observation from Peter Rubenstein's lab, coincidentally another classmate from Harvard, about the release of nonmuscle actin from cultured embryonic skeletal muscle cells (Rubenstein 1982). A small amount of Hsp110 or possibly a cognate form of this protein was also detected among the proteins released from control cells. Elimination of the medium washes also eliminated the release of these extracellular proteins. We were encouraged by the highly reproducible nature of the gel patterns from independent experiments that immediately suggested that we were not dealing with the release of cytosolic protein from a variable number of necrotic cells in these cultures. The release was not dependent on heat shock: similar proteins were released from control cells as well. Further, the release was not continuous but appeared to be stimulated by a simple medium change, i.e. a change in the extracellular environment of the cultured cells. A thumbnail calculation indicated that, only a small amount of the newly synthesized Hsps were released from cells, roughly 1–2 %. The release was very rapid, on the order of seconds not minutes and thus it was unlikely to involve the 'classical' pathway of ER-Golgi-plasma membrane migration.

In order to formally remove the possibility that the release involved the 'classical' pathway, we showed that the carboxylic ionophore monensin, which inhibits protein secretion by vesiculation of the Golgi complex, and colchicine, which blocks transport vesicle movement by inhibiting microtubule assembly, did not block these rapidly released Hsps and actin. They did however block the extracellular appearance of Grp78, which became our positive control protein, since it was known to be produced and to function in the lumen of the ER. Small amounts of these ER proteins were known to escape the ER retention mechanisms so it was not surprising to find small amounts of extracellular Grp78, another member of the Hsp70 family. Other treatments were tested including chasing at 46 °C to block a possible thermolabile component, but to no avail. Also, neither cytochalasin E or the Ca^{++} ionophore A23187 added to the chase medium failed to inhibit the rapid release Hsps and nonmuscle actin. This would become the Achilles heal of the first manuscript submitted to the Journal of Cellular Physiology. We were proposing that the release was specific for the detected set of proteins described above but we had not identified a specific inhibitor of the process, which would have at least operationally defined a new release pathway. The reviewers balked and the editor declined to publish the work until this issue was addressed.

Our original submission did include several experiments aimed at the possibility that the extracellular proteins were released by lysis of a small number of necrotic cells in the cultures. Cells labeled under our standard conditions with ^{35}S-methionine were either treated with 1 % Triton X-100, a nonionic detergent, or dounce homogenized in a low salt buffer. The patterns of proteins released under these conditions were more complex and dominated by the tubulins and vimentin, both known to be released readily from necrotic cells. Neither of these proteins were found under our conditions of rapid release. Therefore, we confidently concluded that the proteins in our gel patterns were not the result of release from necrotic cells in our cultures. After our manuscript was returned from the journal, we spent months considering how to address the reviewers' concerns experimentally. Ultimately the answer had been sitting in front of us the entire time in the guise of small bottles of amino acid analogs that we used to produce analog-substituted proteins to study the induction of Hsps under conditions that did not involve the use of heat as a stressor (Hightower 1980). Analog-substituted proteins often did not fold properly and as a result, they might compete for the rapid release machinery thus blocking release. Alternatively they might not achieve the proper cellular location to engage the release mechanism. We simply added the lysine analog aminoethylcysteine to the radioactive methionine-containing label medium and performed our standard pulse-chase protocol. Synthesis of the ^{35}S-methionine labeled analog-substituted proteins could only occur in metabolically active cells. However, release of these analog-substituted radioactive proteins was blocked, consistent with the idea that the release was from living cells via a specific release pathway that depended on the three-dimensional conformation of the released proteins. We added the new experiment to our manuscript and it was promptly accepted.

The nature of the release mechanism remained elusive. We were well aware from the prior work of both Fred White and the Tytell group that vesicles might be involved. Using ultracentrifugation methods, we found that the Hsps and most of the actin in the releasate remained in the supernatant following centrifugation at 65,000 g (4 °C) for 30 min. We also layered the releasate onto 10–65 % sucrose gradients and subjected them to contrifugation at 65,000 g (4 °C) for 16 h. Again, the rapidly released radioactive proteins remained in the supernatant on the top of the gradient. We did not find evidence of vesicles or large aggregates but such a negative result cannot rule out the possibility that the extracellular Hsps may have been released into labile vesicles that did not survive in our buffers. The presence among the rapidly released proteins of nonmuscle actin, some of which is located in the cellular cortex, suggested the enticing possibility of an active release mechanism involving microfilaments underlying regions of plasma membrane. We cited an immunocytological study (LaThangue 1984) reporting that proteins analogous to Hsc73 and Hsp71 colocalize with actin-containing microfilaments in the ruffling membrane at the leading edge of motile fibroblasts. We proposed that 'perturbations of the cell surface of cultured cells cause the selective release of proteins from such dynamic regions of the cell periphery'. Writing today, we would add exosomes, ectosomes and microparticles to the short list of possibilities.

The specific stimulus associated with the medium washes that triggered release was also frustratingly elusive. Some of the likely possibilities included pH fluctuations, changes in gas tension, disruption of the diffusion boundary at the cell surface, or mechanical stresses such as the flux of media over the cell surface. We recognized at the time that these same factors were under consideration as part of tissue level responses to trauma and we included as a possible *in vivo* correlate the phrase "cellular responses to breaks in tissue homeostasis such as wounding". Recently, I had the pleasure of writing a meeting review with two colleagues describing some of the ongoing research that links cellular stress responses to wound healing (Doshi 2008). It is clear now that the release of Hsps from cells is part of the cellular stress response and that it is a normal release pathway that has been co-opted evolutionarily by Hsps as part of cellular stress responses.

This work is right up there among the most satisfying research endeavors of my professional career from the standpoint of intellectual stimulation and creativity at the bench. In retrospect and if I could magically go back in time to the writing of this paper, I would add a sentence or two to cover the possibility that the released proteins might function as cytokines, or as Calderwood, Asea and colleagues termed them, chaperokines (Asea et al. 2000).

1.4 Extracellular Hsps-Still Hot

In a new age of high throughput and *in silico* research methods we have now firmly established that extracellular detection of Hsps is not restricted to a certain Hsp, cell type or physiological context. Moreover they clearly lack the hydrophobic signal peptide for ER/Golgi release or any uniformly encoded secretory signal sequence. The Stress Observation System (SOS), or the ability of extracellular Hsps to function as cellular communicators in response to stress, was a termed conceived by Antonio De Maio in his aforementioned review (De Maio 2011). Indeed this is a hot area in the chaperone field, and one that has kept researchers busy over the past few decades uncovering these important findings. Key issues still remain surrounding the mechanism of Hsp trafficking from the cytosol including the working destination of these molecules, be it the cell membrane, cell surface or into the extracellular space, and more mysteriously, the function of the extracellular Hsps in mediating cellular responses to stress. Central to their function we must ask if their signaling is intended to be autocrine, paracrine or endocrine. Given the functionally rich history of Hsps in response to stressors and altered physiological states, the answers are most likely not very restrictive.

1.5 Mechanism of Extracellular Hsp Release

The non-classical (non-ER/Golgi) putative secretion mechanisms of Hsps probably occur by a series of events which help them reach their target destination. In a paper by Mambula and Calderwood in 2006, the authors showed that Hsp70 is released from certain tumor cells through a mechanism using the endo-lysosomal compartment (Mambula and Calderwood 2006). Interestingly their data provided evidence that ABC transporter proteins helped facilitate this process as secretion was blocked with their inhibition. In a later review of the subject, Mambula and Calderwood speculated that Hsps are released through this pathway by merging of the endo-lysosomal compartment with the plasma membrane, and subsequent release at the cell surface (Mambula et al. 2007). There are other studies reporting detection of Hsps in the lysosomal compartment following stress in fish, rodents and humans (Mayer et al. 1991; Jethmalani et al. 1997; Polla et al. 2007; Yabu et al. 2011).

Release of Hsps by vesiculation is one of the more commonly proposed mechanisms of release. Heat responsive chaperones may have a higher cortical cell concentration following stress aiding in their release by vesiculation. Vesiculation is typically described as the release of the Hsps through extracellular vesicles. In general, these vesicles originate from the plasma membrane through different mechanisms and would include, exosomes, ectosomes and microparticles. Exosomes are formed through an endocytic process in which the plasma membrane invaginates inward engulfing not only components of the membrane itself (like surface receptors and ligands) but additional cytosolic content (including Hsps). This results in the formation of early endosomes which mature into late endosomes by removal of specific membrane proteins and lipids that are returned to the cell membrane. The late endosomes undergo a secondary invagination resulting in the formation of multivesicular bodies (MVB). These organelles are then able to fuse with the plasma membrane moving their contents into the extracellular microenvironment (Mathivanan et al. 2010). Formation of ectosomes and microparticles occur by way of membrane evagination and involves the help of actin polymerization. Similarly to exosomes, these vesicles also contain both membrane and cytosolic components. What is interesting about both of these processes in extracellular release is that they appear to have a filter for their contents, not just capturing whatever is in the neighborhood but a more specific array of molecules. In a number of different species and physiological conditions, different Hsps have been shown to be released through exosomes. These include the constitutively expressed Hsc70 (Hegmans et al. 2004; Buschow et al. 2010), heat inducible Hsp72 (Bausero et al. 2005; Asea et al. 2008; Chalmin et al. 2010; Hurwitz et al. 2010), and Hsp90 (Cheng et al. 2008; Buschow et al. 2010; Mccready et al. 2010), Hsp60 (Gupta and Knowlton 2007), and the small chaperones Hsp27 (Clayton et al. 2005) and AlphaB crystallin (Gangalum et al. 2011). Mitochondrial Hsp70 has been detected in ectosomes or microparticles, although its exact role has not been established (Moskovich and Fishelson 2007). Lastly, secretory-like granules were found to play a role in release of Hsp70 following heat shock in A431 (human epithelial carcinoma) cells (Evdonin et al. 2006).

1.6 Hsps on the Cell Surface

When considering the presence of Hsps on the cell surface, translocation is another potential mechanism of transporting these molecules extracellularly. Hsp70 has been found in association with lipid rafts in human colon cancer cell lines. Translocation is enhanced by heat shock and cannot be blocked with secretory inhibitors but can be inhibited with lipid raft disrupting agents (Broquet et al. 2003). This parallels with reports of Hsp70 being able to incorporate into artificial lipid bilayers and play a role in channel conductance (Arispe and De Maio 2000; Vega et al. 2008). The ability of Hsp70 to incorporate into the lipid membrane is thought to be dependent on membrane fluidity and composition (Arispe et al. 2004; Gehrmann et al. 2008; Horvath et al. 2008; Sugawara et al. 2009). As Hsps are also involved in shuttling other transmembrane proteins across these lipid membranes, it seems logical that these chaperones could then remain at the cell surface essentially having piggybacked their substrate. It is now well established that a number of different Hsps can be found on the surface of tumor cells in both stressed and unstressed conditions. For example, high levels of cytosolic Hsp72 accumulate in response to certain therapies in tumor cells and can then become expressed on the surface of these tumor cells (Gehrmann et al. 2005). Hsp72 is expressed on the surface of many cancer cells lines in response to stress (Multhoff et al. 1995). Hsp72 is also expressed on the surface of tumor cells under physiological conditions (Botzler et al. 1996; Multhoff et al. 1997). Hsp70B', an Hsp70 family member expressed in response to extreme stress, is only found on the surface of tumor cells in response to proteasome inhibitors and not mild heat shock (Noonan et al. 2008). Interestingly, in normal cells, Hsp70 is associated with cell surface receptors while in tumors it is integrated into the cell membrane. This distinction becomes important in terms of technical approaches to distinguishing between these different associations. For details see a recent review (Multhoff and Hightower 2011). In conclusion, we have only scratched the surface in our explorations of the exciting lives of extracellular Hsps, a research area that may well yield new therapies for major human diseases and new insights into human physiology.

References

Arispe N, De Maio A (2000) ATP and ADP modulate a cation channel formed by Hsc70 in acidic phospholipid membranes. J Biol Chem 275:30839–30843

Asea A, Kabingu E, Stevenson MA, Calderwood SK (2000) HSP70 peptide-bearing and peptide-negative preparations act as chaperokines. Cell Stress Chaperones 5:425–431

Arispe N, Doh M, Simakova O, Kurganov B, De Maio A (2004) Hsc70 and Hsp70 interact with phosphatidylserine on the surface of PC12 cells resulting in a decrease of viability. FASEB J 18:1636–1645

Asea A, Jean-Pierre C, Kaur P, Rao P, Linhares IM, Skupski D, Witkin SS (2008) Heat shock protein-containing exosomes in mid-trimester amniotic fluids. J Reprod Immunol 79:12–17

Bausero MA, Gastpar R, Multhoff G, Asea A (2005) Alternative mechanism by which IFN-gamma enhances tumor recognition: active release of heat shock protein 72. J Immunol 175:2900–2912

Botzler C, Issels R, Multhoff G (1996) Heat-shock protein 72 cell-surface expression on human lung carcinoma cells in associated with an increased sensitivity to lysis mediated by adherent natural killer cells. Cancer Immunol Immunother 43:226–230

Broquet AH, Thomas G, Masliah J, Trugnan G, Bachelet M (2003) Expression of the molecular chaperone Hsp70 in detergent-resistant microdomains correlates with its membrane delivery and release. J Biol Chem 278:21601–21606

Buschow SI, van Balkom BW, Aalberts M, Heck AJ, Wauben M, Stoorvogel W (2010) MHC class II-associated proteins in B-cell exosomes and potential functional implications for exosome biogenesis. Immunol Cell Biol 88:851–856

Chalmin F, Ladoire S, Mignot G et al (2010) Membrane-associated Hsp72 from tumor-derived exosomes mediates STAT3-dependent immunosuppressive function of mouse and human myeloid-derived suppressor cells. J Clin Invest 120:457–471

Cheng CF, Fan J, Fedesco M et al (2008) Transforming growth factor alpha (TGFalpha)-stimulated secretion of HSP90alpha: using the receptor LRP-1/CD91 to promote human skin cell migration against a TGFbeta-rich environment during wound healing. Mol Cell Biol 28:3344–3358

Clayton A, Turkes A, Navabi H, Mason MD, Tabi Z (2005) Induction of heat shock proteins in B-cell exosomes. J Cell Sci 118:3631–3638

De Maio A (2011) Extracellular heat shock proteins, cellular export vesicles, and the stress observation system: a form of communication during injury, infection, and cell damage. Cell Stress Chaperones 16:235–249

Doshi BM, Perdrizet GA, Hightower LE (2008) Wound healing from a cellular stress response perspective. Cell Stress Chaperones 13:393–399

Evdonin AL, Martynova MG, Bystrova OA, Guzhova IV, Margulis BA, Medvedeva ND (2006) The release of Hsp70 from A431 carcinoma cells is mediated by secretory-like granules. Eur J Cell Biol 85:443–455

Gangalum RK, Atanasov IC, Zhou ZH, Bhat SP (2011) Alpha B-crystallin is found in detergent-resistant membrane microdomains and is secreted via exosomes from human retinal pigment epithelial cells. J Biol Chem 286:3261–3269

Gehrmann M, Schonberger J, Zilch T, Rossbacher L, Thonigs G, Eilles C, Multhoff G (2005) Retinoid- and sodium-butyrate-induced decrease in heat shock protein 70 membrane-positive tumor cells is associated with reduced sensitivity to natural killer cell lysis, growth delay, and altered growth morphology. Cell Stress Chaperones 10:136–146

Gehrmann M, Liebisch G, Schmitz G, Anderson R, Steinem C, De Maio A, Pockley G, Multhoff G (2008) Tumor-specific Hsp70 plasma membrane localization is enabled by the glycosphingolipid Gb3. PLoS One 3:e1925

Guidon PT Jr, Hightower LE (1986) Purification and initial characterization of the 71-kilodalton rat heat-shock protein and its cognate as fatty acid binding proteins. Biochemistry 25:3231–3239

Gupta S, Knowlton AA (2007) HSP60 trafficking in adult cardiac myocytes: role of the exosomal pathway. Am J Physiol Heart Circ Physiol 292:H3052–H3056

Hegmans JP, Bard MP, Hemmes A et al (2004) Proteomic analysis of exosomes secreted by human mesothelioma cells. Am J Pathol 164:1807–1815

Hightower LE (1980) Cultured animal cells exposed to amino acid analogues or puromycin rapidly synthesize several polypeptides. J Cell Physiol 102:407–427

Hightower LE, Guidon PT Jr (1989) Selective release from cultured mammalian cells of heat-shock (stress) proteins that resemble glia-axon transfer proteins. J Cell Physiol 138:257–266

Hightower LE, Guidon PT Jr, Whelan SA, White CN (1985) Stress responses in avian and mammalian cells. In: Atkinson BG, Walden DB (ed) Changes in eukaryotic gene expression response to environmental stress. Academic, New York, pp 197–210

Horvath I, Multhoff G, Sonnleitner A, Vigh L (2008) Membrane-associated stress proteins: more than simply chaperones. Biochim Biophys Acta 1778:1653–1664

Hurwitz MD, Kaur P, Nagaraja GM, Bausero MA, Manola J, Asea A (2010) Radiation therapy induces circulating serum Hsp72 in patients with prostate cancer. Radiother Oncol 95:350–358

Jethmalani SM, Henle KJ, Gazitt Y, Walker PD, Wang SY (1997) Intracellular distribution of heat-induced stress glycoproteins. J Cell Biochem 66:98–111

LaThangue NB (1984) A major heat-shock protein defined by a monoclonal antibody. EMBO J 3:1871–1879

Leeyoon D, Easton D, Murawski M, Burd R, Subjeck JR (1995) Identification of a major subfamily of large hsp70-like proteins through the cloning of the mammalian 110-kDa heat shock protein. J Biol Chem 270:15725–15733

Mambula SS, Calderwood SK (2006) Heat shock protein 70 is secreted from tumor cells by a nonclassical pathway involving lysosomal endosomes. J Immunol 177:7849–7857

Mambula SS, Stevenson MA, Ogawa K, Calderwood SK (2007) Mechanisms for Hsp70 secretion: crossing membranes without a leader. Methods 43:168–175

Mathivanan S, Ji H, Simpson RJ (2010) Exosomes: extracellular organelles important in intercellular communication. J Proteomics 73:1907–1920

Mayer RJ, Lowe J, Landon M, McDermott H, Tuckwell J, Doherty F, Laszlo L (1991) Ubiquitin and the lysosome system: molecular immunopathology reveals the connection. Biomed Biochim Acta 50:333–341

McCready J, Sims JD, Chan D, Jay DG (2010) Secretion of extracellular hsp90alpha via exosomes increases cancer cell motility: a role for plasminogen activation. BMC Cancer 10:294

Moskovich O, Fishelson Z (2007) Live cell imaging of outward and inward vesiculation induced by the complement c5b-9 complex. J Biol Chem 282:29977–29986

Multhoff G, Hightower LE (2011) Distinguishing integral and receptor-bound heat shock protein 70 (Hsp70) on the cell surface by Hsp70-specific antibodies. Cell Stress Chaperones 16:251–255

Multhoff G, Botzler C, Wiesnet M, Muller E, Meier T, Wilmanns W, Issels RD (1995) A stress-inducible 72-kDa heat-shock protein (HSP72) is expressed on the surface of human tumor cells, but not on normal cells. Int J Cancer 61:272–279

Multhoff G, Botzler C, Jennen L, Schmidt J, Ellwart J, Issels R (1997) Heat shock protein 72 on tumor cells: a recognition structure for natural killer cells. J Immunol 158:4341–4350

Noonan EJ, Fournier G, Hightower LE (2008) Surface expression of Hsp70B' in response to proteasome inhibition in human colon cells. Cell Stress Chaperones 13:105–110

Polla BS, Gabert F, Peyrusse BM, Jacquier-Sarlin MR (2007) Increased proteolysis of diphtheria toxin by human monocytes after heat shock: a subsidiary role for heat-shock protein 70 in antigen processing. Immunology 120:230–241

Ritossa FM (1962) A new puffing pattern induced by a temperature shock and DNP in Drosophila. Experientia 18:571–573

Rubenstein P, Ruppert T, Sandra A (1982) Selective isoactin release from cultured embryonic sketetal muscle cells. J Cell Biol 92:164–169

Schwalbe JC, Hightower LE (1982) Maturation of the envelope glycoproteins of Newcastledisease virus on cellular membranes. J Virol 41:947–957

Sugawara S, Kawano T, Omoto T, Hosono M, Tatsuta T, Nitta K (2009) Binding of Silurus asotus lectin to Gb3 on Raji cells causes disappearance of membrane-bound form of HSP70. Biochim Biophys Acta 1790:101–109

Tytell M, Greenberg SG, Lasek RJ (1986) Heat shock-like protein is transferred from glia to axon. Brain Res 363:161–164

Vega VL, Rodriguez-Silva M, Frey T et al (2008) Hsp70 translocates into the plasma membrane after stress and is released into the extracellular environment in a membrane-associated form that activates macrophages. J Immunol 180:4299–4307

Velazquez JM, Di Domenico BJ, Lindquist S (1980) Intracellular location of heat shock proteins in *Drosophila*. Cell 20:679–689

White FP (1979) Subcellular particles involved in the translocation of proteins in rat brain. J Neurobiol 10:591–607

White FP (1980) The synthesis and possible transport of specific proteins by cells associated with brain capillaries. J Neurochem 35:88–94

Yabu T, Imamura S, Mohammed MS, Touhata K, Minami T, Terayama M, Yamashita M (2011) Differential gene expression of HSC70/HSP70 in yellowtail cells in response to chaperone-mediated autophagy. FEBS J 278:673–685

Chapter 2
Discovery of the Extracellular Agonist Actions of Molecular Chaperones and Protein-Folding Catalysts

Brian Henderson

Abstract Surprisingly, the history of the agonist actions of extracellular molecular chaperones can be traced back to the 1970s, with the cytokine macrophage migration inhibitory factor (MIF) and chaperonin (Hsp)10. The next cell stress protein to be identified as a molecular chaperone was the peptidylprolyl isomerase, cyclophilin A, in 1992. It is only later in the 1990s that the major signalling cell stress proteins—chaperonin (Hsp)60 and Hsp70 are found to have agonist activities. There are still ongoing discoveries of stress proteins with agonist actions and the latest such proteins are a new group of molecular chaperones—the extracellular/circulating molecular chaperones which include clusterin and α-acid1-glycoprotein.

2.1 Introduction

The phenomenon of the human system called science is endlessly fascinating for the paradoxes it encompasses. Science, at its heart, is the creation of the story of the Universe/Multiverse we live in, with its various disciplines focused on different parts of this larger picture. The basic unit of science is the testable hypothesis and the discoveries that the hypothesis machine provides us with are never complete and always need to be altered or even radically changed. Starting with the initial discovery of the heat shock response (Ritossa et al. 1962), the evolution of the phenomenon of the cell stress response and the discovery of the process of protein chaperoning (Laskey et al. 1978), it was the basic assumption that the proteins involved in the stress response were only active within the cell. This paradigm of molecular chaperones and protein-folding catalysts (PFCs—collectively cell stress proteins) being found only within the cell, has seriously, and negatively, influenced the new paradigm that has emerged since the 1990s, that molecular chaperones and PFCs can be secreted from the cell interior and function as cell surface receptors or cell signalling soluble agonists (See Chap. 1 for a personal history of the discovery of the secretion of Hsp70). This failure, by the cell stress protein community, largely to ignore the biology of extracellular molecular chaperones and PFCs, is curious, as in 1977, 1 year before Laskey's coining of the term chaperone for the function

B. Henderson (✉)
Department of Microbial Diseases, UCL-Eastman Dental Institute,
University College London, 256 Gray's Inn Road, London WC1X 8LD, UK
e-mail: brian.henderson@ucl.ac.uk

B. Henderson, A. G. Pockley (eds.), *Cellular Trafficking of Cell Stress Proteins in Health and Disease*, Heat Shock Proteins 6, DOI 10.1007/978-94-007-4740-1_2,

of nucleoplasmin, a group of Australian scientists identified an immunosuppressive factor in the serum of women in the first trimester of pregnancy (Morton et al. 1977; Noonan et al. 1979). Unfortunately, the molecular nature of this factor was not identified until 1994, when it was shown to be human chaperonin (Cpn)10 (Cavanagh and Morton 1994). However, this result seems to have been overlooked and the growing numbers of publications, arising from around the early 1990s, on the cell signalling actions of recombinant cell stress proteins, which would have been strongly supported by the physiological actions of Cpn10, were largely ignored or criticised on the grounds that they were due to bacterial contaminants (e.g. Tsan and Gao 2009). This criticism is still extant at the time of writing, even though there are a growing number of reports of the actions of extracellular cell stress proteins that could not possibly be due to bacterial contamination—either because the actions have nothing to do with the activity of pro-inflammatory bacterial contaminants or because the proteins under study are made in eukaryotic systems or are, indeed, totally synthetic proteins/peptides (Henderson and Pockley 2010; Henderson et al. 2010). It is hoped that reviews like this one, which provide a historical perspective, will encourage a fairer response to the study of the extracellular actions of molecular chaperones.

2.2 Secretion of Molecular Chaperones

Cell stress proteins are soluble signalling mediators—is a proposition that can only be accepted if there is evidence that these proteins are capable of being secreted. Again, the cell stress community has been guilty of ignoring key information, such as the early work from Tytel and Hightower that specific cell stress proteins are released from viable cells (Tytell et al. 1986; Hightower and Guidon 1989—see Chap. 1 for full details). The major problem has been a lack of understanding that both in bacteria (Holland 2010) and in eukaryotic cells (Nickel and Rabouille 2009), there are a plethora of protein secretion pathways in addition to the classical signal peptide secretion mechanism. Good evidence now exists that eukaryotic molecular chaperones can be secreted via one or other of these newly discovered secretion pathways (Table 2.1). In contrast, we know almost nothing about the release pathways that are involved in the secretion of bacterial molecular chaperones. Clearly much more work is needed to determine if the secretion of the many cell stress proteins found in the body fluids is due to a novel system for maintaining homeostasis, or if it is involved in tissue and cell pathology. However, it is now perfectly clear that a number of the major cell stress proteins are normally secreted and therefore their presence in body fluids and their actions on cells is not some artefact of the scientific process but is a manifestation of normal biological processes in both bacteria and eukaryotic cells.

Both bacteria and eukaryotes have a common set of cell stress proteins. It would appear, from the current literature, that eukaryotic cells have evolved to secrete numerically more of these proteins than bacteria have (Fig. 2.1). However, this may only represent the particular personal focus on the release of these proteins by both of these cell Kingdoms, with less attention being paid to bacterial cell stress proteins.

Table 2.1 The known secretion pathways for cell stress proteins

Protein	Secretion pathway	Reference
Thioredoxin	Novel pathway with some similarities to that of IL-1β	Rubartelli et al. 1990, 1992; Tassi et al. 2009
HSPB5	Exosomal secretion pathway	Gangalum et al. 2010
Peroxiredoxin	Brefeldin-insensitive non-classical pathway	Chang et al. 2006
PPIs	Unique vesicle-associated process	Suzuki et al. 2006
Cpn/Hsp60	Exosomal secretion pathway	Gupta and Knowlton 2007
Hsp70	Exosomal or vesicle-dependent secretion pathway	Lancaset and Febbraio 2005; Zhan et al. 2009
Hsp70	Non-classical pathway involving lysosomes	Mambula and Calderwood 2006
HspB1	Classic secretion pathway	Evdonin et al. 2009
BiP	Brefeldin-inhibited secretion pathway	Xiao et al. 1999
Hsp90	Possible exosomal pathway	Cheng et al. 2008

PPIs peptidyprolyl isomerases

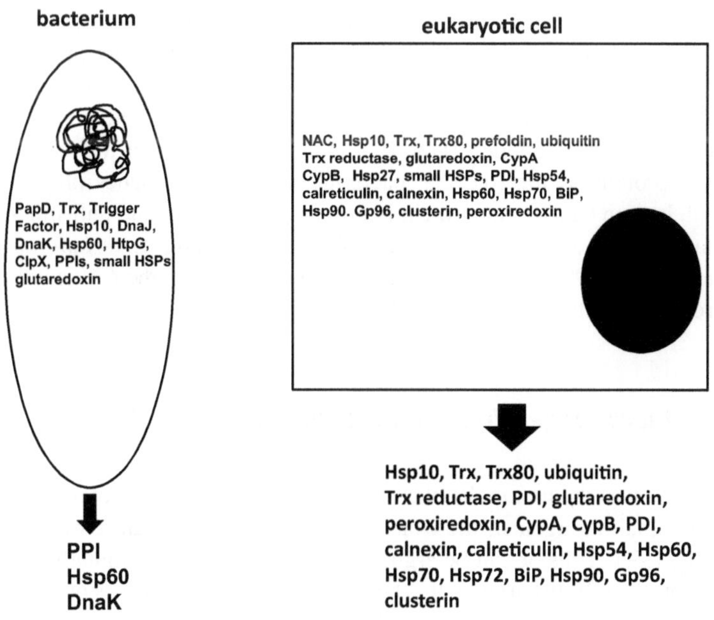

Fig. 2.1 The secretion of the total population of molecular chaperones and PFCs by bacteria and eukaryotic cells

2.3 Identification of the Cell Signalling Agonist Actions of Cell Stress Proteins

2.3.1 Chaperonin 10 (HSPE1)

As has already been discussed, the first evidence that cell stress proteins have the ability to act as cell signalling agonists was the report that a serum factor found in early pregnancy, and in consequence termed, early pregnancy factor (EPF), was an immunosuppressive factor (Morton et al. 1977). The nature of this factor was not clarified in the initial studies (e.g. Clarke et al. 1978) and it was only in 1979 that the immunosuppressive actions of this protein were defined (Noonan et al. 1979). However, it took until 1994 before the identity of EPF was confirmed. Using platelets as the source of EPF, led to the isolation of a pure form of this protein and its unequivocal demonstration as chaperonin (Cpn) 10, the co-chaperone for chaperonin (Cpn)60 (Cavanagh and Morton 1994). About a decade later it was shown that administration of recombinant human Cpn10 to rats could inhibit experimental inflammation (e.g. Zhang et al. 2003). Perhaps not surprisingly, this has led on to the Australian biopharmaceutical company AC Bio testing recombinant human Cpn10 for the treatment of a variety of human diseases including rheumatoid arthritis (Vanags et al. 2006) and multiple sclerosis (Broadley et al. 2009). One of the curiosities of the agonist actions of cell stress proteins is that it is impossible to predict what the next moonlighting action of these proteins will be. Proteins with more than one unique function are termed moonlighting proteins (Jeffery et al. 1999). It has recently been reported that Cpn10 is an erythropoietin-inducible secreted protein with effects on endothelial cell differentiation (Dobocan et al. 2009). Another interesting facet of cell stress proteins is the role that bacterial protein homologues play in disease. Thus with human Cpn10, the protein is immunosuppressive and anti-inflammatory. A similar situation is seen with the homologue from *Mycobacterium tuberculosis* (e.g. Ragno et al. 1996). In contrast the Cpn10 protein from the *Chlamydiae* appear to be pro-inflammatory (Zhou et al. 2011; Jha et al. 2011).

2.3.2 Macrophage Migration Inhibitory Factor (MIF)

While Cpn10 was the first molecular chaperone and cell stress protein to be discovered as a signalling agonist, the exact identification of the nature of EPF was not made until 1994. Thus, it can be argued that Cpn10 is not the signalling cell stress protein prototype. If this argument is heeded, then which cell stress protein takes the prize of being the first to be found with cell signalling activity? The surprising answer is that we can now take the history of cell stress proteins as immunomodulators back to 1971, and the discovery of the cytokine—macrophage migration inhibitory factor (MIF—e.g. Bartfield and Atoynatan 1971). This is one of the earliest and most confusing of cytokine molecules, which has been implicated in the pathology of diseases including: sepsis, pneumonia, diabetes, rheumatoid arthritis, inflammatory bowel disease, psoriasis and cancer and for which novel low molecular mass inhibitors have been developed (Al-abed and Van Patten 2011). Whilst being around since the early 1970s, it was only in the late 1990s that the other guise of this protein was identified with the discovery that MIF is a thiol protein oxidoreductase (Kleeman et al. 1998; Potolicchio et al. 2003). A later study confirmed the influence MIF has on protein folding (Cherepkova et al. 2006). MIF is a secreted protein and it appears to induce its myriad effects by binding to CD74 on target cells, and such binding is now known to play a role in cancer (Shachar and Haran 2011). Unfortunately, nothing is really known about the connection between the protein-folding action of MIF and its myriad extracellular agonist functions. However, MIF serves to remind the reader that a potent cytokine can also function as a protein folding protein and should reinforce the hypothesis that molecular chaperones and PFCs can also function as cell signalling agonists with cytokine activity.

2.3.3 Thioredoxin

Again, the reader could accuse the writer that both examples of cell stress proteins so far discussed were only found to be molecular chaperones many years after their initial discovery as cytokine-like factors and that the earliest discovery of a molecular chaperone as a cell signalling agonist has not yet been defined. The example of the first molecular chaperone to be defined as a cell signalling agonist must then be human thioredoxin (Trx). This story can be said to start with the finding that lymphocyte activation is dependent on cell surface thiol oxidation status (Noelle and Lawrence 1981). Some years later it was reported that HTLV-1-transformed T lymphocytes secreted a factor, termed adult T cell leukaemia (**ATL**)-**d**erived **f**actor (ADF) that enhanced the expression of the p55 subunit of the IL-2 receptor. ADF synthesis was increased by classic T cell activators such as mitogens and phorbol ester. ADF was then shown to be human Trx (Tagaya et al. 1989). ADF was shown to be an autocrine growth factor for lymphocytes and to synergise with both IL-1 and IL-2 (Wakasugi et al. 1990). As has been shown in Table 2.1, Trx was the first protein folding catalyst to be shown to be secreted by a novel secretion pathway whose complete elucidation still eludes us. Interestingly, the secretion of key inflammatory cytokines such as IL-1 (Tassi et al. 2009) and the chemokine MCP-1 (Chen et al. 2010) is controlled by Trx. In addition to functioning to enhance the growth of T and B lymphocytes, Trx has also been shown to be secreted in greater amounts from T regulatory cells (T regs) and this is associated with a decreased level of T reg apoptosis (Mougiakakos et al. 2011).

Since the initial discovery of ADF/thioredoxin it has been established that Trx is a naturally secreted protein whose levels in the circulation are regulatable and with a wide range of important biological function (Holmgren and Lu 2010). This chapter is not the place to review the actions of thioredoxin and while elevated levels can be associated with poor outcome in patients with AIDS (Nakamura et al. 2001) it is generally found that administration of this protein promotes beneficial effects in a variety of experimental animal disease states (Nakamura 2008). It is expected that this protein, or homologues of the protein, will find a place as therapeutics for a range of human diseases.

2.3.4 Peptidylprolyl Isomerases

These enzymes, which can be classified into the cyclophilins, FK506-binding proteins (FKBPs) and parvulins, are responsible for catalysing peptide bond *cis/trans* isomerisation (Schiene-Fischer et al. 2011). The cyclophilins (Cyps) and FKBs are intracellular proteins that bind to immunosuppressants, such as cyclosporine (Göthel and Marahiel 1999) and transduce their immunosuppressive actions. The first evidence that the cyclophilins had any signalling actions was the report that exposure of the mouse macrophage cell line, RAW267.4 to lipopolysaccharide (LPS), resulted in the release of cyclophilin (Cyp)A. The purified CypA exhibited pro-inflammatory activity *in vivo* and acted as a chemoattractant for human neutrophils and monocytes. This activity was blocked by cyclosporine A, but not by the structural analogue, cyclosporine H, which does not bind cyclophilin. The finding that CypA is released in larger amounts by LPS-stimulated than by non-stimulated macrophages and that the activity it exhibits is pro-inflammatory is identical to the situation with the secretion of true pro-inflammatory cytokines such as IL-1β and TNFα (Sherry et al. 1992). Thus this protein seems like it should be added to the group of pro-inflammatory cytokines.

Up to this date, all secreted cell stress proteins with agonist actions had been eukaryotic proteins. In the same year as mouse CypA was shown to be a pro-inflammatory factor it was reported that the essential *Legionella pneumophila* virulence protein, Mip (macrophage infectivity promoter), involved in the invasion of macrophages by this bacterium (Engleberg et al. 1989), was also a peptidylprolyl isomerase (Fischer et al. 1992). However, Mip was not a cyclophilin, but an FKBP (Engleberg et al. 1989). Other bacteria have also been found to utilise homologues of Mip to invade cells including *Chlamydia trachomatis* (Lundemose et al. 1993) and, coming more up to date, *Burkholderia pseudomallei* (Norville et al. 2011). In addition to being bacterial invasins, one of the peptidylprolyl isomerases of *Helicobacter pylori* is both a major immunogen in patients with gastric ulcers (Attanasov et al. 2002) and a signalling protein promoting major changes in the behaviour of both epithelial cells (Basak et al. 2005) and macrophages (Pathak et al. 2006).

Coming back to the human cyclophilin, if this protein had any role in disease then it would be found in relevant body fluids. This hypothesis was confirmed by the finding of elevated levels of CypA in the synovial fluid of patients with the inflammatory disease, rheumatoid arthritis, compared to those with the non-inflammatory condition, osteoarthrosis (Billich et al. 1997). Similarly, serum levels of peptidyl prolyl isomerase activity were elevated in patients with sepsis (Tegeder et al. 1997). Again, coming up to date, it has recently been shown that knock-out of the gene encoding CypA renders mice unresponsive to acetaminophen-induced liver injury (Dear et al. 2011).

One of the major problems in the study of the biology of secreted cell stress proteins is identifying the nature of the agonist receptor and many such agonist proteins have no clearly identified receptor. This failure to identify single receptors for proteins such as Hsp60, Hsp70 and BiP has exacerbated the criticism that the effects of these proteins are artefactual. It is, however, satisfying, to record that secreted CypA and CypB have a well identified receptor in the protein, CD147 (Yurchenko et al. 2001). It is now established that the binding of CypA to CD147 is independent of the peptidyl prolyl isomerase active site (Song et al. 2011). It is reported that CypA and the chemokine CXCL2 have a cooperative influence in terms of neutrophil chemotaxis thus showing that the cell stress proteins can interact with cytokine networks (Heine et al. 2011). There is now significant evidence for the hypothesis that cyclophilin-CD147 interactions are pathogenic and this agonist-receptor pairing is now seen as an important therapeutic target for a range of human disease (Yurchenko et al. 2010).

2.3.5 Chaperonin (Hsp)60 (HSPD1)

Chaperonin (Cpn)60, generally known as heat shock protein (Hsp)60 in the human context, is one of the best studied secreted molecular chaperone agonists and one of the most heavily criticised, particularly the signalling activity of the human homologue. However, the discovery that this protein had signalling actions was made using the Cpn60.2 protein from the major human pathogen, *Mycobacterium tuberculosis* (Friedland et al. 1993). This paper revealed that exposure of the human macrophage cell line, THP-1 to recombinant *M. tuberculosis* Cpn60.2 induced the formation of a number of pro-inflammatory cytokines, including TNFα. The inference of this paper was that *M. tuberculosis* Cpn60.2, a protein better known at the time as Hsp65, was an activator of macrophages. This interpretation has had deleterious consequences for the field of secreted/signalling molecular chaperone biology. It was assumed that as this Cpn60 protein induced macrophage activation, then all homologues would do the same. The importance of this interpretation lies in the fact that the potent Gram-negative cell wall component, LPS, also stimulates 'macrophage activation'. This left the Cpn60 protein open to the objection that its effect on macrophages was not due to an inherent biological activity, but to contamination with the LPS coming from the *E. coli* strains in which the Cpn60 proteins were expressed (Tsan and Gao 2009).

Macrophage activation is an oft-used term. In recent years, attempts have been made to subdivide this 'activation' into a number of distinct states (Gordon and Martinez 2010). Currently, macrophage activation, depending on which review is read, can be divided into: (i) classical (induced by gamma-interferon); (ii) innate (induced by bacterial components, principally LPS and; (iii) various forms of so-called alternative activation states (Gordon and Martinez 2010). Both classical and innate activation involve the upregulation of genes encoding for proteins involved in bacterial recognition and bacterial killing. These would include, Fc gamma family proteins, MHC class II proteins and free radical inducing proteins. In addition, the classical and innate activators are major inducers of pro-inflammatory cytokines. It was therefore fascinating to find that when Ralph van Furth's group measured the effects of *M. tuberculosis* Cpn60.2 on monocytes that the major signs of macrophage activation, apart from cytokine synthesis, were missing (Peetermans et al. 1994). Thus, unlike LPS, the *M. tuberculosis* Cpn60.2 protein did not induce, for example, the synthesis of MHC class II proteins, required if the macrophage was going to present antigen to T lymphocytes. This Cpn60.2 protein is therefore inducing an activation state in macrophages which has not really been defined. However, what it is not doing, is mimicking the actions of LPS. Van Furth also showed that the *M. tuberculosis* Cpn60.2 protein activated human vascular endothelial cells in a manner distinct from that of LPS or human pro-inflammatory cytokines (Verdegaal et al. 1996). These early findings have been completely ignored by those workers who seek to criticise the study of the signalling actions of chaperonin 60 proteins.

Since these early studies of the signalling actions of the *M. tuberculosis* Cpn60.2 protein it has been shown that the Cpn60 proteins from *Aggregatibacter actinomycetemcomitans, E. coli, Mycobacterium leprae, Rhizobium leguminosarum, Chlamydia pneumoniae* and *C. trachomatis* and *Helicobacter pylori* can all stimulate macrophages and other cells to secrete cytokines (Henderson and Martin 2011). However, it was not until 1999, that it was shown that human Hsp60 would also induce both human and mouse macrophages to release pro-inflammatory cytokines (Chen et al. 1999). There is now a substantial literature on this protein which has been recently reviewed by Henderson and Pockley (2010).

In addition to stimulating monocytes, *M. tuberculosis* actually invades these cells and survives within them by subverting their anti-bacterial responsiveness (Rohde et al. 2007). It has been found that *M. tuberculosis* actually secretes large amounts of Cpn60.2 which adheres to the bacterial surface and acts as an adhesin for binding to macrophages. Binding is to the cell surface receptor CD43 (Hickey et al. 2009, 2010), a protein that is known to control the rate of intracellular growth of *M. tuberculosis* (Randhawa et al. 2008). Thus this Cpn60.2 protein plays multiple roles in allowing the bacterium to interact with macrophages. Further information on the secretion and role of circulating human Cpn60 will be provided by Pockley in Chap. 3.

2.3.6 Hsp70

The abbreviation Hsp70, now refers to at least twelve separate proteins, which are designated by a new nomenclature which was briefly described at the beginning of the book (Hageman et al. 2011). It was known since 1989 that Hsp70 could be secreted by cells (Hightower and Guidon 1989—reviewed in detail in Chap. 1) yet it was only in the twenty-first century that the first demonstration that Hsp70 could act as a stimulating agonist with human monocytes was published (Asea et al. 2000).

As with Hsp60, the nature of the receptor transducing the signal from Hsp70 has proved difficult to identify. Early studies, using cells transfected with Toll-like receptors (TLRs), suggested that TLR2/TLR4 were the receptors for Hsp70 (Asea et al. 2002). However, later studies from this group failed to replicate these findings and suggested that Hsp70 bound principally to the scavenger receptor LOX-1—a non-signalling cell surface protein (Thériault et al. 2005). A later study increased the number of scavenger receptors able to bind human Hsp70 (Thériault et al. 2006). Another proposed receptor for human Hsp70 is CD40 (Becker et al. 2002). Much more information on receptors for Hsp70 is to be found in Chap. 13 from Stuart Calderwood's group.

Some years after the discovery that human Hsp70 stimulated monocyte activation it was found that the Hsp70 (DnaK) protein from *M. tuberculosis* was also able to stimulate human monocytes to produce a range of key chemokines important in the pathogenesis of tuberculosis (Wang et al. 2001). The mycobacterial protein bound to CD40. This appeared to be the same as reported for the human homologue (Becker et al. 2002). However, the binding site for CD40 on the human and *M. tuberculosis* Hsp70 proteins is different. Thus the human protein binds via the amino-terminal nucleotide-binding domain in its ADP state (Becker et al. 2002). In contrast, the *M. tuberculosis* Hsp70 protein uses its C-terminal peptide binding domain to bind to CD40 (Wang et al. 2002). This suggests that the use of the CD40 receptor by the bacterial and eukaryotic Hsp70 proteins is an example of convergent evolution. In addition to binding to CD40, the *M. tuberculosis* Hsp70 protein also binds to the chemokine receptor and HIV-co-receptor, CCR5 (Floto et al. 2006).

2.3.7 Hsp27

Thus far all the cell stress proteins described have had the ability to stimulate leukocyte cytokine synthesis. In 2000, Carol Miller-Graziano and co-workers described the ability of human Hsp27 to promote human monocyte production of the anti-inflammatory cytokine, IL-10, relative to the production of modest levels of the pro-inflammatory cytokine, TNFα (De et al. 2000). This protein also blocks, by a unique mechanism, the differentiation of monocytes into dendritic cells (Laudanski et al. 2007). What is particularly interesting, is the actions of macrophages within tumours. Such tumor-associated macrophages (TAMs) are known to have major effects on tumour progression by producing factors that promote angiogenesis, remodel tissue and inhibit anti-tumoural immune responses (Muhktar et al. 2011). Miller-Graziano has reported that breast tumours produce very high levels of Hsp27, much of which is secreted. When exposed to Hsp27 monocytes are induced to become macrophages with immunomodulatory actions, having low levels of classic macrophage activation markers (HLA-DR, CD86 etc). These cells induce severe unresponsiveness and anergy in T lymphocytes and have very low tumouricidal activity but marked pro-angiogenic activity. These are the classic markers of the TAM and important in tumour growth (Banerjee et al. 2011). Thus it would appear that Hsp27, by controlling monocyte activation, can be a pro-tumoural protein.

1971	**Identification of macrophage migration inhibitory factor (MIF)**
1977	**EPF discovered as immunosuppressive factor in pregnancy**
1989	**Thioredoxin found to be a secreted lymphocyte modulator**
1992	**Cyclophilin A identified as pro-inflammatory secreted macrophage product**
1992	***Legionella pneumophila* virulence factor, Mif, involved in invasion of macrophages, shown to be a peptidyl prolyl isomerase**
1993	**First bacterial chaperone, chaperonin 60, shown to be an inducer of cytokine synthesis**
1994	**EPF shown to be chaperonin 10**
1999	**MIF shown to be a chaperone**
1999	**Human Hsp60 shown to stimulate macrophage cytokine synthesis**
2000	**Human Hsp70 shown to stimulate macrophage cytokine synthesis**
2001	**Hsp27, first example of a monocyte-deactivating chaperone**
2001	**BiP, a human ER chaperone also found to have anti-inflammatory mode of action**

Fig. 2.2 Timeline of the discovery of the signalling activity of cell stress proteins

To provide the reader with a visual representation of the history of the cell stress protein as a signalling agonist a time line has been provided (Fig. 2.2).

2.4 Conclusions

To paraphrase an ancient saying. 'always expect the unexpected'. This is the situation in the field of cell stress proteins, with a growing number of such proteins exhibiting a myriad of biological actions over-and-above the facility to promote correct protein folding. These proteins can now be defined as moonlighting proteins—that is—proteins with more than a single unique function (Jeffery 1999). Currently around

16 human cell stress proteins and 4 bacterial proteins have been found to moonlight, largely as immune modulators with agonist properties. It is predicted that all molecular chaperones and protein folding catalysts will have the ability to be expressed on the surface of cells and/or to be secreted and thus to signal. This (the author postulates) allows the key homeostatic mechanism, the cell stress system, to broadcast to other cells and tissues that there is some form of stressor in the vicinity allowing nearby cells to prepare for it. It will be of interest to see if this hypothesis can survive a full-blown Popperian test.

References

Al-Abed Y, Van Patten S (2011) MIF as a disease target: ISO-1 as a proof-of-concept therapeutic. Future Med Chem 3:45–63

Asea A, Kraeft SK, Kurt-Jones EA, Stevenson MA, Chen LB, Finberg RW, Koo GC, Calderwood SK (2000) HSP70 stimulates cytokine production through a CD14-dependant pathway, demonstrating its dual role as a chaperone and cytokine. Nat Med 6:435–442

Asea A, Rehli M, Kabingu E, Boch JA, Bare O, Auron PE, Stevenson MA, Calderwood SK (2002) Novel signal transduction pathway utilized by extracellular HSP70: role of toll-like receptor (TLR) 2 and TLR4. J Biol Chem 277:15028–1534

Atanassov C, Pezennec L, d'Alayer J, Grollier G, Picard B, Fauchère JL (2002) Novel antigens of *Helicobacter pylori* correspond to ulcer-related antibody pattern of sera from infected patients. J Clin Microbiol 40:547–552

Banerjee S, Lin CF, Skinner KA, Schiffhauer LM, Peacock J, Hicks DG, Redmond EM, Morrow D, Huston A, Shayne M, Langstein HN, Miller-Graziano CL, Strickland J, O'Donoghue L, De AK (2011) Heat shock protein 27 differentiates tolerogenic macrophages that may support human breast cancer progression. Cancer Res 71:318–327

Bartfeld H, Atoynatan T (1971) Activity and properties of macrophage migration inhibitory factor produced by mixed lymphocyte cultures. Nat New Biol 230:246–247

Basak C, Pathak SK, Bhattacharyya A, Pathak S, Basu J, Kundu M (2005) The secreted peptidyl prolyl cis, trans-isomerase HP0175 of *Helicobacter pylori* induces apoptosis of gastric epithelial cells in a TLR4- and apoptosis signal-regulating kinase 1-dependent manner. J Immunol 174:5672–5680

Becker T, Hartl FU, Wieland F (2002) CD40, an extracellular receptor for binding and uptake of Hsp70-peptide complexes. J Cell Biol 158:1277–1285

Billich A, Winkler G, Aschauer H, Rot A, Peichl P (1997) Presence of cyclophilin A in synovial fluids of patients with rheumatoid arthritis. J Exp Med 185:975–980

Broadley SA, Vanags D, Williams B, Johnson B, Feeney D, Griffiths L, Shakib S, Brown G, Coulthard A, Mullins P, Kneebone C (2009) Results of a phase IIa clinical trial of an anti-inflammatory molecule, chaperonin 10, in multiple sclerosis. Mult Scler 15:329–336

Cavanagh AC, Morton H (1994) The purification of early-pregnancy factor to homogeneity from human platelets and identification as chaperonin 10. Eur J Biochem 222:551–560

Chang JW, Lee SH, Lu Y, Yoo YJ (2006) Transforming growth factor-beta1 induces the non-classical secretion of peroxiredoxin-I in A549 cells. Biochem Biophys Res Commun 345:118–123

Chen B, Guan D, Cui ZJ, Wang X, Shen X (2010) Thioredoxin 1 downregulates MCP-1 secretion and expression in human endothelial cells by suppressing nuclear translocation of activator protein 1 and redox factor-1. Am J Physiol Cell Physiol 298:C1170–C1179

Cheng CF, Fan J, Fedesco M, Guan S, Li Y, Bandyopadhyay B, Bright AM, Yerushalmi D, Liang M, Chen M, Han YP, Woodley DT, Li W (2008) Transforming growth factor alpha (TGFalpha)-stimulated secretion of HSP90 alpha: using the receptor LRP-1/CD91 to promote human skin cell migration against a TGFbeta-rich environment during wound healing. Mol Cell Biol 28:3344–3358

Chen W, Syldath U, Bellmann K, Burkart V, Kolb H (1999) Human 60-kDa heat-shock protein: a danger signal to the innate immune system. J Immunol 162:3212–3219

Cherepkova OA, Lyutova EM, Eronina TB, Gurvits BY (2006) Accelerated protein aggregation induced by macrophage migration inhibitory factor under heat stress conditions. Biochemistry (Mosc) 71:140–145

Clarke FM, Morton H, Clunie GJ (1978) Detection and separation of two serum factors responsible for depression of lymphocyte activity in pregnancy. Clin Exp Immunol 32:318–323

De AK, Kodys KM, Yeh BS, Miller-Graziano C (2000) Exaggerated human monocyte IL-10 concomitant to minimal TNF-alpha induction by heat-shock protein 27 (Hsp27) suggests Hsp27 is primarily an antiinflammatory stimulus. J Immunol 165:3951–3958

Dear JW, Simpson K, Nicolai MP, Catterson JH, Street J, Huizinga T, Craig D, Dhaliwal K, Webb S, Bateman DN, Webb DJ (2011) Cyclophilin A Is a damage-associated molecular pattern molecule that mediates acetaminophen-induced liver injury. J Immunol 187:3347–3352

Dobocan MC, Sadvakassova G, Congote LF (2009) Chaperonin 10 as an endothelial-derived differentiation factor: role of glycogen synthase kinase-3. J Cell Physiol 219:470–476

Engleberg NC, Carter C, Weber DR, Cianciotto NP, Eisenstein BI (1989) DNA sequence of *mip*, a *Legionella pneumophila* gene associated with macrophage infectivity. Infect Immun 57:1263–1270

Evdonin A, Kinev A, Tsupkina N, Guerriero V, Raynes DA, Medvedeva N (2009) Extracellular HspBP1 and Hsp72 synergistically activate epidermal growth factor receptor. Biol Cell 101:351–360

Fischer G, Bang H, Ludwig B, Mann K, Hacker J (1992) Mip protein of Legionella pneumophila exhibits peptidyl-prolyl-cis/trans isomerase (PPlase) activity. Mol Microbiol 6:1375–1383

Floto RA, MacAry PA, Boname JM, Mien TS, Kampmann B, Hair JR, Huey OS, Houben EN, Pieters J, Day C, Oehlmann W, Singh M, Smith KG, Lehner PJ (2006) Dendritic cell stimulation by mycobacterial Hsp70 is mediated through CCR5. Science 314:454–458

Friedland JS, Shattock R, Remick DG, Griffin GE (1993) Mycobacterial 65-kD heat shock protein induces release of proinflammatory cytokines from human monocytic cells. Clin Exp Immunol 91:58–62

Gangalum RK, Atanasov IC, Zhou ZH, Bhat SP (2010) AlphaB-crystallin is found in detergent-resistant membrane microdomains and is secreted via exosomes from human retinal pigment epithelial cells. J Biol Chem 286:3261–32699

Gordon S, Martinez FO (2010) Alternative activation of macrophages: mechanism and functions. Immunity 32:593–604

Göthel SF, Marahiel MA (1999) Peptidyl-prolyl cis-trans isomerases, a superfamily of ubiquitous folding catalysts. Cell Mol Life Sci 55:423–436

Gupta S, Knowlton AA (2007) HSP60 trafficking in adult cardiac myocytes: role of the exosomal pathway. Am J Physiol Heart Circ Physiol 292:H3052–3056

Hageman J, van Waarde MA, Zylicz A, Walerych D, Kampinga HH (2011) The diverse members of the mammalian HSP70 machine show distinct chaperone-like activities. Biochem J 435:127–142

Heine SJ, Olive D, Gao JL, Murphy PM, Bukrinsky MI, Constant SL (2011) Cyclophilin A cooperates with MIP-2 to augment neutrophil migration. J Inflamm Res 4:93–104

Henderson B, Pockley AG (2010) Molecular chaperones and protein-folding catalysts as intercellular signaling regulators in immunity and inflammation. J Leukoc Biol 88:445–462

Henderson B, Martin A (2011) Bacterial virulence in the moonlight: multitasking bacterial moonlighting proteins are virulence determinants in infectious disease. Infect Immun 79:3476–3491

Henderson B, Calderwood SK, Coates AR, Cohen I, van Eden W, Lehner T, Pockley AG (2010) Caught with their PAMPs down? The extracellular signalling actions of molecular chaperones are not due to microbial contaminants. Cell Stress Chaperones 15:123–141

Hickey TB, Thorson LM, Speert DP, Daffé M, Stokes RW (2009) *Mycobacterium tuberculosis* Cpn60.2 and DnaK are located on the bacterial surface, where Cpn60.2 facilitates efficient bacterial association with macrophages. Infect Immun 77:3389–3401

Hickey TB, Ziltener HJ, Speert DP, Stokes RW (2010) *Mycobacterium tuberculosis* employs Cpn60.2 as an adhesin that binds CD43 on the macrophage surface. Cell Microbiol 12:1634–1647

Hightower LE, Guidon PT (1989) Selective release from cultured mammalian cells of heat-shock (stress) proteins that resemble glia-axon transfer proteins. J Cell Physiol 138:257–266
Holland IB (2010) The extraordinary diversity of bacterial protein secretion mechanisms. Methods Mol Biol 619:1–20
Holmgren A, Lu J (2010) Thioredoxin and thioredoxin reductase: current research with special reference to human disease. Biochem Biophys Res Commun 396:120–124
Jeffery CJ (1999) Moonlighting proteins. Trends Biochem Sci 24:8–11
Jha R, Vardhan H, Bas S, Salhan S, Mittal A (2011) Chlamydia trachomatis heat shock proteins 60 and 10 induce apoptosis in endocervical epithelial cells. Inflamm Res 60:69–78
Kleemann R, Kapurniotu A, Frank RW, Gessner A, Mischke R, Flieger O, Jüttner S, Brunner H, Bernhagen J (1998) Disulfide analysis reveals a role for macrophage migration inhibitory factor (MIF) as thiol-protein oxidoreductase. J Mol Biol 280:85–102
Lancaster GI, Febbraio MA (2005) Exosome-dependent trafficking of HSP70: a novel secretory pathway for cellular stress proteins. J Biol Chem 280:23349–2355
Laskey RA, Honda BM, Mills AD, Finch JT (1978) Nucleosomes are assembled by an acidic protein which binds histones and transfers them to DNA. Nature 275:416–420
Laudanski K, De A, Miller-Graziano C (2007) Exogenous heat shock protein 27 uniquely blocks differentiation of monocytes to dendritic cells. Eur J Immunol 37:2812–2824
Lundemose AG, Kay JE, Pearce JH (1993) *Chlamydia trachomatis* Mip-like protein has peptidyl-prolyl cis/trans isomerase activity that is inhibited by FK506 and rapamycin and is implicated in initiation of chlamydial infection. Mol Microbiol 7:777–783
Mambula SS, Calderwood SK (2006) Heat shock protein 70 is secreted from tumor cells by a nonclassical pathway involving lysosomal endosomes. J Immunol 177:7849–7857
Morton H, Rolfe B, Clunie GJ (1977) An early pregnancy factor detected in human serum by the rosette inhibition test. Lancet 1:394–397
Mougiakakos D, Johansson CC, Jitschin R, Böttcher M, Kiessling R (2011) Increased thioredoxin-1 production in human naturally occurring regulatory T cells confers enhanced tolerance to oxidative stress. Blood 117:857–861
Mukhtar RA, Nseyo O, Campbell MJ, Esserman LJ (2011) Tumor-associated macrophages in breast cancer as potential biomarkers for new treatments and diagnostics. Expert Rev Mol Diagn 11:91–100
Nakamura H (2008) Extracellular functions of thioredoxin. Novartis Found Symp 291:184–192
Nakamura H, De Rosa SC, Yodoi J, Holmgren A, Ghezzi P, Herzenberg LA, Herzenberg LA (2001) Chronic elevation of plasma thioredoxin: inhibition of chemotaxis and curtailment of life expectancy in AIDS. Proc Natl Acad Sci U S A 98:2688–26893
Nickel W, Rabouille C (2009) Mechanisms of regulated unconventional protein secretion. Nat Rev Mol Cell Biol 10:148–155
Noelle RJ, Lawrence DA (1981) Modulation of T-cell function. II. Chemical basis for the involvement of cell surface thiol-reactive sites in control of T-cell proliferation. Cell Immunol 60:453–469
Noonan FP, Halliday WJ, Morton H, Clunie GJ (1979) Early pregnancy factor is immunosuppressive. Nature 278:649–651
Norville IH, Harmer NJ, Harding SV, Fischer G, Keith KE, Brown KA, Sarkar-Tyson M, Titball RW (2011) A *Burkholderia pseudomallei* Mip-like protein has rapamycin inhibitable peptidyl-prolyl isomerase activity and has pleiotropic effects on virulence. Infect Immun 79:4299–4307
Pathak SK, Basu S, Bhattacharyya A, Pathak S, Banerjee A, Basu J, Kundu M (2006) TLR4-dependent NF-kappaB activation and mitogen- and stress-activated protein kinase 1-triggered phosphorylation events are central to *Helicobacter pylori* peptidyl prolyl cis-, trans-isomerase (HP0175)-mediated induction of IL-6 release from macrophages. J Immunol 177:7950–7958
Peetermans WE, Raats CJ, Langermans JA, van Furth R (1994) Mycobacterial heat-shock protein 65 induces proinflammatory cytokines but does not activate human mononuclear phagocytes. Scand J Immunol 39:613–617

Potolicchio I, Santambrogio L, Strominger JL (2003) Molecular interaction and enzymatic activity of macrophage migration inhibitory factor with immunorelevant peptides. J Biol Chem 278:30889–30895

Ragno S, Winrow VR, Mascagni P, Lucietto P, Di Pierro F, Morris CJ, Blake DR (1996) A synthetic 10-kD heat shock protein (hsp10) from *Mycobacterium tuberculosis* modulates adjuvant arthritis. Clin Exp Immunol 103:384–390

Randhawa AK, Ziltener HJ, Stokes RW (2008) CD43 controls the intracellular growth of *Mycobacterium tuberculosis* through the induction of TNF-alpha-mediated apoptosis. Cell Microbiol 10:2105–2117

Ritossa F (1962) A new puffing pattern induced by temperature shock and DNP in drosophila. Cell Mol Life Sci 18:571–573

Rohde K, Yates RM, Purdy GE, Russell DG (2007) *Mycobacterium tuberculosis* and the environment within the phagosome. Immunol Rev 219:37–54

Rubartelli A, Cozzolino F, Talio M, Sitia R (1990) A novel secretory pathway for interleukin-1 beta, a protein lacking a signal sequence. EMBO J 9:1503–1510

Rubartelli A, Bajetto A, Allavena G, Wollman E, Sitia R (1992) Secretion of thioredoxin by normal and neoplastic cells through a leaderless secretory pathway. J Biol Chem 267:24161–24164

Shachar I, Haran M (2011) The secret second life of an innocent chaperone: the story of CD74 and B cell/chronic lymphocytic leukemia cell survival. Leuk Lymphoma 52:1446–1454

Schiene-Fischer C, Aumüller T, Fischer G (2011) Peptide bond cis/trans Isomerases: a biocatalysis perspective of conformational dynamics in proteins. Top Curr Chem [Epub ahead of print]

Sherry B, Yarlett N, Strupp A, Cerami A (1992) Identification of cyclophilin as a proinflammatory secretory product of lipopolysaccharide-activated macrophages. Proc Natl Acad Sci U S A 89:3511–3515

Song F, Zhang X, Ren XB, Zhu P, Xu J, Wang L, Li YF, Zhong N, Ru Q, Zhang DW, Jiang JL, Xia B, Chen ZN (2011) Cyclophilin A (CyPA) induces chemotaxis independent of its peptidylprolyl cis-trans isomerase activity: direct binding between CyPA and the ectodomain of CD147. J Biol Chem 286:8197–8203

Suzuki J, Jin Z-G, Meoli DF, Matoba T, Berk BC (2006) Cyclophilin A is secreted by a vesicular pathway in vascular smooth muscle cells. Circ Res 98:811–817

Tagaya Y, Maeda Y, Mitsui A, Kondo N, Matsui H, Hamuro J, Brown N, Arai K, Yokota T, Wakasugi H et al (1989) ATL-derived factor (ADF), an IL-2 receptor/Tac inducer homologous to thioredoxin; possible involvement of dithiol-reduction in the IL-2 receptor induction. EMBO J 8:757–764

Tassi S, Carta S, Vené R, Delfino L, Ciriolo MR, Rubartelli A (2009) Pathogen-induced interleukin-1beta processing and secretion is regulated by a biphasic redox response. J Immunol 183:1456–1462

Tegeder I, Schumacher A, John S, Geiger H, Geisslinger G, Bang H, Brune K (1997) Elevated serum cyclophilin levels in patients with severe sepsis. J Clin Immunol 17:380–386

Thériault JR, Mambula SS, Sawamura T, Stevenson MA, Calderwood SK (2005) Extracellular HSP70 binding to surface receptors present on antigen presenting cells and endothelial/epithelial cells. FEBS Lett 579:1951–1960

Thériault JR, Adachi H, Calderwood SK (2006) Role of scavenger receptors in the binding and internalization of heat shock protein 70. J Immunol 177:8604–8611

Tsan MF, Gao B (2009) Heat shock proteins and immune system. J Leukoc Biol 85:905–910

Tytell M, Greenberg SG, Lasek RJ (1986) Heat shock-like protein is transferred from glia to axon. Brain Res 363:61–64

Vanags D, Williams B, Johnson B, Hall S, Nash P, Taylor A, Weiss J, Feeney D (2006) Therapeutic efficacy and safety of chaperonin 10 in patients with rheumatoid arthritis: a double-blind randomised trial. Lancet 368:855–863

Verdegaal EME, Zegveld ST, van Furth R (1996) Heat shock protein 65 induces CD62e, CD106 and CD54 on cultured human endothelial cells and increases their adhesiveness for monocytes and granulocytes. J Immunol 157:369–376

Wakasugi N, Tagaya Y, Wakasugi H, Mitsui A, Maeda M, Yodoi J, Tursz T (1990) Adult T-cell leukemia-derived factor/thioredoxin, produced by both human T-lymphotropic virus type I- and Epstein-Barr virus-transformed lymphocytes, acts as an autocrine growth factor and synergizes with interleukin 1 and interleukin 2. Proc Natl Acad Sci U S A 87:8282–8286

Wang Y, Kelly CG, Karttunen JT, Whittall T, Lehner PJ, Duncan L, MacAry P, Younson JS, Singh M, Oehlmann W, Cheng G, Bergmeier L, Lehner T (2001) CD40 is a cellular receptor mediating mycobacterial heat shock protein 70 stimulation of CC-chemokines. Immunity 15:971–983

Wang Y, Kelly CG, Singh M, McGowan EG, Carrara AS, Bergmeier LA, Lehner T (2002) Stimulation of Th1-polarizing cytokines, C-C chemokines, maturation of dendritic cells, and adjuvant function by the peptide binding fragment of heat shock protein 70. J Immunol 169:2422–2429

Xiao G, Chung TF, Pyun HY, Fine RE, Johnson RJ (1999) KDEL proteins are found on the surface of NG108-15 cells. Brain Res Mol Brain Res 72:121–128

Yurchenko V, O'Connor M, Dai WW, Guo H, Toole B, Sherry B, Bukrinsky M (2001) CD147 is a signaling receptor for cyclophilin B. Biochem Biophys Res Commun 288:786–788

Yurchenko V, Constant S, Eisenmesser E, Bukrinsky M (2010) Cyclophilin-CD147 interactions: a new target for anti-inflammatory therapeutics. Clin Exp Immunol 160:305–317

Zhan R, Leng X, Liu X, Wang X, Gong J, Yan L, Wang L, Wang Y, Wang X, Qian LJ (2009) Heat shock protein 70 is secreted from endothelial cells by a non-classical pathway involving exosomes. Biochem Biophys Res Commun 387:229–233

Zhang B, Walsh MD, Nguyen KB, Hillyard NC, Cavanagh AC, McCombe PA, Morton H (2003) Early pregnancy factor treatment suppresses the inflammatory response and adhesion molecule expression in the spinal cord of SJL/J mice with experimental autoimmune encephalomyelitis and the delayed-type hypersensitivity reaction to trinitrochlorobenzene in normal BALB/c mice. J Neurol Sci 212:37–46

Zhou Z, Wu Y, Chen L, Liu L, Chen H, Li Z, Chen C (2011) Heat shock protein 10 of Chlamydophila pneumoniae induces proinflammatory cytokines through Toll-like receptor (TLR) 2 and TLR4 in human monocytes THP-1. In Vitro Cell Dev Biol Anim 47:541–549

[illegible]

Chapter 3
Molecular Chaperones and Protein-Folding Catalysts in Biological Fluids

A. Graham Pockley

Abstract Although yet to be fully accepted by the wider scientific community, it is clear that heat shock (stress) proteins can be released from a number of different cell types via mechanisms that do not involve overt cell death. These proteins have now been found in a number of biological fluids, in which they have the potential to elicit a range of essential functions. This chapter summarises elements of the current literature relating to the presence of heat shock proteins that have been found in biological fluids and highlights the role(s) of these, as appropriate. Although much progress has been made, we need to better structure our studies in order to provide more informative insights into the functions of these proteins and perhaps consider their presence in a broader context which involves the application of Systems Biology principles. It is likely that heat shock protein profiling will provide more insight into the physiological significance of these multifunctional proteins, the sequence conservation of which illustrates their importance to organismal regulation and homeostasis.

Although the initial observations that heat shock (stress) proteins can be released from viable cells were made some time ago (see full discussion of this topic in Chap. 1), the concept that these proteins can have functions beyond the classically considered intracellular chaperone activities has taken some time to be established. Some might argue that it has yet to be definitively proven, and the aim of this book is to consolidate the available literature and knowledge in this area in order to strengthen this concept and allow the field to develop.

The pioneering work of Tytell and colleagues, which demonstrated the transfer of glia-axon transfer proteins, including members of the 70 and 100 KDa heat shock protein families, from adjacent glial cells into the squid giant axon (Tytell et al. 1986) and Hightower and Guidon, which showed that heat treatment induced the release of the constitutively expressed and inducible members of 70 kDa heat shock protein from rat embryonal cells (Hightower and Guidon 1989) in the absence of necrosis, suggested that these proteins might have additional extracellular functions. Larry Hightower, one of the early pioneers in the field, has provided an illuminating historical context to the discovery of the release of heat shock proteins and the

A. G. Pockley(✉)
John van Geest Cancer Research Centre, Nottingham Trent University,
Clifton Lane, Nottingham, NG11 8NS, UK
e-mail: graham.pockley@ntu.ac.uk

B. Henderson, A. G. Pockley (eds.), *Cellular Trafficking of Cell Stress Proteins in Health and Disease*, Heat Shock Proteins 6, DOI 10.1007/978-94-007-4740-1_3,

frustrations that accompanied this discovery at the beginning of this volume (see Chap. 1).

The key barrier then, and to a certain extent today, relates to the fact that the mechanisms that are involved in the release of heat shock proteins from viable cells have yet to be definitively described. This is especially relevant, as these proteins do not possess the N-terminal signal peptide sequences that are typically required for secretion. However, 'non-classical' secretion of proteins that lack such sequences has been observed for a number of proteins such as fibroblast growth factors 1 and 2, IL-1, as well as viral proteins. The mechanisms that are involved in such secretory events have been considered elsewhere (Chimini and Rubartelli 2005). The paucity of information in this area should not therefore cast doubt on the validity of the findings that such proteins are released/secreted by viable cells.

Although the heat shock protein secretory mechanisms are not yet fully elucidated, active processes that involve a number of potential pathways such as lipids (Broquet et al. 2003), exosomes (Bausero et al. 2005; Gastpar et al. 2004; Lancaster and Febbraio 2005a) and secretory lysosomal endosomes (Mambula and Calderwood 2006) have been indicated. The spontaneous release of soluble Hsp70 by viable human colon and pancreatic carcinoma cells tumour sublines differentially expressing Hsp70 on their cell surface is low, whereas detergent-soluble vesicles actively released by tumours contain high amounts of Hsp70/Bag-4 and Hsp70/Hsp40 (Gastpar et al. 2005). The biochemical and biophysical properties of these vesicles (as defined by density, acetylcholine esterase activity, protein composition) indicate that they are exosomes (Gastpar et al. 2005). The mechanisms that are potentially involved in the secretion/release of heat shock proteins from cells have been considered elsewhere in this volume (Chaps. 1, 5, 6, 7) and so will not be expanded upon here.

Another aspect of this area which has, to some extent at least, inhibited the wider adoption of the concept that heat shock proteins might act as physiologically-relevant intercellular signalling molecules are the differing opinions of the scientific disciplines. These have been clearly illustrated by Larry Hightower's historical experience which he highlights elsewhere in this volume. He recalls that an immunologist 'did not care how the Hsps were released', but was 'interested in what they did afterward', whereas a prominent biochemist and cell biologist said that he 'would never convince colleagues of the biological relevance of extracellular Hsps until he had determined the molecular mechanism of their release from cells'.

It is also necessary to accept that proteins might have multiple functions, the manifestations of which are dictated by the context in which the proteins are encountered. For instance, can proteins such as heat shock proteins exhibit distinct profiles of physiological activities when in the intracellular and extracellular environments? If so, and this indeed appears to be the case, then this would argue against the concept of 'one protein, one function'. Although in some quarters this remains a difficult concept to accept, it is now some time ago that Campbell and Scanes first proposed the term "protein moonlighting" to describe the capacity of certain proteins to exhibit more than one biological function, specifically the apparent immunological functions of "endocrine peptides" (Campbell and Scanes 1995). A number of prokaryote and eukaryote proteins have since been shown to exhibit such moonlighting functions, and this concept has been expanded upon by a number of proponents (Jeffery 1999, 2003, 2009; Henderson and Pockley 2010).

The author's own experience in this area relates to his initial identification of Hsp60 and Hsp70 in the peripheral circulation of normal individuals (Pockley et al. 1998, 1999). He too experienced significant difficulties when it came to publishing this work, despite the seminal studies that had already 'established' the principle that heat shock proteins could be released as part of a physiological mechanism by the Tytell (Tytell et al. 1986) and Hightower (Hightower and Guidon 1989) laboratories some years earlier. Although probably not fully appreciated at the time, the concept that heat shock proteins can exist in the peripheral circulation had already been established by a study which had reported the presence of a protein ('early pregnancy factor') in the serum of women in the first trimester of pregnancy in 1977 (Morton et al. 1977—see also Chap. 2). This protein was demonstrated to have immunosuppressive properties 2 years later (Noonan et al. 1979) and was identified as being heat shock protein 10 (Hsp10) in 1994 (Cavanagh and Morton 1994 see also Chap. 2). It is therefore the case that Hsp10, a 10 kDa monomer which caps the Hsp60 oligomer and facilitates protein folding (Richardson et al. 1998) is also present at low levels in non-pregnant individuals (Shamaei-Tousi et al. 2007). Hsp10 inhibits the secretion of several inflammatory mediators (Johnson et al. 2005) and can act as an immunosuppressive protein with the capacity to attenuate a variety of human inflammatory diseases (Broadley et al. 2009; Van Eden 2008; Vanags et al. 2006; Williams et al. 2008). The finding that that circulating levels of Hsp10 in patients with periodontal disease are lower than in matched, disease-free, controls and that levels only return to normal after effective therapy suggest that circulating Hsp10 levels are controlled by local levels of inflammation (Shamaei-Tousi et al. 2007). This molecular chaperone therefore appears to be a homeostatic controller of inflammation, in addition to being an integral component of the intracellular molecular chaperone machinery.

Returning to Hsp60 and Hsp70, the discovery of these proteins in the peripheral circulation of overtly normal individuals led to a certain degree of confusion, as these proteins had become considered as being pro-inflammatory molecules when present in the extracellular environment. Indeed, one of the major issues for investigators studying the immunobiology of extracellular stress proteins is the apparently contradictory evidence which indicates both pro- and anti-inflammatory roles for these proteins. It remains the case that the immunological properties of these proteins continue to be discussed in isolation and the time has come for the community to adopt a more systems biology approach to extracellular heat shock proteins in order to better reflect their physiological context and roles. Many studies indicate pro-inflammatory properties for Hsp60 and Hsp70 in their interactions with monocytes, macrophages and dendritic cells (DCs) (Asea et al. 2000, 2002; Chen et al. 1999; Flohé et al. 2003; Kol et al. 2000). Although some speculation that at least some of these inflammatory effects result from the presence of contaminating endotoxin in the recombinant preparations of heat shock proteins that have been used (Gao and Tsan 2003a, b; Gao and Tsan 2004; Tsan and Gao 2009), there is also much evidence to argue against this being the universal explanation for these effects. These have been reviewed elsewhere (Henderson et al. 2010; Henderson and Pockley 2010). It behoves all of us working in this area to ensure that our reagents and experimental design(s) are beyond question when it comes to analyzing and interpreting our data.

There is also a burgeoning body of literature to indicate that Hsp60 and Hsp70 can have profound anti-inflammatory effects. Relatively historic data demonstrate that the induction of T cell reactivity to self Hsp60 and self Hsp70 promotes the development of Th2 type $CD4^+$ T cells producing the regulatory cytokines IL-4 and IL-10 and down-regulates disease in a number of experimental models of inflammatory disease (Kingston et al. 1996; Quintana et al. 2004; Tanaka et al. 1999; Van Eden et al. 2005; Wendling et al. 2000). Furthermore, DNA vaccines encoding for these proteins inhibit experimental arthritis and diabetes in non-obese diabetic mice (Quintana et al. 2003, 2004). The recognition of conserved (self) epitopes on these highly conserved molecules dominantly down-regulates the inflammatory capacity of the non-conserved (non-self) epitopes (Anderton et al. 1995) and the net outcome of any immune response therefore hinges on the relative strengths of these antagonistic events (reviewed in (Pockley et al. 2008). Human Hsp60 can act as a co-stimulator and activator of $CD4^+CD25^+$ regulatory T cell populations by interacting with Toll-like receptor 2 (TLR2) (Zanin-Zhorov et al. 2006). The treatment of such cells with Hsp60 enhances their ability to regulate the activities of $CD8^+$ T cell populations via mechanisms that involve direct cell-cell contact and the secretion of the immunoregulatory cytokines IL-10 and TGF-β (Zanin-Zhorov et al. 2006). It remains to be demonstrated that other heat shock proteins can elicit similar effects, although evidence that Hsp70 can elicit similar anti-inflammatory effects is highlighted later. The anti-inflammatory potential of Hsp60 and Hsp60-derived peptides has also been demonstrated in studies that have used these to modulate the rejection of murine skin allografts (Birk et al. 1999; Luna et al. 2007). The interactions of Hsp60 with the innate and adaptive immune systems and their immunoregulatory consequences have been reviewed and considered by Quintana and Cohen (Quintana and Cohen 2011 see also Chap. 3).

The availability of commercial and in-house assays for measuring heat shock proteins has led to a dramatic increase in the number of studies that have measured circulating levels of these proteins in various disease states. Our own initial studies and those of others demonstrated an association of Hsp60 levels with the presence of early atherosclerosis (Pockley et al. 2000; Xu et al. 2000), although the precise relationship between the two has yet to be definitely defined and it is always difficult to distinguish 'cause' from 'effect'. For example, is atherosclerosis driving circulating Hsp60 levels or is Hsp60 driving atherosclerosis? This is a problem which permeates this field and more work needs to be undertaken in order to better understand the physiological roles of circulating heat shock proteins and their impact on disease processes. The concept that Hsp60 drives atherosclerosis via an autoimmune type response has been recently reviewed (Grundtman et al. 2011; Grundtman and Wick 2011). In addition to its apparent involvement in the triggering and driving of cardiovascular disease, it should also be noted that the induction of anti-Hsp60 immunity can also attenuate cardiovascular disease (Li et al. 2011a; Maron et al. 2002; Van Puijvelde et al. 2007).

Currently, the primary focus of studies which measure circulating heat shock proteins is the identification of these as potential biomarkers of disease. In this regard, myocardial infarction has been shown to increase circulating levels of Hsp60 in an experimental mouse model (Li et al. 2011b) and elevated Hsp60 to predict adverse events after acute myocardial infarction (Novo et al. 2011). The potential of Hsp60 as a biomarker of disease has also been illustrated by the findings that levels correlate with the time required for remission from flare-ups in patients with juvenile idiopathic arthritis (Wu et al. 2011). Hsp60 has also been reported to be present in the saliva of 10 % of control and 93 % of patients with type 2 diabetes (Yuan et al. 2011) and in patients with periodontitis (Alfakry et al. 2011; Rizzo et al. 2012). The literature in this and associated areas is expanding at a rapid rate and the author leaves the reader to track down the work in his/her specific area. An interesting recent development has been the identification of elevated levels of Hsp60 in the circulation of patients with colorectal cancer and the proposition that be a potential marker for this disease (Hamelin et al. 2011).

The other stress proteins for which there is an established literature relating to their presence in biological fluids are members of the 70 kDa family of molecules, specifically the constitutive member (Hsc70, HSPA8, (Kampinga et al. 2009), the inducible form (Hsp70, HSPA1A) and glucose-regulated protein 78 (grp78, Binding Immunoglobulin Protein (BiP), HspA5). As indicated previously, Hsc70 and Hsp70 were the proteins that were the focus of the initial studies by Tytell (Tytell et al. 1986) and Hightower (Hightower and Guidon 1989). Michael Tytell first reported the transfer of glia-axon transfer proteins, which include Hsp70, Hsc70 and Hsp100, from adjacent glial cells into the squid giant axon (Tytell et al. 1986). This finding prompted the suggestion that the release of such proteins might be a mechanism by which glial cells, which are capable of generating effective stress protein-mediated resistance to physical and metabolic insults, can protect adjacent neuronal cells which exhibit a deficient response to stress.

Stress proteins have been reported to be released from a wide range of viable cells such as human islet cells, cultured rat cortical astrocytes, a human neuroblastoma cell line, a human keratinocyte-derived cell line, cultured vascular smooth muscle cells, murine and human prostate cancer cells and human peripheral blood mononuclear cells, to highlight but a few (Frostegård and Pockley 2005) and the reader is referred to the current literature for the latest developments. Hsc70 has also been reported to be released from leukaemic cells (Barreto et al. 2003).

Remaining with Hsc70, it would appear that this molecule might play a role in, arguably, the most important biological process for the survival of species, namely reproduction. Proteomic analysis of porcine oviductal fluid has revealed that epithelial cells in the oviductal lumen secrete a number of molecules in response to the presence of spermatozoa, most notable of which are heat shock (stress) proteins (HSPs) (Georgiou et al. 2005). Heat shock proteins have also been identified in soluble fractions of pig and cow oviductal apical plasma membranes (sAPM) and in the human apical epithelium (Boilard et al. 2004; Lachance et al. 2007; Lloyd et al. 2009). These are potentially important findings, as the oviduct and oviductal sperm storage play key roles in reproduction by providing a secure reservoir in which spermatozoa can attain full fertilizing properties. Hsc70 appears to interact with components of the sperm cell surface membrane (Boilard et al. 2004; Lloyd et al. 2009) and exposure to Hsc70 prolongs the survival of boar, bull and ram sperm (Elliott et al. 2009; Lloyd et al. 2009). Ongoing studies are attempting to elucidate the mechanisms that are involved in these protective effects and their potential impact of reproductive success and potential.

As indicated previously, the demonstration by Hightower and Guidon that heat treatment broadened the spectrum of proteins released from cultured rat embryo cells, from a small set of proteins including the constitutively-expressed member of the 70 kDa family of molecules, Hsc70, to include its inducible counterpart, Hsp70, and Hsp110 extended the profile of heat shock proteins that might have a physiological role in the extracellular environment (Hightower and Guidon 1989). Subsequent work has demonstrated that exogenous Hsp70 has also been shown to have a range of cytoprotective properties, although in some instances the purity of the preparations that were being investigated is unclear and the precise contribution of Hsp70 and Hsc70 to the reported effects must be treated with some degree of caution. Notwithstanding this, these studies have clearly demonstrated that these proteins can protect heat-stressed cynomolgus macaque aortic cells (Johnson et al. 1990) and serum-deprived rabbit arterial smooth muscle cells (Johnson and Tytell 1993) from cytotoxic damage, the latter by a mechanism which involves cell association, but not internalisation. Although the mechanisms accounting for this cytoprotective effect remain unclear, the fact that extracellular Hsp70 can increase intracellular Hsp70 levels might be a contributing factor (Berberian et al. 1990). Extracellular Hsp70 has also been shown to have a number of neuroprotective properties and a capacity to inhibit motor and sensory neuron degeneration (Robinson et al. 2005; Tidwell et al. 2004; Tytell 2005).

Although the literature is not as extensive, Hsp70 has also been shown to have a number of anti-inflammatory effects, including the down-regulation of inflammatory disease in experimental models by mechanisms which involve the generation of Th2 cells producing the regulatory cytokine IL-10 (Kingston et al. 1996; Tanaka et al. 1999; Van Eden et al. 2005; Wendling et al. 2000). Mycobacterial Hsp70 inhibits the maturation of DCs and induces their production of IL-10 (Detanico et al. 2004) and has been shown to prolong experimental melanoma and regular skin allograft models via a mechanism which is dependent on $CD4^+CD25^+$ regulatory T cell populations (Borges et al. 2010).

The literature relating to the presence of Hsp70 in the peripheral circulation broadly tracks with that which relates to the presence of Hsp60 in the peripheral circulation (see Chap. 18 for more details). A number of investigators have used their own in-house or commercial assays to measure the levels of this protein in a number of circumstances. The author's first demonstration that Hsp70 is present in the circulation of normal, healthy volunteers (Pockley et al. 1998) was followed up by studies demonstrating that circulating levels of Hsp70 (and indeed Hsp60) decline with age (Rea et al. 2001). Interestingly, the circulating form of Hsp70 appears to protect against atherosclerosis, at least in individuals with established hypertension

(Pockley et al. 2003) and Hsp70 has been associated with a lowered risk of cardiovascular disease (Zhu et al. 2003). The finding that serum Hsp70 levels decline with age has been confirmed by the Asea laboratory (Terry et al. 2006) and is consistent with the observations that ageing cells are at a reduced capacity to respond to stress and synthesize heat shock proteins (Faassen et al. 1989; Fargnoli et al. 1990; Heydari et al. 1995; Nitta et al. 1994; Pahlavani et al. 1988). Elevated circulating levels of Hsp70 have also been associated with infection and inflammation (Molvarec et al. 2009; Njemini et al. 2003, 2004) and increased Hsp70 levels in the serum and urine have been reported in patients with chronic kidney disease (Lebherz-Eichinger et al. 2012). Circulating levels of Hsp70 also appear to be associated with normal and pathological pregnancies (Molvarec et al. 2007; Molvarec et al. 2010; Tamasi et al. 2010). In addition to the above, circulating Hsp70 levels are increased in response to exercise (Lancaster and Febbraio 2005b) and also experimentally-induced psychological stress (Fleshner et al. 2004). This latter finding might reflect a common 'stress' response in that during the course of a study investigating levels of circulating Hsp60 in the plasma of 229 healthy British civil servants taking part in the Whitehall II study, a prospective study which aimed to identify risk factors for coronary heart disease (Marmot et al. 1991), Henderson and colleagues identified a significant association between elevated levels of Hsp60, low socioeconomic status and social isolation in males and females, as well as with psychological distress in women (Lewthwaite et al. 2002). The literature relating to circulating Hsp70 levels in health, ageing and disease has recently been reviewed elsewhere (Njemini et al. 2011).

Although Hsp70 has been proposed as a novel biomarker for the early detection of pancreatic cancer (Dutta et al. 2012) or a favourable outcome following radiofrequency ablation of tumors (Haen et al. 2011), the specificity of such measurements might be questionable given the number of other factors that can influence circulating Hsp70 levels.

As an advisory note, care must be taken when comparing and contrasting levels of Hsp70 in extracellular fluids that have been determined using different assays, as in-house and commercial assays, for reasons that remain unclear, produce markedly different data (Njemini et al. 2005). This finding is also the author's unpublished observation.

Before concluding, it is important to make reference to another stress protein which was originally thought to be an autoantigen which was involved in the triggering and progression of autoimmune disease, but which has since been characterized as a potent immunomodulatory molecule which has clinical potential, namely glucose-regulated protein 78 (grp78, Binding Immunoglobulin Protein, BiP) (Shields et al. 2011). BiP is a fine example of a moonlighting protein. It was identified as being essential for the assembly of immunoglobulin molecules (Haas and Wabl 1983) and is required for the translocation of nascent polypeptides across the endoplasmic reticulum membrane and protects cells against ER stress (Gething 1999). It can also be expressed on the cell surface and acts as regulator of coagulation (Bhattacharjee et al. 2005) and cell proliferation (Davidson et al. 2005; Misra et al. 2006). This is reviewed in much more detail in Chap. 14. In addition to these, it has been shown to be a potent immunoregulator (Shields et al. 2011). BiP is present in the circulation of healthy individuals and at lower levels in patients with rheumatoid arthritis (Shields et al. 2011). It is also found in synovial and oviductal fluid (Corrigall et al. 2004; Marin-Briggiler et al. 2010). In contrast to the heat shock proteins that have been considered above, the secretion of BiP is likely to be via a classic route as it possesses the C-terminus ER retention signal (lysine, aspartic acid, glutamic acid, leucine (KDEL) amino acid sequence) which is common to proteins that reside in the ER. The multiple activities of BiP and its potential as a therapeutic agent for the management of inflammatory disease have been eloquently and comprehensively considered elsewhere and so will not be expanded upon here.

In summary, it is clear that many, if not all, of the stress proteins that are released from cells under normal physiological conditions could be considered as moonlighting proteins as they appear to possess a range of biological functions, the nature of which depend on the context in which they are encountered. This chapter has focussed attention on those proteins that are known to be secreted by cells and therefore to be present under normal circumstances in biological fluids. However, it should not be forgotten that there are a number of other heat shock (stress) proteins that are not as well established as being in the extracellular environment under normal conditions which have also been shown to have contrasting effects. A good example of such proteins is glucose-regulated protein 94 (gp96, HSPC4). Gp96 is a 94–96 kDa member of the Hsp90 family of molecular chaperones/stress proteins which resides within the lumen of the endoplasmic reticulum. In addition to being an intracellular chaperone (Gething and Sambrook 1992; Young et al. 1993), the administration of gp96, which has been purified from tumors, has been shown to induce tumor-specific cytolytic T cells and a protective tumor-specific immunity which is defined by peptides with which the administered gp96 is associated (Binder and Srivastava 2005; Chandawarkar et al. 1999; Udono and Srivastava 1994). In contrast, no protective effect is observed when high doses ($2 \times 10\,\mu$g intradermally) of tumor-derived gp96 are administered to mice (Chandawarkar et al. 1999). Furthermore, appropriate doses of gp96 which has been purified from normal liver can suppress the onset of diabetes in non-obese diabetic mice and myelin basic protein- or proteolipid protein-induced autoimmune encephalomyelitis (EAE) in SJL mice (Chandawarkar et al. 2004), as well as prolonging the survival of murine skin allografts (Kovalchin et al. 2006) and rat cardiac allografts (Slack et al. 2007). The mechanisms that underlie these effects might involve the induction, activation and/or recruitment of as yet unidentified immunoregulatory T cell populations (Chandawarkar et al. 2004; Kovalchin et al. 2006). In our hands, gp96 could not be shown to be an activator of DCs, but did appear to activate $CD3^+$ T cells *in vitro* (Mirza et al. 2006), and, contrastingly, lead to a state of peripheral T cell hyporesponsiveness following *in vivo* administration to rats bearing cardiac transplants (Slack et al. 2007).

It is clear that heat shock proteins can be released into the extracellular environment in order to elicit a range of, one presumes, essential physiological functions given the high degree of phylogenetic conservation which is a feature of these proteins. It is essential that we continue to promote this field and via sound scientific endeavor convince the wider scientific community that this is an area which should provide a wealth of valuable and clinically-relevant leads that can be developed into effective therapeutic strategies. We need to take a broader view of the biology of these proteins and ensure that they and their functions are considered in the context of the Systems Biology of the organism rather than in artificial isolation.

References

Alfakry H, Paju S, Sinisalo J et al (2011) Periodontopathogen- and host-derived immune response in acute coronary syndrome. Scand J Immunol 74:383–389

Anderton SM, Van Der Zee R, Prakken B, Noordzij A, Van Eden W (1995) Activation of T cells recognizing self 60-kD heat shock protein can protect against experimental arthritis. J Exp Med 181:943–952

Asea A, Kraeft S-K, Kurt-Jones EA et al (2000) Hsp70 stimulates cytokine production through a CD14-dependent pathway, demonstrating its dual role as a chaperone and cytokine. Nature Med 6:435–442

Asea A, Rehli M, Kabingu E et al (2002) Novel signal transduction pathway utilized by extracellular HSP70. Role of Toll-like receptor (TLR) 2 and TLR4. J Biol Chem 277:15028–15034

Barreto A, Gonzalez JM, Kabingu E, Asea A, Fiorentino S (2003) Stress-induced release of HSC70 from human tumors. Cell Immunol 222:97–104

Bausero MA, Gastpar R, Multhoff G, Asea A (2005) Alternative mechanism by which IFN-γ enhances tumor recognition: active release of heat shock protein 72. J Immunol 175:2900–2912

Berberian P, Johnson A, Bond M (1990) Exogenous 70kD heat shock protein increases survival of normal and atheromatous arterial cells. FASEB J 4:A1031

Bhattacharjee G, Ahamed J, Pedersen B et al (2005) Regulation of tissue factor–mediated initiation of the coagulation cascade by cell surface grp78. Arterioscler Thromb Vasc Biol 25:1737–1743

Binder RJ, Srivastava PK (2005) Peptides chaperoned by heat-shock proteins are a necessary and sufficient source of antigen in the cross-priming of $CD8^+$ T cells. Nat Immunol 6:593–599

Birk OS, Gur SL, Elias D et al (1999) The 60-kDa heat shock protein modulates allograft rejection. Proc Nat Acad Sci U S A 96:5159–5163

Boilard M, Reyes-Moreno C, Lachance C et al (2004) Localization of the chaperone proteins GRP78 and HSP60 on the luminal surface of bovine oviduct epithelial cells and their association with spermatozoa. Biol Reprod 71:1879–1889

Borges TJ, Porto BN, Teixeira CA et al (2010) Prolonged survival of allografts induced by mycobacterial Hsp70 is dependent on $CD4^+CD25^+$ regulatory T cells. PLoS One 5:e14264

Broadley SA, Vanags D, Williams B et al (2009) Results of a phase IIa clinical trial of an anti-inflammatory molecule, chaperonin 10, in multiple sclerosis. Multiple Sclerosis 15:329–336

Broquet AH, Thomas G, Masliah J, Trugnan G, Bachelet M (2003) Expression of the molecular chaperone Hsp70 in detergent-resistant microdomains correlates with its membrane delivery and release. J Biol Chem 278:21601–21606

Campbell RM, Scanes CG (1995) Endocrine peptides 'moonlighting' as immune modulators: roles for somatostatin and GH-releasing factor. J Endocrinol 147:383–396

Cavanagh AC, Morton H (1994) The purification of early-pregnancy factor to homogeneity from human platelets and identification as chaperonin 10. Eur J Biochem 222:551–560

Chandawarkar RY, Wagh MS, Kovalchin JT, Srivastava P (2004) Immune modulation with high-dose heat-shock protein gp96: therapy of murine autoimmune diabetes and encephalomyelitis. Int Immunol 16:615–624

Chandawarkar RY, Wagh MS, Srivastava PK (1999) The dual nature of specific immunological activity of tumor-derived gp96 preparations. J Exp Med 189:1437–1442

Chen W, Syldath U, Bellmann K, Burkart V, Kold H (1999) Human 60-kDa heat-shock protein: a danger signal to the innate immune system. J Immunol 162:3212–3219

Chimini C, Rubartelli A (2005) Novel pathways of protein secretion. In: Henderson B, Pockley AG (eds) Molecular chaperones and cell signalling. Cambridge University Press, New York, pp 45–60

Corrigall VM, Bodman-Smith MD, Brunst M, Cornell H, Panayi GS (2004) Inhibition of antigen-presenting cell function and stimulation of human peripheral blood mononuclear cells to express an antiinflammatory cytokine profile by the stress protein BiP: relevance to the treatment of inflammatory arthritis. Arthritis Rheum 50:1164–1171

Davidson DJ, Haskell C, Majest S et al (2005) Kringle 5 of human plasminogen induces apoptosis of endothelial and tumor cells through surface-expressed glucose-regulated protein 78. Cancer Res 65:4663–4672

Detanico T, Rodrigues L, Sabritto AC et al (2004) Mycobacterial heat shock protein 70 induced interleukin-10 production: immunomodulation of synovial cell cytokine profile and dendritic cell maturation. Clin Exp Immunol 135:336–342

Dutta SK, Girotra M, Singla M et al (2012) Serum HSP70: a novel biomarker for early detection of pancreatic cancer. Pancreas 41:530–534

Elliott RM, Lloyd RE, Fazeli A et al (2009) Effects of HSPA8, an evolutionarily conserved oviductal protein, on boar and bull spermatozoa. Reproduction 137:191–203

Faassen AE, O'leary JJ, Rodysill KJ, Bergh N, Hallgren HM (1989) Diminished heat-shock protein synthesis following mitogen stimulation of lymphocytes from aged donors. Exp Cell Res 183:326–334

Fargnoli J, Kunisada T, Fornace AJJ, Schneider EL, Holbrook NJ (1990) Decreased expression of heat shock protein 70 mRNA and protein after heat treatment in cells of aged rats. Proc Natl Acad Sci U S A 87:846–850

Fleshner M, Campisi J, Amiri L, Diamond DM (2004) Cat exposure induces both intra- and extracellular Hsp72: the role of adrenal hormones. Psychoneuroendocrinology 29:1142–1152

Flohé SB, Bruggemann J, Lendemans S et al (2003) Human heat shock protein 60 induces maturation of dendritic cells versus a Th1-promoting phenotype. J Immunol 170:2340–2348

Frostegård J, Pockley AG (2005) Heat shock protein release and naturally-occurring, exogenous heat shock proteins. In: Henderson B, Pockley AG (ed) Molecular chaperones and cell signalling. Cambridge University Press, New York, pp 195–219

Gao B, Tsan MF (2003a) Endotoxin contamination in recombinant human Hsp70 preparation is responsible for the induction of TNFα release by murine macrophages. J Biol Chem 278:174–179

Gao B, Tsan MF (2003b) Recombinant human heat shock protein 60 does not induce the release of tumor necrosis factor α from murine macrophages. J Biol Chem 278:22523–22529

Gao B, Tsan MF (2004) Induction of cytokines by heat shock proteins and endotoxin in murine macrophages. Biochem Biophys Res Commun 317:1149–1154

Gastpar R, Gehrmann M, Bausero MA et al (2005) Heat shock protein 70 surface-positive tumor exosomes stimulate migratory and cytolytic activity of natural killer cells. Cancer Res 65:5238–5247

Gastpar R, Gross C, Rossbacher L, Ellwart J, Riegger J, Multhoff G (2004) The cell surface-localized heat shock protein 70 epitope TKD induces migration and cytolytic activity selectively in human NK cells. J Immunol 172:972–980

Georgiou AS, Sostaric E, Wong CH et al (2005) Gametes alter the oviductal secretory proteome. Mol Cell Proteomics 4:1785–1796

Gething MJ (1999) Role and regulation of the ER chaperone BiP. Semin Cell Dev Biol 10:465–472

Gething MJ, Sambrook J (1992) Protein folding in the cell. Nature 355:33–45

Grundtman C, Kreutmayer SB, Almanzar G, Wick MC, Wick G (2011) Heat shock protein 60 and immune inflammatory responses in atherosclerosis. Arterioscler Thromb Vasc Biol 31:960–968

Haas IG, Wabl M (1983) Immunoglobulin heavy chain binding protein. Nature 306:387–389

Grundtman C, Wick G (2011) The autoimmune concept of atherosclerosis. Curr Opin Lipidol 22:327–334

Haen SP, Gouttefangeas C, Schmidt D et al (2011) Elevated serum levels of heat shock protein 70 can be detected after radiofrequency ablation. Cell Stress Chaperones 16:495–504

Hamelin C, Cornut E, Poirier F et al (2011) Identification and verification of heat shock protein 60 as a potential serum marker for colorectal cancer. FEBS J 278:4845–4859
Henderson B, Pockley AG (2010) Molecular chaperones and protein folding catalysts as intercellular signaling regulators in immunity and inflammation. J Leuk Biol 88:445–462
Henderson B, Calderwood SK, Coates AR et al (2010) Caught with their PAMPs down? The extracellular signalling actions of molecular chaperones are not due to microbial contaminants. Cell Stress Chaperones 15:123–141
Heydari AR, Conrad CC, Richardson A (1995) Expression of heat shock genes in hepatocytes is affected by age and food restriction in rats. J Nutrition 125:410–418
Hightower LE, Guidon PT (1989) Selective release from cultured mammalian cells of heat-shock (stress) proteins that resemble glia-axon transfer proteins. J Cell Physiol 138:257–266
Jeffery CJ (1999) Moonlighting proteins. Trends Biochem Sci 24:8–11
Jeffery CJ (2003) Moonlighting proteins: old proteins learning new tricks. Trends Genet 19:415–417
Jeffery CJ (2009) Moonlighting proteins—an update. Mol Biosyst 5:345–350
Johnson AD, Berberian PA, Bond MG (1990) Effect of heat shock proteins on survival of isolated aortic cells from normal and atherosclerotic cynomolgus macaques. Atherosclerosis 84:111–119
Johnson AD, Tytell M (1993) Exogenous Hsp70 becomes cell associated, but not internalised by stressed arterial smooth muscle cells. In Vitro Cell Dev Biol 29A:807–812
Johnson BJ, Le TT, Dobbin CA et al (2005) Heat shock protein 10 inhibits lipopolysaccharide-induced inflammatory mediator production. J Biol Chem 280:4037–4047
Kampinga HH, Hageman J, Vos MJ et al (2009) Guidelines for the nomenclature of the human heat shock proteins. Cell Stress Chaperones 14:105–111
Kingston AE, Hicks CA, Colston MJ, Billingham MEJ (1996) A 71-kD heat shock protein (hsp) from Mycobacterium tuberculosis has modulatory effects on experimental rat arthritis. Clin Exp Immunol 103:77–82
Kol A, Lichtman AH, Finberg RW, Libby P, Kurt-Jones EA (2000) Heat shock protein (HSP) 60 activates the innate immune response: CD14 is an essential receptor for HSP60 activation of mononuclear cells. J Immunol 164:13–17
Kovalchin JT, Mendonca C, Wagh MS, Wang R, Chandawarkar RY (2006) In vivo treatment of mice with heat shock protein, gp96, improves survival of skin grafts with minor and major antigenic disparity. Transplant Immunol 15:179–185
Lachance C, Bailey JL, Leclerc P (2007) Expression of Hsp60 and Grp78 in the human endometrium and oviduct, and their effect on sperm functions. Hum Reprod 22:2606–2614
Lancaster GI, Febbraio MA (2005a) Exosome-dependent trafficking of HSP70: a novel secretory pathway for cellular stress proteins. J Biol Chem 280:23349–23355
Lancaster GI, Febbraio MA (2005b) Mechanisms of stress-induced cellular HSP72 release: implications for exercise-induced increases in extracellular HSP72. Exerc Immunol Rev 11:46–52
Lebherz-Eichinger D, Ankersmit HJ, Hacker S et al (2012) HSP27 and HSP70 serum and urine levels in patients suffering from chronic kidney disease. Clin Chim Acta 413:282–286
Lewthwaite J, Owen N, Coates A, Henderson B, Steptoe A (2002) Circulating human heat shock protein 60 in the plasma of British civil servants. Circulation 106:196–201
Li J, Zhao X, Zhang S, Wang S, Du P, Qi G (2011a) ApoB-100 and HSP60 peptides exert a synergetic role in inhibiting early atherosclerosis in immunized ApoE-null mice. Protein Pept Lett 18:733–740
Li Y, Si R, Feng Y et al (2011b) Myocardial ischemia activates an injurious innate immune signaling via cardiac heat shock protein 60 and Toll-like receptor 4. J Biol Chem 286:31308–31319
Lloyd RE, Elliott RM, Fazeli A, Watson PF, Holt WV (2009) Effects of oviductal proteins, including heat shock 70 kDa protein 8, on survival of ram spermatozoa over 48 h in vitro. Reprod Fertil Dev 21:408–418
Luna E, Postol E, Caldas C et al (2007) Treatment with encapsulated Hsp60 peptide (p277) prolongs skin graft survival in a murine model of minor antigen disparity. Scand J Immunol 66:62–70
Mambula SS, Calderwood SK (2006) Heat shock protein 70 is secreted from tumor cells by a nonclassical pathway involving lysosomal endosomes. J Immunol 177:7849–7857

Marin-Briggiler CI, Gonzalez-Echeverria MF, Munuce MJ et al (2010) Glucose-regulated protein 78 (Grp78/BiP) is secreted by human oviduct epithelial cells and the recombinant protein modulates sperm-zona pellucida binding. Fertil Steril 93:1574–1584

Marmot MG, Smith GD, Stansfeld S et al (1991) Health inequalities among British civil servants: the Whitehall II study. Lancet 337:1387–1393

Maron R, Sukhova G, Faria A-M et al (2002) Mucosal administration of heat shock protein-65 decreases atherosclerosis and inflammation in aortic arch of low-density lipoprotein receptor-deficient mice. Circulation 106:1708–1715

Mirza S, Muthana M, Fairburn B, Slack LK, Hopkinson K, Pockley AG (2006) The stress protein gp96 is not an activator of resting rat bone marrow-derived dendritic cells, but is a co-stimulator and activator of $CD3^+$ T cells. Cell Stress Chaperones 11:364–378

Misra UK, Deedwania R, Pizzo SV (2006) Activation and cross-talk between Akt, NF-κB, and unfolded protein response signaling in 1-LN prostate cancer cells consequent to ligation of cell surface-associated GRP78. J Biol Chem 281:13694–13707

Molvarec A, Rigo J Jr, Lazar L et al (2009) Increased serum heat-shock protein 70 levels reflect systemic inflammation, oxidative stress and hepatocellular injury in preeclampsia. Cell Stress Chaperones 14:151–159

Molvarec A, Rigo J Jr, Nagy B et al (2007) Serum heat shock protein 70 levels are decreased in normal human pregnancy. J Reprod Immunol 74:163–169

Molvarec A, Tamasi L, Losonczy G, Madach K, Prohaszka Z, Rigo J Jr (2010) Circulating heat shock protein 70 (HSPA1A) in normal and pathological pregnancies. Cell Stress Chaperones 15:237–247

Morton H, Rolfe B, Clunie GJ (1977) An early pregnancy factor detected in human serum by the rosette inhibition test. Lancet 1:394–397

Nitta Y, Abe K, Aoki M, Ohno I, Isoyama S (1994) Diminished heat shock protein 70 mRNA induction in aged rats after ischemia. Am J Physiol 267:H1795–H1803

Njemini R, Lambert M, Demanet C, Mets T (2003) Elevated serum heat-shock protein 70 levels in patients with acute infection: use of an optimized enzyme-linked immunosorbent assay. Scand J Immunol 58:664–669

Njemini R, Demanet, C, Mets T (2004) Inflammatory status as an important determinant of heat shock protein 70 serum concentrations during aging. Biogerontology 5:31–38

Njemini R, Demanet C, Mets T (2005) Comparison of two ELISAs for the determination of Hsp70 in serum. J Immunol Methods 306:176–182

Njemini R, Bautmans I, Onyema OO, Van Puyvelde K, Demanet C, Mets T (2011) Circulating heat shock protein 70 in health, aging and disease. BMC Immunol 12:24

Noonan FP, Halliday WJ, Morton H, Clunie GJ (1979) Early pregnancy factor is immunosuppressive. Nature 278:649–651

Novo G, Cappello F, Rizzo M et al (2011) Hsp60 and heme oxygenase-1 (Hsp32) in acute myocardial infarction. Transl Res 157:285–292

Pahlavani MA, Cheung TH, Chesky JA, Richardson A (1988) Influence of exercise on the immune function of rats of various ages. J App Physiol 64:1997–2001

Pockley AG, Shepherd J, Corton J (1998) Detection of heat shock protein 70 (Hsp70) and anti-Hsp70 antibodies in the serum of normal individuals. Immunol Invest 27:367–377

Pockley AG, Bulmer J, Hanks BM, Wright BH (1999) Identification of human heat shock protein 60 (Hsp60) and anti-Hsp60 antibodies in the peripheral circulation of normal individuals. Cell Stress Chaperones 4:29–35

Pockley AG, Wu R, Lemne C, Kiessling R, De Faire U, Frostegård J (2000) Circulating heat shock protein 60 is associated with early cardiovascular disease. Hypertension 36:303–307

Pockley AG, Georgiades A, Thulin T, De Faire U, Frostegård J (2003) Serum heat shock protein 70 levels predict the development of atherosclerosis in subjects with established hypertension. Hypertension 42:235–238

Pockley AG, Muthana M, Calderwood SK (2008) The dual immunoregulatory role of stress proteins. Trends Biochem Sci 3:71–79

Quintana FJ, Carmi P, Mor F, Cohen IR (2003) DNA fragments of the human 60-kDa heat shock protein (HSP60) vaccinate against adjuvant arthritis: identification of a regulatory HSP60 peptide. J Immunol 171:3533–3541
Quintana FJ, Carmi P, Mor F, Cohen IR (2004) Inhibition of adjuvant-induced arthritis by DNA vaccination with the 70-kd or the 90-kd human heat-shock protein: immune cross-regulation with the 60-kd heat-shock protein. Arthritis Rheum 50:3712–37120
Rea IM, McNerlan S, Pockley AG (2001) Serum heat shock protein and anti-heat shock protein antibody levels in aging. Exp Gerontology 36:341–352
Quintana FJ, Cohen IR (2011) The HSP60 immune system network. Trends Immunol 32:89–95
Richardson A, Landry SJ, Georgopoulos C (1998) The ins and outs of a molecular chaperone machine. Trends Biochem Sci 23:138–143
Rizzo M, Cappello F, Marfil R et al (2012) Heat-shock protein 60 kDa and atherogenic dyslipidemia in patients with untreated mild periodontitis: a pilot study. Cell Stress Chaperones 17:399–407
Robinson MB, Tidwell JL, Gould T et al (2005) Extracellular heat shock protein 70: a critical component for motoneuron survival. J Neurosci 25:9735–9745
Shamaei-Tousi A, D'aiuto F, Nibali L et al (2007) Differential regulation of circulating levels of molecular chaperones in patients undergoing treatment for periodontal disease. PLoS One 2:e1198
Shields AM, Panayi GS, Corrigall VM (2011) Resolution-associated molecular patterns (RAMP): RAMParts defending immunological homeostasis? Clin Exp Immunol 165:292–300
Slack LK, Muthana M, Hopkinson K et al (2007) Administration of the stress protein gp96 prolongs rat cardiac allograft survival, modifies rejection-associated inflammatory events and induces a state of peripheral T cell hyporesponsiveness. Cell Stress Chaperones 12:71–82
Tamasi L, Bohacs A, Tamasi V et al (2010) Increased circulating heat shock protein 70 levels in pregnant asthmatics. Cell Stress Chaperones 15:295–300
Tanaka S, Kimura Y, Mitani A et al (1999) Activation of T cells recognizing an epitope of heat-shock protein 70 can protect against rat adjuvant arthritis. J Immunol 163:5560–5565
Terry DF, Wyszynski DF, Nolan VG et al (2006) Serum heat shock protein 70 level as a biomarker of exceptional longevity. Mech Ageing Dev 127:862–868
Tidwell JL, Houenou LJ, Tytell M (2004) Administration of Hsp70 in vivo inhibits motor and sensory neuron degeneration. Cell Stress Chaperones 9:88–98
Tsan MF, Gao B (2009) Heat shock proteins and immune system. J Leukoc Biol 85:905–910
Tytell M (2005) Release of heat shock proteins (Hsps) and the effects of extracellular Hsps on neural cells and tissues. Int J Hyperthermia 21:445–455
Tytell M, Greenberg SG, Lasek RJ (1986) Heat shock-like protein is transferred from glia to axon. Brain Res 363:161–164
Udono H, Srivastava PK (1994) Comparison of tumor-specific immunogenicities of stress-induced proteins gp96, hsp90 and hsp70. J Immunol 152:5398–5403
Van Eden W (2008) XToll, a recombinant chaperonin 10 as an anti-inflammatory immunomodulator. Curr Opin Investig Drugs 9:523–533
Van Eden W, Van Der Zee R, Prakken B (2005) Heat shock proteins induce T-cell regulation of chronic inflammation. Nat Immunol 5:318–330
Van Puijvelde GH, Van Es T, Van Wanrooij EJ et al (2007) Induction of oral tolerance to HSP60 or an HSP60-peptide activates T cell regulation and reduces atherosclerosis. Arterioscler Thromb Vasc Biol 27:2677–2683
Vanags D, Williams B, Johnson B et al (2006) Therapeutic efficacy and safety of chaperonin 10 in patients with rheumatoid arthritis: a double-blind randomised trial. Lancet 368:855–863
Wendling U, Paul L, Van Der Zee R, Prakken B, Singh M, Van Eden W (2000) A conserved mycobacterial heat shock protein (hsp) 70 sequence prevents adjuvant arthritis upon nasal administration and induces IL-10-producing T cells that cross-react with the mammalian self-hsp70 homologue. J Immunol 164:2711–2717
Williams B, Vanags D, Hall S et al (2008) Efficacy and safety of chaperonin 10 in patients with moderate to severe plaque psoriasis: evidence of utility beyond a single indication. Arch Dermatol 144:683–685

Wu CT, Ou LS, Yeh KW, Lee WI, Huang JL (2011) Serum heat shock protein 60 can predict remission of flare-up in juvenile idiopathic arthritis. Clin Rheumatol 30:959–965

Xu Q, Schett G, Perschinka H et al (2000) Serum soluble heat shock protein 60 is elevated in subjects with atherosclerosis in a general population. Circulation 102:14–20

Young D, Romain E, Moreno C, O'brien R, Born W (1993) Molecular chaperones and the immune system response. Phil Trans Royal Soc Lond 339:363–367

Yuan J, Dunn P, Martinus RD (2011) Detection of Hsp60 in saliva and serum from type 2 diabetic and non-diabetic control subjects. Cell Stress Chaperones 16:689–693

Zanin-Zhorov A, Cahalon L, Tal G, Margalit R, Lider O, Cohen IR (2006) Heat shock protein 60 enhances $CD4^+CD25^+$ regulatory T cell function via innate TLR2 signaling. J Clin Invest 116:2022–2032

Zhu J, Quyyumi AA, Wu H et al (2003) Increased serum levels of heat shock protein 70 are associated with low risk of coronary artery disease. Arterioscler Thromb Vasc Biol 23:1055–1059

Chapter 4
Hsp27 Phosphorylation Patterns and Cellular Consequences

Sergiy Kostenko and Ugo Moens

Abstract The human small heat shock protein 27 (HSP27 or HSPB1) is a multifunctional protein that participates in a variety of cellular processes such as controlling protein folding, F-actin-dependent processes, cytoprotection/anti-apoptosis, differentiation, cell proliferation, and gene expression. The structural and physiological properties of HSP27 are partially controlled by phosphorylation and several protein kinases that mediate phosphorylation have been identified. While phosphorylation of serine residues 15, 78 and 82 has been most extensively studied, other *in vivo* phosphorylation modifications have been identified. Here we review the different phosphorylations of HSP27 and consider the consequences of phosphorylation on HSP27's conformation, subcellular localization, and cellular roles. We also address the phosphorylation pattern under pathogenic conditions and discuss the possible implications of HSP27 phosphorylation in human disease.

4.1 Introduction

The heat shock proteins (HSPs) or cell stress proteins (CSPs) are a large family of proteins found in all three Kingdoms of life. In humans there are now recognised to be five families of HSPs: HSPA, HSPB, HSPC, HSPH, and DNAJ (see beginning of book) consisting of 13, 11, 5, 4, and 40 genes, respectively (Kampinga et al. 2009). DNAJ pseudogenes and proteins containing a J domain have been identified so that the DNAJ subfamily may contain additional members (Vos et al. 2008; Kampinga et al. 2009). There is also the HSPD, HSPE and CCT proteins which will not be discussed (Kampinga et al. 2009). The members of the human HSPB or small HSP (sHSP) family are characterised by predicted small molecular masses ranging from 16.3 to 28.4 kDa and a highly conserved region of ~90 residues referred to as the α-crystallin domain with β-sheet structure (Gusev et al. 2002; Haslbeck et al. 2005;

U. Moens (✉) · S. Kostenko
Institute of Medical Biology, Faculty of Health Sciences,
University of Tromsø, NO-9037 Tromsø, Norway
e-mail: ugo.moens@uit.no; sergiyk@uit.no

B. Henderson, A. G. Pockley (eds.), *Cellular Trafficking of Cell Stress Proteins in Health and Disease*, Heat Shock Proteins 6, DOI 10.1007/978-94-007-4740-1_4,

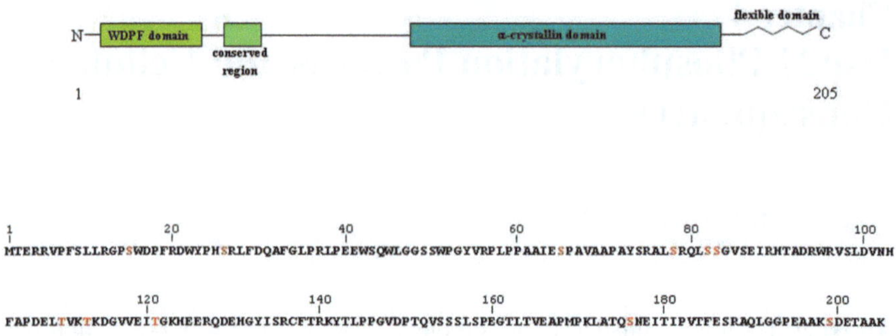

Fig. 4.1 Functional organization of the human HSP27. The WDPF domain, the conserved N-terminal region, and the α-crystallin domain are shown. The *zigzag line* corresponds to the flexible domain in the C-terminal part of the protein. The *lower panel* gives the primary structure of human HSP27. The one-letter amino acid code is used. The in vivo identified phosphorylation sites Ser-15, Ser-26, Ser-65, Ser-78, Ser-82, Ser-83, Thr-110, Thr-113, Thr-121, Ser-176 and Ser-199 are indicated in *red*

Sun and MacRae 2005; Arrigo 2007; Vos et al. 2008). Somewhat less conserved is the N-terminal WDPF domain, owing its name to the presence of the amino acid residues W (tryptophan), D (aspartic acid), P (proline), and F (phenylalanine). The amino-terminal part of the protein also contains the partially conserved sequence PSRLFDQXFGEXLL, while the carboxy-terminal region consists of a flexible motif. The sHSPs are often found in oligomeric complexes involving one or more family members (Lelj-Garolla and Mauk 2005; Haslbeck et al. 2005). One of the best studied small HSPs is HSP27 (HSPB1 according to the new nomenclature; Kampinga et al. 2009) and the functional structure and amino acid composition of human HSP27 is given in Fig. 4.1. The WDPF region of HSP27 is crucial for oligomerization, but other sequences in the N-terminal region are required (Kappe et al. 2003). The α-crystallin domain is also essential for oligomerization of HSP27, while the flexible region has been suggested to participate in the interaction with target proteins, to be involved in oligomerization, and to be important for solubility (reviewed in Gusev et al. 2002; Lelj-Garolla and Mauk 2005). The role of the partially conserved sequence remains unknown (Gusev et al. 2002). HSP27 can oligomerize into large aggregates up to 800 kDa, but it can also form hetero-oligomers with other sHSP family members (e.g. HSPB6 = HSP20, HSPB4 = αA-crystallin). Unphosphorylated HSP27 forms large multimers, ranging in size from 12 to 35-mer, which are responsible for protection against cellular stress, while phosphorylation results in conformational changes and formation of smaller di- and tetrameric units (Kato et al. 1994; Lambert et al. 1999; Rogalla et al. 1999; Bova et al. 2000; Lelj-Garolla and Mauk 2005; Hayes et al. 2009). This posttranslational change also modulates the direct interaction with proteins such as other HSPs, actin, HSF1, and oestrogen receptor-β (see further; Ferns et al. 2006; Sun et al. 2006; Al-Madhoun et al. 2007; Brunet Simioni et al. 2009; Bukach et al. 2009).

Table 4.1 Predicted phosphoacceptor sites in HSP27(HSPB1) and proven *in vitro* and *in vivo* phosphorylation sites. The kinases known to mediate phosphorylation of HSP27 are shown (Kostenko and Moens 2009). (See text for details)

predicted phosphosites	context	in vitro phosphorylation	in vitro kinase	in vivo phosphorylation	in vivo kinase
Thr-2	MTERRV			N	
Ser-9	RVPFSLLRG			N	
Ser-15	AAIESPAVA	Y	PKA*,PKB,PKC,MK2,MK3,MK5	Y	PKG,MK2,MK3
Ser-26	WYPHSRLFD			Y	not identified
Tyr-54	SWPGYVRPL			N	
Ser-65	LRGPSWDPF			Y	not identified
Tyr-73	AAPAYSRAL			N	
Ser-78	SRALSRQLS	Y	PKA,PKC,PKG,RSK2,MK2,MK3,MK5	Y	RSK2,MK2,MK3,MK5
Ser-82	SRQLSSGVS	Y	PKA,PKB,PKD,PKG,RSK2,MK2,MK3,MK5	Y	PKB,PKD,PKG,RSK2,MK2,MK3,MK5
Ser-83	RQLSSGVSE			Y	not identified
Ser-86	SSGVSEIRH	Y	PKA	N	
Thr-91	EIRHTADRW			N	
Ser-98	RWRVSLDVN			N	
Thr-110	PDELTVKTK			Y	not identified
Thr-113	LTVKTKDGV			Y	not identified
Thr-121	VVEITGKHE			Y	not identified
Tyr-133	DEHGYISRC			N	
Ser-135	HGYISRCFT			N	
Thr-139	SRCFTRKYT			N	
Thr-143	TRKYTLPPG	Y	PKA,PKG	N	
Ser-156	QVSSSLSPE			N	
Ser-158	SSSLSPEGT			N	
Thr-162	SPEGTLTVE			N	
Thr-174	PKLATQSNE			N	
Ser-176	LATQSNEIT			Y	not identified
Ser-199	EAAKSDETA			Y	not identified

*Abbreviations: PKA= protein kinase A/cAMP-dependent protein kinase; PKB= protein kinase B or AKT, PKC= protein kinase C; PKD= protein kinase D; PKG= cGMP-dependent protein kinase; MK= mitogen-activated protein kinase-activated protein kinase

HSP27 can be detected in most cells and tissues examined, but the expression levels may vary with some cells expressing undetectable or relatively low levels, while others express HSP27 abundantly. Expression levels also vary during development, growth cycle progression, differentiation, and with pathophysiological conditions of the cell (Kostenko and Moens 2009; Uhlen et al. 2010). A plethora of different stimuli, but also several pathological conditions can alter the phosphorylation levels of HSP27 (reviewed in Kostenko and Moens 2009). This chapter reviews the protein kinases and phosphatases that control the phosphorylation of human HSP27 and focuses on the physiological consequences of HSP27 phosphorylation under normal and pathogenic conditions.

4.2 Phosphorylation/Dephosphorylation of HSP27

Analyzing the primary sequence of HSP27 protein (205 amino acids, Accession: CAG38728.1; GI: 49168466) reveals that serine and threonine are statistically over-represented with respectively 21 (10.2 %) and 14 (6.8 %) residues, while only 5 (2.4 %) tyrosine residues are present. Different phosphorylation prediction algorithm programs propose 14 serines, 9 threonines and 3 tyrosines to have a high probability to be phosphor-acceptor sites for a variety of protein kinases (Table 4.1; Blom et al. 1999; Huang et al. 2005; Liding et al. 2007; Gnad et al. 2011). While most studies have examined the phosphorylation status of Ser-15, -78 and -82 (reviewed in Kostenko and Moens 2009), few have addressed the entire phosphorylation pattern of HSP27. Phosphoproteome profiling studies of human cells have identified Ser-15, -26, -65, -78, -82, -83, -176 and -199, and Thr-110, -113, -121 as genuine *in vivo* phosphorylation sites (Beausoleil et al. 2004; Yang et al. 2006; Olsen et al. 2010). However, not all protein kinases responsible for these phosphorylation events are known. Ser-15, -78 and -82 can be phosphorylated by the mitogen-activated protein kinase-activated protein kinases-2 and -3 (MK2 and MK3), while MK5 can phosphorylate Ser-78 and Ser-82 *in vivo* (reviewed in Kostenko and Moens 2009). RSK2, another mitogen-activated protein kinase-activated protein kinase, also phosphorylates these sites both *in vitro* and *in vivo* (Kang et al. 2010). MK2 is the major HSP27 kinase that mediates phosphorylation induced by stress, while MK5 seems to act as an HSP27 kinase in response to activation of the cAMP/PKA pathway (Kostenko et al. 2009; Shiryaev et al. 2011). cGMP-dependent protein kinase (PKG) can phosphorylate HSP27 *in vivo* at residues Ser-15 and Ser-82, while Ser-82 is also phosphoacceptor site for protein kinase B (PKB/AKT) and protein kinase D (PKD) (reviewed in Kostenko and Moens 2009). ATM (Ataxia telangiectasia mutated), a protein kinase involved in double-strand break and oxidative stress responses, can induce phosphorylation of HSP27 at Ser-78, but this post-translational modification is mediated by the $p38^{MAPK}$-MK2 pathway (Cosentino et al. 2011). PKC may act as a Ser-82 kinase in SH-SY5Y cells (Dokas et al. 2011). The nature of the protein kinases that phosphorylate the other sites remains unknown.

Time-dependent studies reveal that stimulus-induced HSP27 phosphorylation is often transient, so that HSP27 phosphatases are also implicated in controlling the phosphorylation status of this protein. Protein phosphatases PP2A, PP2B and PP1 can dephosphorylate phosphoHSP27 *in vitro* although with different stoichiometries, with PP2A being the most potent HSP27 phosphatase (Gaestel et al. 1992; Cairns et al. 1994; Levin and Santell 1991; Loktionova and Kabakov 1998; Tar et al. 2006). Cell culture studies confirm a role for PP2A, but not PP2B and PP1 as HSP27 phosphatase (Cairns et al. 1994; Berrou and Bryckaert 2009; Chandrika et al. 2010). It is not known whether specific protein phosphatases preferentially dephosphorylate particular phosphorylated residues, but dephosphorylation of HSP27 may be signal-dependent as Berrou and Bryckaert showed that platelet-derived growth factor caused an increase in PP2A activity which was concomitant with HSP27 dephosphorylation (Berrou and Bryckaert 2009). Knockdown of the catalytic subunit of PP2A in human HER-2/neu positive BT474 and SKBR3 breast cancer cell lines correlated with increased HSP27 phosphorylation at Ser-78. This indicates that PP2A may be responsible for *in vivo* dephosphorylation of phosphoSer-78 (Wong et al. 2010). Studies in human vascular smooth muscle cells with low concentrations of okadaic acid that only inhibit PP2A or use of the specific PP2A inhibitor, fostriecin, demonstrated that PP2A mediated low density lipoprotein (LDL)-induced dephosphorylation of phosphoSer-82 HSP27 (Garcia-Arguinzonis et al. 2010).

HSP27 is involved in a remarkable variety of cellular functions in addition to its protein folding (chaperone activity), protecting cells against different types

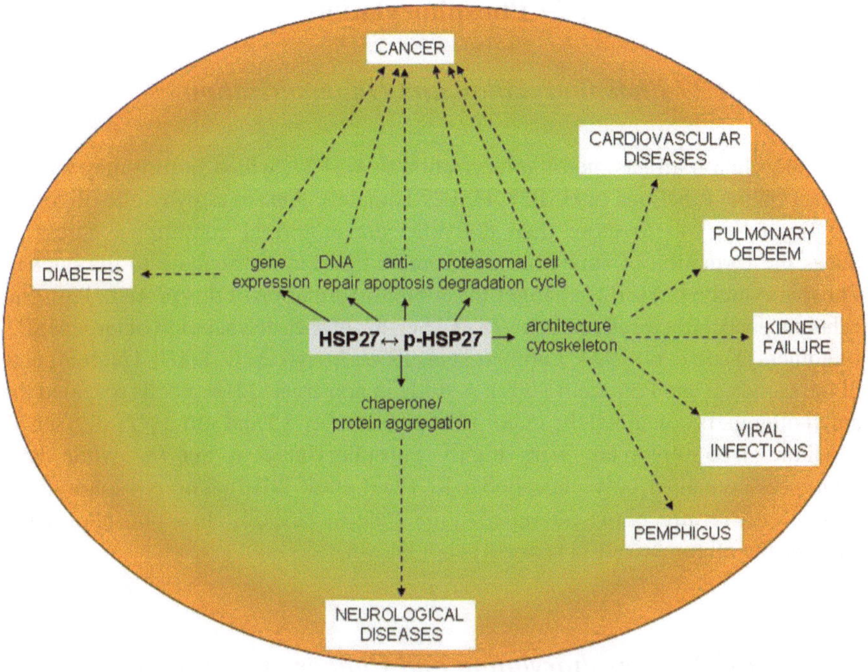

Fig. 4.2 The role of phosphoHSP27 in normal functions and its possible implication in diseases. HSP27 will be present in non-phosphorylated and phosphorylated forms depending on the cellular conditions. The *central* part is of the figure is in *green* and symbolizing the normal function of HSP27/phosphoHSP27. The *outer* part is in *red*, symbolizing pathogenic conditions. Aberrant phosphorylation of HSP27 can impair its normal functions and be a causal factor in the development of diseases. The *stippled lines* indicate that perturbed phosphorylation of HSP27 may favour the progress of a normal cell into a diseased cell

of stress. These involve: controlling cytoskeletal architecture, cell growth, differentiation and intracellular redox state (Arrigo 2007; Lanneau et al. 2007; Kostenko and Moens 2009). Phosphorylation affects both the structure of HSP27 complexes and the functions of HSP27. Studies with phosphomimicking or non-phosphorylatable HSP27 variants have been pivotal in elucidating the importance of phosphorylation in modulating HSP27's functions. The phosphomimicking mutants are referred to HSP27-3D when serine residues (Ser or S) 15, 78 and 82 are replaced by aspartic acid (Asp or D), while non-phosphorylatable HSP27 mutants are HSP27-3A and HSP27-3G where Ser-15, -78 and -82 are replaced by alanine (Ala or A) or glycine (Gly or G), respectively. The roles of phosphoHSP27 under normal cellular conditions is reviewed in the first part of this chapter and summarized (Fig. 4.2). The second part of the chapter will discuss the possible implications of aberrant HSP27 phosphorylation in the development of diseases.

4.3 Function of HSP27 Phosphorylation

4.3.1 HSP27 Phosphorylation and Oligomerization

Phosphorylation affects the oligomerization of HSP27, which in turn may have an impact on the functions of HSP27. HSP27 is mainly present as cytosolic oligomers ranging from 140 to more than 800 kDa under normal conditions. Stress induces the formation of small sized oligomers that can redistribute to the nucleus. Phosphorylation of HSP27 changes the complex formation of this protein. Phosphorylation, especially at residues Ser-78 and Ser-82, causes dissociation of large HSP27 oligomers (Kato et al. 1994; Lambert et al. 1999; Arrigo 2001, 2007; Thériault et al. 2004; Gendron and Petrucelli 2009). A study in non-treated HeLa cells revealed that small oligomers contained the majority of phosphoSer-15 and some phosphoSer-82. Medium sized oligomers were highly phosphorylated at Ser-78, while large oligomers contained only phosphoSer-82 (Paul et al. 2010). The phosphorylation pattern clearly affects the oligomerization state of HSP27. How phosphorylation governs the oligomerization process is not known.

4.3.2 HSP27 Phosphorylation and Chaperone Function

HSP27 can act as an ATP-independent chaperone by binding to unfolded proteins and preventing their aggregation (Knauf et al. 1994). Multimeric HSP27 complexes form the chaperone competent state and help refolding of denaturated proteins. Moreover, such complexes prevent the aberrantly folded proteins generated by heat stress from forming toxic aggregates by storing them. Stored misfolded proteins are then either refolded or degraded (Jakob et al. 1993; Kampinga et al. 1994; Ehrnsperger et al. 1997; Rogalla et al. 1999; Sun and MacRae 2005; Paul et al. 2010). The phosphorylation independent chaperone function of HSP27 seems to be disputed. HSP27 mutants mimicking phosphorylation show effective chaperone function activity *in vitro* and *in vivo* (Panasenko et al. 2003; Bryantsev et al. 2007a). Results from other groups suggest that the oligomeric forms of unphosphorylated HSP27 are a storage state for proteins during stress conditions, while the smaller (phosphorylated) HSP27 subunits in cooperation with other chaperones (e.g. HSP70) allow the trapped proteins to regain their native structure when physiological conditions are restored (Ehrnsperger et al. 1997; Bryantsev et al, 2007a; Hayes et al. 2009). *In vitro* studies monitoring the chaperone activity of recombinant wild-type, MK2-phosphorylated HSP27, and HSP-3D on insulin and α-lactalbumin aggregation demonstrated that phosphorylated HSP27 is a more effective molecular chaperone than the phosphomimic HSP27-3D mutant, which is slightly better than the unphosphorylated protein (Hayes et al. 2009). The reasons for the disagreement about whether phosphorylation of HSP27 affects its property as a chaperone are unclear, but may be explained by the use of different substrates, HSP from different species, different purification protocols, source (bacterial expressed, tissue-derived), different phosphorylation status (single, double or triple), and minor changes in the recombinant HSP27 due to additional N- or C-tagged sequences (Hayes et al. 2009).

4.3.3 HSP27 Phosphorylation and F-actin Dynamics and F-actin-Dependent Processes

One of the best studied impacts of HSP27 on its function is actin filament dynamics. In its unphosphorylated form, HSP27 inhibits actin polymerization by binding to the barbed end of actin filaments. Phosphorylation of HSP27 reduces its affinity for the barbed ends of actin, but may allow small oligomers of phosphorylated HSP27 to bind to the sites of F-actin, thus stabilizing them (Zhu et al. 1994a; Lavoie et al. 1995; Huot et al. 1996; Mounier and Arrigo 2002; Dominguez and Holmes 2011). Phosphorylation of HSP27 favours F-actin remodelling, and as such, the phosphorylation state of HSP27 determines its role in actin filament dynamics and actin-dependent processes such as pinocytosis, lamellipodia formation and cell motitity (Lavoie et al. 1993; Piotrowicz et al. 1998; Pichon et al. 2004; Pivovarova et al. 2005; Nomura et al. 2007; Pivovarova et al. 2007; Doshi et al. 2010; Chen et al. 2010; Dokas et al. 2011). Accordingly, prevention of phospho-HSP27 dephosphorylation by the PP2A inhibitor fostriecin stimulated F-actin polymerization in vascular smooth muscle cells (Garciá-Arguinzonis et al. 2010). The phosphorylation-dependent role of HSP27 in stress-induced F-actin rearrangement has focused on the serine 15, 78 and 82 residues. Mutant HSP27 in which these residues were replaced by non-phosphorylatable Ala or by phosphomimicking Asp confirmed that prevention of phosphorylation of HSP27 at these sites blocked stress-induced F-actin remodelling and F-actin dependent processes, while overexpressing the HSP27-3D mutant or microinjection of phosphorylated HSP27 resulted in F-actin rearrangement (Schneider et al. 1998; Hedges et al. 1999; Landry and Huot 1999). The effect of single or double mutants on F-actin dynamics has not been tested meticulously, nor have the consequences of mutations in other putative phosphorylation sites been examined.

4.3.4 HSP27 Phosphorylation and Cytoprotection/Anti-Apoptotic Function

Unphosphorylated and phosphorylated HSP27 exhibit anti-apoptotic effects via different mechanisms. Unphosphorylated HSP27 can block apoptosis by inhibiting activation of pro-caspases or antagonizing Bax-induced mitochondrial injury to reduce the release of cytochrome C, whereas phospho-HSP27 binds to and inhibits Daxx-mediated apoptosis or stimulates the prosurvival AKT pathway (Charette et al. 2000; Concannon et al. 2003; Mosser and Morimoto 2004; Garrido et al. 2006; Lanneau et al. 2007; Havasi et al. 2008; Paul et al. 2010). Several apoptotic inducers increase the phosphorylation status of HSP27 on Ser-15, -78, and -82. However, the phosphorylation pattern and kinetics of individual serine residues varied for the different apoptotic stimuli (Paul et al. 2010). HSP27 has also important anti-oxidant properties related to its ability to maintain glutathione in its reduced form, to decrease reactive oxygen species (ROS), and to neutralize the toxic effects of oxidized proteins (reviewed in Arrigo 2007 and in Lanneau et al. 2007). In unstressed cells, HSP27 decreases the basal level of intracellular ROS. Oxidative stress leads to a drastic changes in HSP27 phosphorylation and reduces the size of the HSP27 oligomers to about 200 kDa. Wild-type and HSP27-3A were able to generate large aggregates of >200 kDa, while HSP27-3D and HSP27-3G formed small oligomers of <200 kDa. Only the large aggregates reduced ROS production, enhanced glutathione levels, and provided cellular protection against oxidative stress. This observation suggests that large unphosphorylated oligomers of HSP27 represent the active form of the protein which modulates ROS levels and protect cells against oxidative stress (Mehlen et al. 1997; Arrigo 2007). Hydrogen peroxide-treatment of control Chinese hamster CCL39 cells or cells overexpressing HSP27-3A caused F-actin fragmentation and cell death, while cells over-expressing wild-type HSP27 showed increased resistance against fragmentation of F-actin and cell survival (Huot et al. 1996). These results confirm that phosphorylation of HSP27 is important for cells to resist oxidative stress. Hippocampal HiB5 cells overexpressing HSP-2D (where Ser-15 is not substituted by Asp) or HSP27-3D displayed higher actin stability and survival rate against heat shock than cells overexpressing HSP27-2A or HSP27-3A (Geum et al. 2002). It is therefore likely that phosphorylation of HSP27 is important for maintenance of the actin stress fibres in response to heat shock and thus enhances thermoresistance (Geum et al. 2002). Actin stability and survival rate in cells overexpressing HSP27-2D or HSP27-3D were not quantified, so that the contribution of Ser-15 phosphorylation cannot be evaluated. Ectopic expression of wild-type or HSP27-3A protected MCF-7 cells against doxorubicin-induced cell death, while HSP27-3D did not increase cell survival, indicating that survival of doxorubicin-treated MCF-7 cells requires unphosphorylated HSP27 (Fujita et al. 2011). The neuroprotective action of HSP27 in C6 glioma cells and in human neuroblastoma SK-N-SH cells also depends on its phosphorylation (Wyttenbach et al. 2002; Tanabe et al. 2008). However, results from another group demonstrated that HSP27-S15D and HSP27-S78D/S82D, but not HSP27-3D could not protect rat neuronal 13.S.1.24 cells against oxidative stress. This suggests that the neuroprotective effect of phosphoHSP27 may be cell-specific and is impaired when all three sites are phosphorylated (Rogalla et al. 1999).

A recent study showed that the anti-apoptotic protein Bcl-2, targeted at the endoplasmic reticulum, triggered HSP27 phosphorylation at Ser-15, -78 and -82 and stimulated long term survival of the colon cancer cell lines HCT116 and SW480. Inhibition of the MEK, p38 MAPK and Akt protein kinases prevented HSP27 phosphorylation and long term survival induced by endoplasmic reticulum-targeted Bcl-2, while silencing of HSP27 reduced cell survival. These results suggest that Bcl-2 targeted to the endoplasmic reticulum can activate different protein kinases that phosphorylate HSP27 and that phosphorylation of HSP27 is important in mediating long term survival (Chandrika et al. 2010). The exact pathway by which endoplasmic reticulum targeted Bcl-2 induces HSP27 phosphorylation and cell survival is not known. The group of Arrigo examined the anti-apoptotic effect of Ser-15, -78, and -82 phosphorylation in HeLa cells treated with four different apoptosis inducers (Paul et al. 2010). Etoposide, Fas antibody, staurosporine and cytochalasin D all provoked phosphorylation of the three sites, but with different stoichiometry and kinetics. For example, etoposide strongly increased phosphorylation of Ser-15, but had little effect on Ser-82, whereas Fas antibodies treatment resulted in similarly robust phosphorylation of all three residues. However, phosphorylation of Ser-82 was very obvious after 1h, while phosphorylation of Ser-78 was pronounced after 3 h, and phosphorylation of Ser-15 only increased after 3 h. The phosphorylation pattern of the different oligomeric structures also changed, but the significance is not yet clear. Transfection studies with HSP27-3A and HSP27-3D confirmed the essential role of phosphorylation in the anti-apoptotic activity of HSP27, but the precise contribution of each phosphoserine needs to be further determined. Moreover, the different stimuli caused distinct localization of phosphoHSP27 (Paul et al. 2010). In conclusion, HSP27 has multiple and complex strategies to exert anti-apoptotic activities. These strategies include changes in localization, oligomerization and phosphorylation.

4.3.5 HSP27 Phosphorylation and Differentiation

Increased HSP27 phosphorylation has been observed during differentiation of a variety of cell types, including HL-60, MCF-7, PC12, myelomonocytic leukaemia cells, keratinocytes and erythroid progenitors, while inhibition of HSP27 accumulation results in failure to differentiate (Michishita et al. 1991; Spector et al. 1994; Horman et al. 1997; Uddin et al. 2004; Shi et al. 2008b; Hong et al. 2009; Bertrand-Vallery et al. 2010; de Thonel et al. 2010; Robitaille et al. 2010; Jonak et al. 2011). Two important questions that remain unanswered are: Is phosphorylation of HSP27 functionally involved in the differential process and if so, what is the mechanism by which phosphoHSP27 contributes to differentiation? Recent studies have shed some light on the possible role of HSP27 phosphorylation in the differentiation process. Rat pheochromocytoma PC12 cells stably overexpressing glial-cell-line-derived neurotropic factor (GDNF) receptor α1 and the RET receptor tyrosine kinase (PC12-GFRα1-RET cells) were induced to differentiate into neuron-like cells by exposing them to glial cell-derived neurotrophic factor (GDNF) (Hong et al. 2009). Phosphoproteome studies revealed that GDNF-triggered neurite outgrowth was accompanied with increased HSP27 phosphorylation. Western blot analysis with HSP27 and phosphoSer-82 specific antibodies confirmed increased phosphorylation of HSP27 at this site after GDNF administration, while no differences in total HSP27 levels were observed between untreated and treated cells. SiRNA-mediated knock-down of HSP27, as well as overexpressing HSP27-3A strongly reduced GDNF-induced neurite outgrowth, while overexpressing wild-type or HSP27-3D did not influence neurite outgrowth caused by GDNF. In the absence of GDNF, overexpression of wild-type HSP27, HSP27-3A or HSP27-3D had no effect on the morphology of the PC12-GFRα1-RET cells. Inhibition of the MEK/ERK or the $p38^{MAPK}$ pathway, but not the JNK pathway, negatively interfered with GDNF-induced neurite outgrowth (Hong et al. 2009). These studies suggest that phosphorylation of HSP27 plays a key role in GDNF-induced neurite outgrowth. Phosphorylation is mediated, at least partially by the MEK/ERK and $p38^{MAPK}$ signalling pathways, but the HSP27 kinases were not characterized. Whether phosphorylation of all three (Ser-15, -78, and -82), a particular one, a combination, or additional residues are required for GDNF-induced differentiation of PC12 cells remains to be investigated. Another example that illustrates the involvement of phosphoHSP27 in differentiation is haematopoiesis. Differentiation of blood progenitor cells coincides with nuclear accumulation of phosphoHSP27 (de Thonel et al. 2010). Because phosphoHSP27 is translocated to the nucleus upon differentiation, it may play a role in nuclear processes such as transcription and post transcriptional events (see further).

4.3.6 HSP27 Phosphorylation and Bone Metabolism

Osteocalcin is involved in bone mineralization and its synthesis is induced by bone morphogenetic proteins (BMP) and the thyroid hormone triiodothyronine (T_3). Studies using stably transfected osteoblast-like MC3T3-E1 cells that overexpress either wild-type, 3A or 3D HSP27 showed that unphosphorylated (i.e. wild-type or HSP27-3A), but not phosphorylated (i.e. HSP27-3D) suppressed BMP and T_3-induced osteocalcin expression. On the other hand, matrix mineralization was more extensive in cells overexpressing wild-type or non-phosphorylatable HSP27 than in cells expressing HSP27-3D. Thus unphosphorylated HSP27 appears to act as a positive regulator of bone calcification by down regulation osteocalcin synthesis and upregulating matrix mineralization (Kato et al. 2011). Osteocalcin acts also on pancreatic β-cells, adipocytes and muscular tissue to increase insulin production, adiponectin synthesis, and insulin sensitivity, respectively. Thus aberrant HSP27 phosphorylation levels may lead to perturbed osteocalcin synthesis, which contributes to the pathogenesis of diseases such as diabetes (Villafan-Bernal et al. 2011).

4.3.7 HSP27 Phosphorylation and Ageing

The cytoskeleton changes during ageing, suggesting that HSP27 may be involved. Indeed, researchers noticed that HSPB1 levels decreased in cultured human dermal fibroblast as a function of ageing and that phosphoHSP27 levels decreased in gastrointestinal smooth muscle of older rats (Bitar and Patil 2004; Orejuela et al. 2007). Similar observations were done in ageing brain (Kreuger-Naug et al. 2002; Calderwood et al. 2009). Minor increases in total and phosphorylated HSPB1 levels have also been reported in brain (8 % increase) and skeletal muscle (3 % increase) from elderly people compare to adults, while HSPB1 levels increased by 180 % in foreskin (Lu et al. 2004; Welle et al. 2004; Yamaguchi et al. 2007; Craig et al. 2010;

Laimer et al. 2010). The physiological role of increased HSP27 phosphorylation in ageing cells is incompletely understood, but phosphoHSP27 may increase cytoskeletal stability. Ageing cells have probably been more exposed to stress and hence actin disrupting and modifying conditions. It is possible that increased HSP27 expression and phosphorylation may protect or repair actin filaments in aged cells. On the other hand, actin dynamics could be used as biosensor to maintain the cell in a population. HSP27/phosphoHSP27 levels could be part of this biosensor (Gourlay and Ayscough 2005).

4.3.8 HSP27 Phosphorylation and Cell Proliferation

Two observations indicate that phosphorylation of HSP27 may regulate the growth rate of cells: phosphoHSP27 levels accumulate as cells reached growth saturation and overexpression of HSP27-3D, but not HSP27-3A or HSP-3G inhibits cell proliferation (Arata et al. 1997; Salinthone et al. 2007; Matsushima-Nishiwaki et al. 2008; Trott et al. 2009). The molecular basis for HSP27-induced growth arrest is not exactly known, but several mechanisms may be operational. Overexpression of HSP27-D in human microvascular endothelial cell line HMEC-1 resulted in differential expression of 65 identifiable protein spots compared to control cells by liquid chromatography/mass spectrometry. Fifty-six peptide spots were down-regulated; most of them representing proteins that participate in regulation of the cytoskeleton and in the ubiquitin-proteasome pathway. Among these were stathmin, a protein that regulates the dynamics of microtubules, cofilin and vimentin. There is evidence that all three proteins are implicated in cell cycle control (Tzivion et al. 2000; Rubin and Atweh 2004; Hsu et al. 2010). Overexpression of HSP27-3D in human hepatocarcinoma HuH7 cells resulted in significant reduction in cyclin D1 levels caused by phosphoHSP27-mediated inhibition of the MEK/ERK pathway, which may explain cell cycle arrest (Matsushima-Nishiwaki et al. 2008). Another study has shown that HSP27 can directly interact with p53 and enhance transcription of the cyclin-dependent protein kinase inhibitor $p21^{Cip\text{-}1/Waf\text{-}1}$ and cause G_2/M cell cycle arrest of cardiac H9c2 cells (Venkatakrishnan et al. 2008). The p53-HSP27 interaction was more pronounced in heat shock treated cells than in control cells, but whether this was because of the increased levels of total HSP27 protein and/or due to enhanced phosphorylation of HSP27 was not tested. In line with the results obtained by Venkatakrishnan, O'Callaghan-Sunol and her colleagues found that depletion of HSP27 led to growth arrest of HCT116 colon carcinoma through the activation of the $p53/p21^{Cip\text{-}1/Waf\text{-}1}$ pathway (O'Callaghan-Sunol et al. 2007). Taken together, the phosphorylation status of HSP27 is an important regulator of the rate of cell proliferation. Hence identifying nuclear target proteins for phospoHSP27 may increase our understanding how phosphoHSP27 may participate in cell cycle regulation. One such target protein may be the cyclin-dependent kinase inhibitor $p27^{Kip1}$ which is triggered for degradation in cells overexpressing wild-type or HSP27-3D to promote S phase progression (Parcellier et al. 2006).

4.3.9 HSP27 Phosphorylation and DNA Repair

ATM promotes repair of double strand breaks, but also acts in cellular response to oxidative stress. In a recent study, Cosentino et al. (2011) demonstrated that ATM mediates HSP27 phosphorylation at Ser-78 and this promotes binding of phosphoHSP27 to glucose-6-phosphate dehydrogenase (G6PD). Interaction of phosphoHSP27 with G6PD induces the activity of G6PD, resulting in stimulation of the pentose phosphate pathway and nucleotide production which are required for the repair of double strand breaks (Cosentino et al. 2011).

4.3.10 HSP27 Phosphorylation and Subcellular Localization

Under normal conditions, HSP27 is predominantly present in the cytoplasm. A variety of stimuli that induce phosphorylation of HSP27, also cause accumulation of phosphoHSP27 in the nucleus (Wong et al. 2000; Geum et al. 2002; Al-Madhoun et al. 2007; Bryantsev et al. 2007b; Leja-Szpak et al. 2007; Shi et al. 2008a; Hong et al. 2009; de Thonel et al. 2010; Fujita et al. 2011). In resting HeLa cells, wild-type HSP27, HSP27-3A, and HSP27-3D mutant are homogenously localized in the cytoplasm. However, under heat shock wild type HSP27 and HSP27-3D, but not HSP27-A were partially translocated to the nucleus, suggesting that phosphorylation of Ser-15, Ser-78 or/and Ser-82 is not sufficient for HSP27 to enter the nucleus under normal conditions, but stimulates nuclear import after heat shock. The truncated HSP27(93-205) mutant, lacking the N-terminal 92 amino acids, was detected partially in the cytoplasm and the nucleus in untreated cells, but predominantly in the nucleus after heat shock. The C-terminal truncated HSP27 (1-173) mutant did not translocate to the nucleus after heat shock (Al-Madhoun et al. 2007). These results may indicate that phosphorylation of Ser-15, -78, and -82 are required, but not sufficient to trigger nuclear import of HSP27. The dissimilarity in subcellular distribution of HSP27 (1-173) and HSP27 (93-205) may suggest that residues between 173 and 205 may be involved in regulating the cytoplamic-nucleus allocation of HSP27. This region encompasses Ser-176 and Ser-199 which have been shown to be *in vivo* phosphoacceptor sites (see Table 4.1). The phosphorylation-dependent translocation of HSP27 may also be affected by interacting proteins. The oestrogen receptor-β (ER-β) can interact with HSP27 and this association is stimulated when HSP27 is phosphorylated at Ser-15, -78, and -82 (Al-Madhoun et al. 2007). Under control conditions, ERβ was largely confined to the nucleus. Only HSP27-3A accumulated in the nucleus after treatment of HeLa cells with 17β-oestradiol, while ER-β and wild-type HSP27, as well as HSP27-3D colocalized in discrete cytoplasmic speckle-like structures after exposure to 17β-oestradiol (Al-Madhoun et al. 2007). HSP27 resides predominantly in the cytoplasm of hippocampal HiB cells under normal conditions, while heat shock triggers nuclear import. The double mutant HSP27 S78A/S82A remained in the cytoplasm after heat shock, while HSP27 S78D/S82D was in the nucleus even in the absence of heat shock. These results indicate that phosphorylation of

Ser-78 and -82 are sufficient for nuclear localization in these cells. Whether phosphorylation of Ser-15 has an additional effect on HSP27's nuclear distribution was not tested (Geum et al. 2002). Bryantsev and co-workers showed that phosphorylation in itself was not sufficient to redistribute HSP27 from the cytoplasm to the nucleus as HSP27-3D resided in the cytoplasm in non-stressed normal rat kidney cells and Rat-1 fibroblasts (Bryantsev et al. 2007a, b). The size of nuclear pores varies with cell-type and with the condition of the cell and this may explain the discrepancy in results obtained by the two groups (reviewed in Elad et al. 2009 and Webster et al. 2009). Brynatsev et al. (2007a) observed nuclear accumulation of wild-type HSP27 and HSP27-3D upon heat shock. Surprisingly, HSP27-3G and HSP27-3A mutants behaved differently in subcellular localization upon heat shock. HSP27-3G translocated to the nucleus, whereas HSP27-3A remained in the cytoplasm after heat shock treatment of Rat-1 and mouse fibroblasts L929 (Bryantsev et al. 2007a). Moreover, Bryantsev et al. (2007b) found that HSP27-3D, but not HSP27-3A was selectively recruited to mitotic interchromatin granules in the absence of stress. Likewise, HSP27-3D fused to a nuclear localization signal peptide, but not wild-type and HSP27-3A were recruited to nuclear speckles in unstressed interphase cells. These findings strongly suggest that that oligomeric size and maybe additional events beside phosphorylation control the subcellular redistribution of HSP27 upon stress (Bryantsev et al. 2007b). Osteoblast-like MC3T3-E1 cells stably overexpressing wild-type HSP27 or HSP27-3A display HSP27 scattered in the cytosol, while HSP27 localized to the endoplasmic reticulum in cells stably overexpressing HSP27-3D (Kato et al. 2011). The exact reason is not known, but phosphorylated HSP27 may act as a functional regulator of the endoplasmic reticulum or protein modification processes occurring at this site.

The biological relevance of nuclear import of HSP27 is not completely understood. As outlined above, phosphorylated HSP27 can retain ERβ in the cytoplasm and therefore HSP27 may act as a co-repressor of oestrogen signalling. Nuclear translocation of phosphoHSP27 may also function to protect nuclear structures to prevent apoptotic cell death. Indeed, during the recovery period after heat shock, HSP27 gradually redistributed to the cytoplasm and there was an increment of apoptotic cells (Geum et al. 2002). Recruitment of phosphoHSP27 to nuclear speckles may support specific activities such as mRNA processing, release of storage mRNA during recovery after stress or proteasomal degradation of unfolded proteins that fail to refold (Bryantsev et al. 2007a, b).

4.3.11 HSP27 Phosphorylation and Vascular Functions

Several observations support a phosphorylation-dependent role for HSP27 in the biology of platelets. Zhu et al. (1994b) observed that HSP27 formed a specific complex with platelet factor XIII. It suggests that HSP27 might regulate FXIII activity within platelets and contributes to the stabilization of fibrin clot (Zhu et al. 1994b). In the resting platelet, HSP27 and platelet factor XIII are mainly cytoplasmic. Activation by agonist such as thrombin leads to phosphorylation of HSP27 followed by rapid translocation of HSP27 and FXIII to the peripheral submembranous region, which immediately precedes the association of HSP27 with the platelet cytoskeleton (Zhu et al. 1994a). Butt et al. (2001) identified HSP27 as a substrate of cGMP-dependent protein kinase (cGK) in intact platelets. They suggested that HSP27 participates in the cGK initiated inhibition of platelet aggregation and that cGK-mediated phosphorylation of HSP27 might influence association between platelet factor XIII and HSP27 (Butt et al. 2001). Another platelet activator, adenosine diphosphate (ADP), induces serotonin (5-HT), platelet-derived growth factor (PDGF) -AB and CD40 ligand (CD40 L) release, which play pivotal roles in the platelet secretion and aggregation (Kato et al. 2008; Doi et al. 2009; Kato et al. 2010). In addition, ADP has been reported to induce HSP27 phosphorylation in human platelets via activation of p38 MAPK and p44/p42 MAPK (Kato et al. 2008). Inhibition of p44/p42 MAPK is correlated with suppression of HSP27 phosphorylation level and inhibition of 5-HT, PDGF-AB and sCD40L secretion but not with aggregation of platelets (Kato et al. 2010). These findings suggest that HSP27 phosphorylation is sufficient for granule secretion from human platelets but not for platelet aggregation.

Additionally, HSP27 plays a pivotal role in vascular endothelial growth factor (VEGF) -dependent angiogenesis. VEGF induces phosphorylation of HSP27 at Ser-82 through PKC-dependent activation of PKD and independent of p38 MAPK (Evans et al. 2008, 2010). The exact role of phosphoHSP27 in VEGF-induced angiogenesis is not known, but probably phosphoHSP27 is involved in cytoskeletal dynamics.

4.3.12 HSP27 Phosphorylation and Gene Expression

Phosphorylation of HSP27 may have effects on transcriptional, post-transcriptional, translational and post-translational level. Non-phosphorylated HSP27 exerts a negative feedback on transcription of heat shock factor-1 (HSF-1)-regulated genes by inhibiting the transcriptional activity of HSF-1. Non-phosphorylated HSP27 binds HSF-1 and induces sumoylation of HSF-1, which blocks HSF1 transactivation capacity. Phosphorylated HSP27 fails to bind and to induce sumoylation of HSF1 and does not inhibit the transcriptional activity of HSF1. Hence phosphoHSP27 allows transcription of HSF-1 responsive genes (Brunet Simioni et al. 2009). PhosphoHSP27 was shown to bind the transcription factor GATA-1 and to induce its proteasomal degradation in erythroid differentiation (de Thonel et al. 2010). This may influence the transcription of GATA-1-regulated genes. HSP27 also promotes proteasomal degradation of AUF1, a protein involved in mRNA destabilization, in a phosphorylation-dependent manner (Knapinska et al. 2011). The authors demonstrated that HSP27-3D reduced total AUF1 levels, while HSP-3A had no effect. Studies with single, double, and triple phosphomimetic mutants showed that phosphorylation of Ser-15 is sufficient to destabilize AUF1 and promote mRNA degradation. Stabilization of cyclooxygenase 2 mRNA can also be controlled in a phosphoHSP27-dependent manner (Lasa et al. 2000). HSP27 stimulates protein synthesis after heat shock by promoting dephosphorylation (=activation) of eIF2α and by increasing the availability of eIF4E and eIF4G (Doerwald et al. 2006). Whether phosphorylation of HSP27 modulates protein synthesis was not examined, but stress induces HSP27 phosphorylation so that this posttranslational modification may be required. IL-1β treatment of rat C6 glioma cells results in increased expression of IL-6. Overexpression of HSP27-3A did not affect, while HSP27-3D strongly reduced IL1-β induced IL-6 expression. These results indicate that phosphorylation of HSP27 had a negative effect on IL-6 synthesis in response to IL-1β treatment of C6 glioma cells. IL-6 is implicated in the central nerve system in response to injury. It is therefore possible that phosphoHSP27 may exert a neuroprotective role by modulating the expression of IL-6 (Tanabe et al. 2010). The HSP27-3D mutant induced a stronger degradation of $p27^{Kip-1}$ than wild-type or HSP27-3A (Parcellier et al. 2006). All these examples illustrate a role for phosphoHSP27 in gene expression.

4.4 Aberrant HSP27 Phosphorylation and Diseases

Observed changes in HS27 phosphorylation in diseases tissue compared to healthy tissue point to a causal role in pathological processes. In this section some examples of HSP27 phosphorylation patterns in pathophysiological conditions will be given and a possible mechanism for phosphoHSP27 in these processes will be discussed.

4.4.1 Cancer

It is well established that intracellular or/and serum levels of HSP27 are elevated in numerous tumours compared to healthy tissue and that elevated HSP27 levels are correlated with aggressive, metastatic disease and shorter disease-free survival/recurrence following treatment (Ciocca and Calderwood 2005; Tremolada et al. 2005; Calderwood and Ciocca 2008; Khalili et al. 2011 and references therein). Many studies have shown increased phosphorylation levels of HSP27 in different metastatic cancer cells and have indicated that the phosphorylation status of HSP27 correlates with the metastatic potential of cancer cells (Tremolada et al. 2005; Xu et al. 2006; Arrigo 2007; Zhang et al. 2007; Lakshman et al. 2008). However, phosphorylated HSP27 levels can also be inversely correlated with the progression of tumours as e.g. shown in hepatocellular carcinoma (Yasuda et al. 2005). Many tumour cells possess a HSP27 phosphorylation pattern that is different from normal tissue. For example, Ser-82 was more highly phosphorylated than Ser-15 in renal cell carcinoma (Tremolada et al. 2005). Moreover, tumour subtypes may display distinct HSP27 phosphorylation as illustrated in breast cancer. Her-2/neu positive and negative breast cancers have increased pSer-82 levels compared to normal tissue, while no difference was observed for pSer-15. However, pSer-78 levels were significantly higher enhanced in Her-2/neu positive tumours compared to Her-2/neu negative tumours and non-tumour tissues (Zhang et al. 2007). The fact that not all patients amongst different chronic lymphocytic leukaemia patients had increased phosphoHSP27 levels jeopardizes a crucial role for HSP27 in this cancer (O'Hayre et al. 2010).

Little is known about the physiological relevance of altered HSP27 phosphorylation in tumour tissues compared to normal tissues. Inhibition of the p38MAPK/MK2 pathway in human prostate cancer cells reduced HSP27 phosphorylation, matrix metalloproteinase-2 (MMP-2) activation and invasion (Lakshman et al. 2008). These finding suggest that phosphoHSP27 may stimulate invasion through MMP-2 activation. On the other hand, NIH3T3 cells overexpressing wild-type HSP27 or HSP27-3D, but not HSP27-3A in NIH3T3 cells showed reduced MMP-2 expression, cell migration and invasion (Lee et al. 2008). The different results between these two studies may be explained by the use of different cells or because HSP27 was overexpressed in one study.

Recent studies measuring serum levels of HSP27 in breast cancer patients and age-matched healthy individuals revealed significantly higher levels in the cancer patients (Fanelli et al. 1998; Rui et al. 2003; Banerjee et al. 2011). Banerjee et al. (2011) reported that this extracellular HSP27 induced monocytes to secrete both immunosuppressive (e.g. interleukins IL-6 and IL-10, prostaglanding E_2) and proangiogenic molecules such as vascular endothelial growth factor A, IL-8, IL-1β, and tumour necrosis factor (TNF)-α. Soluble HSP27 also triggered increased levels of monocyte chemotactic protein-1, which is responsible for the recruitment of monocytes to the breast tumour site. HSP27 induced differentiation of these monocytes into tumour-associated macrophages that exhibit reduced cytotoxic activity but extremely high neovascularization and tumour promoting properties. Taken together, these findings imply an immunosuppressive and tumour-supporting role for secreted HSP27 in breast cancer and maybe in other cancers because increased serum levels of HSP27 have also been described in hepatocellular and pancreatic carcinomas (Feng et al. 2005; Melle et al. 2007). Whether phosphorylation of HSP27 affects the role of HSP27 in these processes is not known. Banerjee et al. (2011) observed that the phosphoSer-82 level of secreted HSP27 in medium of a primary breast cancer cell and of two established cell lines was 50–75 % lower than recombinant HSP27 (Banerjee et al. 2011). Unfortunately, they did not monitor phosphoSer-82 amplitude of secreted HSP27 in medium of normal human mammary epithelial cell lines or in sera from healthy subjects and breast cancer patients. However, as both recombinant and breast tumour cell-released extracellular HSP27 differentiated monocytes into tumour-associated macrophages, phosphorylation of HSP27, at least at Ser-82, may not be absolutely required for the tumour progression supporting role of secreted HSP27.

Results from the group of Chen demonstrate the importance of differential phosphorylation of HSP27 in metastasis of human head-and-neck squamous cell carcinoma cells (Kang et al. 2010). The authors show that the mitogen-activated protein kinase-activated protein kinase RSK2 phosphorylates HSP27 at Ser-78 and -82 but not -15 *in vivo* and promotes cell invasion by phosphorylating HSP27 to regulate stabilization of actin filaments in human head-and-neck squamous cell carcinoma cell lines. Interestingly, overexpression of HSP27 S78D or S82D did not lead to enhanced cell invasive capability, whereas overexpression of the double mutant HSP27 S78D/S82D significantly potentiated cell invasion. Stable expression of HSP27 S78D/S82D, but not wild-type or phosphodeficient S78A/S82A mutant, partially rescued cell invasion in the head-and-neck squamous cell carcinoma cell line M4e in which RSK2 expression was depleted. These results illustrate that phosphorylation of both sites is required for RSK2-induced invasion of human head-and-neck squamous cell carcinoma cells.

The chemokine CXCL12 secreted by the microenvironment is one of the factors known to promote survival of chronic lymphocytic leukaemia cells (Nishio et al. 2005). Phosphoproteomic analysis showed that phosphoSer-82 HSP27 levels were increased in CXCL12-treated primary chronic lymphocytic leukaemia cells compared to untreated cells (O'Hayre et al. 2010). The enhanced phosphoHSP27 levels may promote cell survival, but may also increase resistance to chemotherapy.

Expression and activity of the androgen receptor (AR) is required for survival and growth of prostate cancer (Lamont and Tindall 2011). In prostate cancer cells, androgen-bound-AR induces HSP27 phosphorylation on Ser-78 and Ser-82 (Ser-15 was not tested) in a p38MAPK-dependent manner and the phosphorylated form of HSP27 becomes more associated with AR. In response to androgen, cytoplasmic AR dissociates from HSP90 and can be chaperoned by phosphorylated HSP27 into the nucleus. Phosphorylated HSP27 stimulated the transcriptional activity of AR. Inhibition of HSP27 phosphorylation by overexpressing HSP27-3A or knockdown of HSP27 shifted the association of AR with HSP90 to MDM2, increased proteasome-mediated AR degradation, decreased AR transcriptional activity, and increased prostate cancer apoptotic rates (Zoubeidi et al. 2007). These findings underscore a contributing, phosphorylation-dependent role for HSP27 in prostate cancer. Another study reported that phosphorylation of HSP27 at Ser 82 by PKD1 in p38 MAPK-dependent fashion mediated repression of AR and induced the translocation of AR from nucleus to cytoplasm (Hassan et al. 2009). The use of different prostate cancer cell lines may explain the discrepancies.

Another mechanism by which HSP27 may promote oncogenesis is through interference with the NF-κB signalling pathway. PhosphoHSP27 functions as a negative regulator of the NF-κB pathway via its association with I-κB kinase. The interaction of HSP27 with I-κB kinase is enhanced by phosphorylation and results in decreased I-κB kinase activity (Park et al. 2003). Because NFκB signalling plays a critical role in inflammation and cancer, aberrant HSP27 phosphorylation may perturb this pathway and as such be a causal factor in cancer (Chen and Greene 2004 and references therein).

4.4.2 Cardiovascular Diseases

Several studies propose a role for phosphoHSP27 in cardiovascular diseases. Phosphorylation of HSP27 at serine residues 15, 78 and 82 increased following cardioplegic arrest and this negatively correlates with the cardiac performance following this intentional cessation of cardiac activity (Clements et al. 2007). Studies in isolated rat cardiomyocytes demonstrated that overexpression of the nonphosphorylatable human HSP27 S15A/S78A/S82A triple mutant completely rescued cardioplegic arrest-induced deficits in myocyte contractility. Although the exogenous human HSP27 S15A/S78A/S82A mutant ablated cardioplegic arrest-triggered impaired contractility, it did not prevent phosphorylation of endogenous ratHSP25. This demonstrates that the beneficial effects of the non-phosphorylatable HSP27 mutant are independent of endogenous HSP27 and may indicate that maintaining a pool of unphosphorylated HSP27 rather than phosphorylation of HSP27 per se, promotes cardiac contractile function (Clements et al. 2011). Another observation supports the view that preservation of non-phosphorylated HSP27 improves cardiac functions. Overexpression of either wild-type or the non-phosphorylatable HSP27 S15A/S78A/S82A mutant partially blocks ischemia-induced contractile deficits in mouse hearts (Hollander et al. 2004). Aberrant phosphoHSP27 levels and migration of smooth muscle cells has been suggested to contribute to multiple cardiac diseases, including atherosclerosis, restenosis and transplant vasculopathy (Hedges et al. 1999; Durán et al. 2007; Gabunia et al. 2011). PhosphoHSP27-induced stimulation of vascular smooth muscle cells migration may be a contributing factor in cardiac disease.

The phosphorylation levels of HSP27 are increased in patients with cardiac graft vasculopathy (de Souza et al. 2005), while they are reduced in atherosclerotic plaques and adjacent artery segments (Park et al. 2006). PhosphoHSP27 levels were also strongly reduced in coronary arteries from patients with ischemic heart disease compared to patients with dilated cardiomyopathy and non-diseased heart donors (Robinson et al. 2010). LDLs stimulate dephosphorylation of HSP27 in vascular smooth muscle cells (Garcia-Arguinzones et al. 2010). The biological consequence is not known, but the authors speculate that it may represent a protective response of the cell to the massive entry of cholesterol and hence play an important mechanistic role in the progression of the atherosclerotic plaques. High levels of LDLs may reduce the concentration of phosphorylated HSP27 in vascular smooth muscle cells, thereby modulating cytoskeletal dynamics and making the cells more susceptible to rupture. Moreover, LDL-triggered HSP27 dephosphorylation may reduce adhesion and migration of vascular smooth muscle cells (Garcia-Arguinzones et al. 2010). PhosphoHSP27 may also protect against vascular disease by stabilizing the actin cytoskeleton within endothelial and/or smooth muscle cells (Robinson et al. 2010).

Another example revealing differential HSP27 phosphorylation in cardiovascular deficiencies derives from a study comparing of the phosphoproteome of aortic samples from patient with aortic aneurysm of which six had bicuspid aortic valve with nine subject that had normal tricuspid aortic valve. Reduced levels of phosphoSer-78 HSP27 were observed in patients with bicuspid aortic valve, but total HSP27 levels were 1.4-fold higher in the subjects with tricuspid aortic valve, while phosphoSer-15 and dually phosphorylated (Ser-78 and -82) HSP27 levels were similar in both groups (Matt et al. 2007). Whether there is a causal relation between altered phosphoHSP27 levels and bicuspid aortic valve remains to be established.

Augmented Hsp27 phosphorylation may also play a causal role in the diabetic heart (Kim et al. 2008a). Whereas in normal heart muscle ~30 % of the energy is provided by carbohydrate and 70 % through fatty acids metabolism, fatty acid is the sole energy source in a diabetic cardiomyocytes. This switch in energy supply is achieved by increased lipoprotein lipase (LPL) activity at the coronary lumen. Phosphorylation of Hsp27 resulted in disruption of Hsp27-PKCδ complexes in cardiomyocytes, allowing PKCδ to activate PKD. Activated PKD stimulated LPL activity at the coronary lumen, causing hydrolysis of triglyceride-rich lipoproteins to fatty acids. The increased Hsp27 phosphorylation is mediated through p38 MAPK and, although not proven by the authors, they suggested MK2 (Kim et al. 2008a, b).

4.4.3 Hypoxia-Induced Injury

Hypoxia increases permeability and gap formation of the pulmonary endothelial barrier leading to pulmonary oedema, and can also expand permeability of the endothelial barrier in other tissues. PhosphoHSP27 may play an important role in these processes as suggested by the work of Liu et al. (2009). Overexpression of HSP27-3D in rat pulmonary microvascular endothelial cells abolishes increased permeability and reduced gap formation induced by hypoxia, despite increased stress fibre formation (Liu et al. 2009). The authors examined the phosphorylation state of myosin light chain-2 (MLC2), myosin phosphatase target subunit (MYPT1) and focal adhesion kinase (FAK) because phosphorylation of these proteins is related to increased permeability of endothelial monolayer. They found that hypoxia augmented the phosphorylation levels of these proteins, but overexpression of HSP27-3D prevented this. These results suggest that targeting HSP27 might be a useful therapeutic strategy in managing endothelial permeability and oedema (Liu et al. 2009).

4.4.4 HSP27 Phosphorylation and Viral Infection

Respiratory syncytial virus (RSV) is the major cause of bronchiolitis and pneumonia in infants, but infection with this virus can also lead to severe respiratory infections in elderly and immunocompromised individuals (Collins and Crowe 2007). RSV infection is characterized by accumulation of fluid in the infected lungs due to increased lung permeability. This fluid extravasation can cause alveolar flooding, running nose, middle ear effusions, and increase in bacterial infections (Collins and Crowe 2007). Singh and co-workers demonstrated that RSV infection was concomitant with increased HSP27 phosphorylation on both Ser-78 and Ser-82 (Ser-15 was not examined), and leads to actin microfilament rearrangement, cell shape modification, and change in cell permeability (Singh et al. 2007). This is in contrast with hypoxia-induced increased cell permeability that was prevented by phospoHSP27 (see previous section). The opposite effect of phosphoHSP27 may be explained by the different stimuli that trigger HSP27 phosphorylation and/or by the cell type.

Epstein-Barr virus, Herpes simplex virus type 1, Sindbis virus and Reovirus infection of cell cultures also trigger HSP27 phosphorylation, but the biological relevance is not known (Nakatsue et al. 1998; Fukagawa et al. 2008; Mathew et al. 2009; Li et al. 2010). Li et al. 2010 showed that Reovirus causes HSP27 phosphorylation by activating the p38MAPK kinase pathway and that inhibition of this pathway increases virus-induced cytopathogenic effect (Li et al. 2010). Nakatsue et al. (1998) speculated that increase in phosphoHSP27 may play a role for delaying Sindbis virus-induced cell death such that phosphoHSP27 may be favourable for viral replication. Infection with herpes simplex virus type 1 induces redistribution of HSP27 with phosphorylated HSP27 excluded from the nucleus, whereas the unphosphorylated form is dispersed throughout the cell. The importance of the difference is subcellular localization during viral infection is not understood. While depletion of total HSP27 reduced viral replication, indicating the importance of HSP27, the exact role of phospoHSP27 remains to be determined (Mathew et al. 2009).

4.4.5 Hsp27 Phosphorylation and the Autoimmune Disease Pemphigus

Pemphigus vulgaris and pemphigus foliaceus are autoimmune diseases of the skin characterized by autoantibodies against desmoglein-3 and -1, respectively, which cause loss of keratinocyte cell-cell adhesion or acantholysis (Berkowitz et al. 2005). These autoantibodies induce phosphorylation of HSP27 in cell culture and skin biopsies from all pemphigus patients display increased phosphoHSP27 levels compared to healthy controls (Berkowitz et al. 2005, 2008a, b). Although studies in cell cultures and animals showed that desmoglein autoantibodies induced reorganization of the actin cytoskeleton and intermediate filament collapse, the exact role of phosphoHSP27 in acantholysis remains unknown (Berkowitz et al. 2005, 2008a, b; Lee et al. 2009). Pretreatment with p38MAPK inhibitor before injection of desmoglein-1 autoantbodies purified from sera of pemphigus foliaceus patients blocked cytoskeletal reorganization in human keratinocyte cell cultures or acantholysis in mouse models (Berkowitz et al. 2005, 2008a, b; Lee et al. 2009), suggesting the MK2 inhibitors could have a beneficial effect for these patients.

4.4.6 Hsp27 Phosphorylation in Kidney Diseases

Abnormal Hsp27 phosphorylation is not only observed in renal cancers, but also other kidney diseases. Enhanced Hsp27 phosphorylation compared to control animals was observed in animal models of nephrotic syndrome and diabetic nephropathy (Smoyer et al. 1996; Dai et al. 2006; Barutta et al. 2008; Park et al. 2008). Increased HSP27 phosphorylation may also be involved in epithelial-to-mesenchymal transition of rat proximal tubular epithelial cells into myofibroblasts in kidney fibrosis, probably through HSP27-induced upregulation of E-cadherin (Vidyasagar et al. 2008). The effect of overexpression of phosphomimicking HSP27 mutant on EMT was not tested and further studies are required to establish whether anomalous phosphoHSP27 levels play a causative role in kidney fibrosis.

4.4.7 HSP27 Phosphorylation and Neurodegenerative Diseases

Perturbed expression and accumulation of the microtubule associated protein tau is a hallmark of many neurodegenerative diseases (Gendron and Petrucelli 2009). Abisambra et al. 2010 showed that wild-type HSP27 was more efficient than HSP27-3D at abrogating tau aggregation *in vitro*. To test the effect *in vivo*, the authors used adeno-associated viral delivery of wild-type HSP27 and HSP27-3D in a transgenic mouse model of tauopathy. These experiments confirmed that wild-type HSP27 facilitated tau clearance better than HSP27-3D and that wild-type HSP27, but not HSP27-3D was able to functionally rescue hippocampal long–term potentiation deficits. These results directs to the importance of the unphosphorylated state of HSP27 to clear its client protein tau. It would have been interesting to test tau aggregation in the presence of HSP27-3A. The authors also demonstrated that HSP27 could block tau aggregation, but could not disaggregate preformed tau fibrils. This property should be considered when designing HSP27-based therapies for tauopathies.

High fat diet exacerbated neurodegeneration in Fisher 344 rats, an animal model of Parkinson's and was associated with decreased phosphoHSP27 levels in the substantia nigra of these animals (Morris et al. 2010). Whether aberrant phospoHSP27 levels in play a causal role in the development of Parkinson's remains to be established, but they may affect proper protein folding and aggregation in neurons and thus contribute to protein misfolding disorders such as Parkinson's disease (Chen and Brown 2007).

4.5 Modulating HSP27 Phosphorylation and Therapy

Specific drugs impeding with HSP27 phosphorylation form an attractive therapeutic target because phosphoHSP27 is involved in a wide variety of malignant and other diseases. Drugs that interfere with HSP27 phosphorylation may include HSP27 kinases inhibitors or molecules that interact with HSP27 thereby preventing its phosphorylation. Cell culture studies with cell permeable MK2 inhibitor peptides showed decreased HSP27 phosphorylation, whereas the synthetic molecule KRIBB3 and derivatives were shown to specifically bind HSP27 and to prevent PMA-induced HSP27 phosphorylation, tumour cell migration at low concentrations (nM range) and microtubule polymerization (Shin et al. 2005; Lopes et al. 2009; Brugnano et al. 2011; Lee et al. 2011). These results indicate that administration of HSP27 kinase inhibitors or peptides that hamper HSP27 phosphorylation have therapeutic potentials. Drugs that stimulate HSP27 phosphorylation may be applied in other clinical conditions. Quercetin and specific quercetin derivates can induce phosphorylation of Hsp27 at Ser-78 by an unknown mechanism (Wang et al. 2009). However, these drugs inhibit expression of Hsp70 and inhibit casein kinase II and CaMK activity, making them unsuitable for therapy. Depletion of HSP27/phosphoHSP27 by treatment with the HSP27 antisense oligonucleotide OGX-427 enhances the cytotoxic effect of radiotherapy on mice with xenografted head-and-neck squamous tumours (Hadchity et al. 2009) and OGX-427 has entered phase I clinical trials for treatment of different cancers (ClinicalTrials.gov). Another example is apigenin, a natural plant flavonoid that induces apoptosis when administered to leukaemia cells. Phosphorylation of HSP27 significantly increased the susceptibility of leukaemia cells to apigenin-induced apoptosis, making this compound a potential drug for tumour cells with increased phosphoHSP27 levels (Gonzalez-Mejia et al. 2010).

4.6 Conclusion and Future Perspective

HSP27 participates in a plethora of cellular processes and phosphorylation of HSP27 can affects its function in these processes. Phosphorylation has mainly been studied on Ser-15, -78 and -82, but additional sites are found to be phosphorylated *in vivo* and the kinases that mediate these phosphorylations have not been identified. The biological relevance of HSP27 phosphorylation at these particular sites, as well as on other sites than Ser-15, -78, -82 remains unexplored. For example, while several studies with triple Ser-15, -78 and -82 into Ala or Asp mutants have shown the importance of multiple phosphorylation of HSP27 at these residues, the contribution of single phosphorylation or combination of double phosphorylations of these amino acids has not always been meticulously investigated. Another challenge for future research is to unveil the exact role of phosphoHSP27 in normal and pathogenic conditions. Many studies have been performed by overexpressing mutant forms of HSP27. Generating transgenic animals in which specific residues of the endogenous *hsp27* gene have been mutated may provide a more authentic clue on the role of HSP27 phosphorylation in cellular processes. On the other hand, HSP27 and phosphoHSP27 protein levels are often increased in stressed and diseased cells, so that overexpression of these mutants actually resembles the pathological conditions. Several decades of intensive research has revealed many exciting functions HSP27. No doubt, HSP27 will not disappoint us as future research will disclose more properties of this protein.

References

Al-Madhoun AS, Chen YX, Haidari L, Rayner K, Gerthoffer W, McBride H, O'Brien ER (2007) The interaction and cellular localization of HSP27 and ERβ are modulated by 17β-estradiol and HSP27 phosphorylation. Mol Cell Endocrinol 270:33–42

Abisambra JF, Blair LJ, Hill SE, Jones JR, Kraft C, Rogers J, Koren J III, Jinwal UK, Lawson L, Johnson AG, Wilcock D, O'Leary JC, Jansen-West K, Muschol M, Golde TE, Weeber EJ, Banko J, Dickey CA (2010) Phosphorylation dynamics regulate Hsp27-mediated rescue of neuronal plasticity deficts in tau transgenic mice. J Neurosci 30:15374–15382

Arata S, Hamaguchi S, Nose K (1997) Inhibition of colony formation of NIH 3T3 cells by the expression of the small molecular weight heat shock protein HSP27: involvement of its phosphorylation and aggregation at the C-terminal region. J Cell Physiol 170:19-26

Arrigo AP (2001) Hsp27: novel regulator of intracellular redox state. IUBMB Life 52:303–307

Arrigo AP (2007) The cellular "networking" of mammalian Hsp27 and its functions in the control of protein folding, redox state and apoptosis. Adv Exp Med Biol 594:14–26

Banerjee S, Lin CFL, Skinner KA, Schiffhauer LM, Peacock J, Hicks DG, Redmond EM, Morrow D, Huston A, Shayne M, Langstein HN, Miller-Graziano CL, Strickland J, O'Donoghue L, De AK (2011) Heat shock protein 27 diferentiates tolerogenic marcrophages that may support human breast cancer progression. Cancer Res 71:318–327

Barutta F, Pinach S, Giunti S, Vittone F, Forbes JM, Chiarle R, Arnstein M, Perin PC, Camussi G, Copper ME, Gruden G (2008) Heat shock protein expression in diabetic nephropathy. Am J Physiol Renal Physiol 295:F1817–F1824

Beausoleil SA, Jedrychowski M, Schwartz D, Elias JE, Villén J, Li J, Cohn MA, Cantley LC, Gygi SP (2004) Large-scale characterization of HeLa cell nuclear phosphoproteins. Proc Natl Acad Sci U S A 101:12130–12135

Berkowitz P, Hu P, Liu Z, Diaz LA, Enghild JJ, Chua MP, Rubenstein DS (2005) Desmosome signaling. Inhibition of p38MAPK prevents pemphigus vulgaris IgG-induced cytoskeleton reorganization. J Biol Chem 280:23778–23784

Berkowitz P, Diaz LA, Hall RP, Rubenstein DS (2008a) Induction of p38MAPK and Hsp27 phosphorylation in pemphigus patient skin. J Invest Dermatol 128:738–740

Berkowitz P, Chua M, Liu Z, Diaz LA, Rubenstein DS (2008b) Autoantibodies in the autoimmune disease pemphigus foliaceus induce blistering via p38 mitogen-activated protein kinase-dependent signalling in the skin. Am J Pathol 173:1628–1636

Berrou E, Bryckaert M (2009) Recruitment of protein phosphatase 2A to dorsal ruffles by platelet-derived growth factor in smooth muscle cells: dephosphorylation of Hsp27. Exp Cell Res 315:836–848

Bertrand-Vallery V, Belot N, Dieu M, Delaive E, Ninane N, Demazy C, Raes M, Salmon M, Poumay Y, Debacq-Chainiaux F, Toussaint O (2010) Proteomic profiling of human keratinocytes undergoing UVB-induced alternative differentiation reveals TRIpartite Motif Protein 29 as a survival factor. PLoS One 5:e10462

Bitar KN, Patil SB (2004) Aging and gastrointestinal smooth muscle. Mech Ageing Dev 125:907–910

Blom N, Gammeltoft S, Brunak S (1999) Sequence- and structure-based prediction of eukaryotic protein phosphorylation sites. J Mol Biol 294:1351–1362

Bova MP, Mchaourab HS, Han Y, Fung, BKK (2000) Subunit exchange of small heat shock proteins. Analysis of oligomer formation of aA-crystallin and hsp27 by fluorescence resonance energy transfer and site-directed truncations. J Biol Chem 275:1035–1042

Brugnano JL, Chan BK, Seal BL, Panitch A (2011) Cell-penetrating peptides can confer biological function: regulation of inflammatory cytokines in human monocytes by MK2 inhibitor peptides. J Control Release 155:128–133

Brunet Simioni M, De Thonel A, Hammann A, Joly AL, Bossis G, Fourmaux E, Bouchot A, Landry J, Piechaczyk M, Garrido C (2009) Heat shock protein 27 is involved in SUMO-2/3 modification of heat shock factor 1 and thereby modulates the transcription factor activity. Oncogene 28:3332–3344

Bryantsev AL, Kurchashova SY, Golyshev SA, Polyakov VY, Wunderink HF, Kanon B, Budagova KR, Kabakov AE, Kampinga HH (2007a) Regulation of stress-induced intracellular sorting and chaperone function of Hsp27 (HspB1) in mammalian cells. Biochem J 407:407–417

Bryantsev AL, Chechenova MB, Shelden EA (2007b) Recruitment of phosphorylated small heat shock protein Hsp27 to nuclear speckles without stress. Exp Cell Res 313:195–209

Bukach OV, Glukhova AE, Seit-Nebi AS, Gusev NB (2009) Heterooligomeric complexes formed by human small heat shock proteins HspB1 (Hsp27) and HspB6 (Hsp20). Biochim Biophys Acta 1794:486–495

Butt E, Immler D, Meyer HE, Kotlyarov A, Laass K, Gaestel M (2001) Heat shock protein 27 is a substrate of cGMP-dependent protein kinase in intact human platelets: phosphorylation-induced actin polymerization caused by HSP27 mutants. J Biol Chem 276:7108–7113
Calderwood SK, Ciocca DR (2008) Heat shock proteins: stress protein with Janus-like properties in cancer. Int J Hyperthermia 24:31–39
Calderwood SK, Murshid A, Prince T (2009) The shock of aging: molecular chaperines and the heat shock response in longevity and aging—a mini-review. Gerontology 55:550–558
Cairns J, Qin S, Philp R, Tan YH, Guy GR (1994) Dephosphorylaton of the small heat shock protein Hsp27 in vivo by protein phosphatase 2A. J Biol Chem 269:9176–9183
Chandrika BB, Maney SK, Lekshmi SU, Retnabhai ST (2010) Endoplasmic reticulum targeted Bcl2 confers long term survival through phosphorylation of heat shock protein 27. Int J Biochem Cell Biol 42:1984–1992
Charette SJ, Lavoie JN, Landry J (2000) Inhibition of Daxx-mediated apoptosis by heat shock protein 27. Mol Cell Biol 20:7602–7612
Chen LF, Greene WC (2004) Shaping the nuclear action of NF-kappaB. Nat Rev Mol Cell Biol 5:392–401
Chen S, Brown IR (2007) Neuronal expression of constitutive heat shock proteins: implications for neurodegenerative diseases. Cell Stress Chaperones 12:51–58
Chen HF, Xie LD, Xu CS (2010) The signal transduction pathways of heat shock protein 27 phosphorylation in vascular smooth muscle cells. Mol Cell Biochem 333:49–56
Ciocca DR, Calderwood SK (2005) Heat shock proteins in cancer: diagnostic, prognostic, predictive, and treatment implications. Cell Stress Chaperones 10:86–103
Clements RT, Sodha NR, Feng J, Mieno S, Boodhwani M, Ramlawi B, Bianchi C, Sellke FW (2007) Phosphorylation and translocation of heat shock protein 27 and alphaB-crystallin in human myocardium after cardioplegia and cardiopulmonary bypass. J Thorac Cardiovasc Surg 134:1461–1470
Clements RT, Feng J, Cordeiro B, Bianchi C, Sellke FW (2011) p38 MAPK-dependent small HSP27 and αB-crystallin phosphorylation in regulation of myocardail function following cardioplegic arrest. Am J Physiol Heart Circ Physiol 300:H1669–H1677
Collins PL, Crowe JE Jr (2007) Respiratory syncytial virus and metapneumovirus. In: Knipe DM, Howley PM (eds) Fields virology. Lippincott Williams & Wilkins, Philadelphia, pp 1601–1646
Concannon CG, Gorman AM, Samali A (2003) On the role of Hsp27 in regulating apoptosis. Apoptosis 8:61–70
Cosentino C, Grieco D, Costanzo V (2011) ATM activates the pentose phosphate pathway promoting anti-oxidant defence and DNA repair. EMBO J 30:546–555
Craig T, Smelick C, de Magalhaes JP (2010) The digital ageing atlas. http://ageing-map.org
Dai T, Natarajan R, Nast CC, LaPage J, Chuang P, Sim J, Tong L, Chamberlin M, Wang S, Adler, SG (2006) Glucose and diabetes: effects on podocyte and glomerular p38MAPK, heat shock protein 25, and actin cytoskeleton. Kidney Int 69:806–814
De Thonel A Vandekerckhove J, Lanneau D, Selvakumar S, Courtois G, Hazoume A, Brunet M, Maurel S, Hammann A, Ribeil JA, Zermati Y, Gabet AS, Boyes J, Solary E, Hermine O, Garrido C (2010) HSP27 controls GATA-1 protein level during erythroid cell differentiation. Blood 116:85–96
de Souza AI, Wait R, Mitchell AG, Banner NR, Dunn MJ, Rose ML (2005) Heat shock protein 27 is associated with freedom from graft vasculopathy after human cardiac transplantation. Cir Res 97:192–198
Doerwald L, van Genesen ST, Onnekink C, Marin-Vinader L, de Lange F, de Jong WW, Lubsen NH (2006) The effect of αB-crystallin and Hsp27 on the availability of translation initiation factors in heat-shocked cells. Cell Mol Life Sci 63:735–743
Doi T, Adachi S, Takai S, Matsushima-Nishiwaki R, Kato H, Enomoto Y, Minamitani C, Otsuka T, Tokuda H, Akamatsu S, Iwama T, Kozawa O, Ogura S (2009) Antithrombin III suppresses ADP-induced platelet granule secretion: inhibition of HSP27 phosphorylation. Arch Biochem Biophys 489:62–67

Dokas LA, Malone AM, Williams FE, Nauli SM, Messer WS Jr (2011) Multiple protein kinases determine the phosphorylated state of the small heat shock protein, HSP27, in SH-SY5Y neuroblastoma cells. Neuropharmacology 61:12–24

Dominguez R, Holmes KC (2011) Actin structure and function. Annu Rev Biophys 40:169–186

Doshi BM, Hightower LE, Lee J (2010) HSPB1, actin filament dynamics, and aging cells. Ann N Y Acad Sci 1197:76–84

Durán MC, Boeri-Erba E, Mohammed S, Martin-Ventura JL, Egido J, Vivanco F Jensen ON (2007) Characterization of HSP27 phosphorylation sites in human atherosclerotic plaque secretome. Methods Mol Biol 357:151–163

Ehrnsperger M, Gräber S, Gaestel M, Buchner J (1997) Binding of non-native protein to Hsp25 during heat shock creates a reservoir of folding intermediates for reactivation. EMBO J 16:221–229

Elad N, Maimon T, Frenkiel-Krispin D, Lim RY, Medalia O (2009) Structural analysis of the nuclear pore complex by integrated approaches. Curr Opin Struct Biol 19:226–232

Evans IM, Britton G, Zachary IC (2008) Vascular endothelial growth factor induces heat shock protein (HSP) 27 serine 82 phosphorylation and endothelial tubulogenesis via protein kinase D and independent of p38 kinase. Cell Signal 20:1375–1384

Evans IM, Bagherzadeh A, Charles M, Raynham T, Ireson C, Boakes A, Kelland L Zachary IC (2010) Characterization of the biological effects of a novel protein kinase D inhibitor in endothelial cells. Biochem J 429:565–572

Fanelli MA, Cuello Carrion FD, Dekker J, Schoemaker J, Ciocca DR (1998) Serological detection of heat shock protein hsp27 in normal and breast cancer patients. Cancer Epidemiol Biomarkers Prev 7:791–795

Feng JT, Liu YK, Song HY, Dai Z, Qin LX, Almofti MR, Fang CY, Lu HJ, Yang PY, Tang ZY (2005) Heat-shock protein 27: a potential biomarker for hepatocellular carcinoma identified by serum proteome analysis. Proteomics 5:4581–4588

Ferns G, Shams S, Shafi S (2006) Heat shock protein 27: its potential role in vascular disease. Int J Exp Path 87:253–274

Fujita R, Ounzain S, Wang AC, Heads RJ, Budhram-Mahadeo VS (2011) Hsp-27 induction requires POU4F2/Brn-3b TF in doxorubicin-treated breast cancer cells, whereas phosphorylation alters its cellular localization following drug treatment. Cell Stress Chaperones 16:427–439

Fukagawa Y, Nishikawa J, Kuramitsu Y, Iwakiri D, Takada K, Imai S, Satake M, Okamoto T, Fujimoto M, Okita K, Nakamura K, Sakaida I (2008) Epstein-Barr virus upregulates phosphorylated heat shock protein 27 kDa in carcinoma cells using the phosphoinositide 3-kinase/Akt pathway. Electrophoresis 29:3192–3200

Gabunia K, Jain S, England RN, Autieri MV (2011) Anti-inflammatory cytokine interleukin-19 inhibits smooth muscle cell migration and activation of cytoskeletal regulators of VSMC motility. Am J Physiol Cell Physiol 300:C896–C906

Gaestel M, Benndorf R, Hayess K, Priemer E, Engel K (1992) Dephosphorylation of the small heat shock protein hsp25 by calcium/calmodulin-dependent (type 2B) protein phosphatase. J Biol Chem 267:21607–21611

Garcia-Arguinzonis M, Padró T, Lugano R, Llorente-Cortes V, Badimon L (2010) Low-density lipoproteins induce heat shock protein 27 dephosphorylation, oligomerization, and subcellular relocalization in human vascular smooth muscle cells. Arterioscler Thromb Vasc Biol 30:1212–1219

Garrido C, Brunet M, Didelot C, Zermati Y, Schmitt E, Kroemer G (2006) Heat shock proteins 27 and 70: anti-apoptotic proteins with tumorigenic properties. Cell Cycle 5:2592–2601

Gendron TF, Petrucelli L (2009) The role of tau in neurodegeneration. Mol Neurodegener 4:13

Geum D, Son GH, Kim K (2002) Phosphorylation-dependent cellular localization and thermoprotective role of heat shock protein 25 in hippocampal progenitor cells. J Biol Chem 277:19913–19921

Gnad F, Gunawardena J, Mann M (2011) Phosida 2011: the posttranslational modification database. Nucleic Acids Res 39:D253–D260

Gonzalez-Mejia ME, Voss OH, Murnan EJ, Doseff AI (2010) Apigenin-induced apoptosis of leukemia cells is mediated by a bimodal and differentially residue-specific phosphorylation of heat-shock protein-27. Cell Death Dis 1:e64

Gourlay CW, Ayscough KR (2005) The actin cytoskeleton in ageing and apoptosis. FEMS Yeast Res 5:1193–1198

Gusev NB, Bogatcheva NV, Marston SB (2002) Structure and properties of small heat shock proteins (sHsp) and their interaction with cytoskeleton proteins. Biochemistry 67:511–519

Hadchity E, Aloy MT, Paulin C, Armandy E, Watkin E, Rousson R, Gleave M, Chapet O, Rodriguez-Lafrasse C (2009) Heat shock protein 27 as a new therapeutic target for radiation sensitization of head and neck squamous cell carcinoma. Mol Ther 17:1387–1394

Haslbeck M, Franzmann T, Weinfurtner D, Buchner J (2005) Some like it hot: the structure and function of small heat-shock proteins. Nat Struct Mol Biol 12:842–846

Hassan S, Biswas MH, Zhang C, Du C, Balaji KC (2009) Heat shock protein 27 mediates repression of androgen receptor function by protein kinase D1 in prostate cancer cells. Oncogene 28:4386–4396

Havasi A, Li Z, Wang Z, Martin JL, Botla V, Ruchalski K, Schwartz JH, Borkan SC (2008) Hsp27 inhibits Bax activation and apoptosis via a phosphatidylinositol 3-kinase-dependent mechanism. J Biol Chem 283:12305–12313

Hayes D, Napoli V, Mazurkie A, Stafford WF, Graceffa P (2009) Phosphorylation dependence of Hsp27 multimeric size and molecular chaperone function. J Biol Chem 284:18801–18807

Hedges JC, Dechert MA, Yamboliev IA, Martin JL, Hickey E, Weber LA, Gerthoffer WT (1999) A role for p38(MAPK)/HSP27 pathway in smooth muscle cell migration. J Biol Chem 274:24211–24219

Hollander JM, Martin JL, Belke DD, Scott BT, Swanson E, Krishnamoorthy V, Dillmann WH (2004) Overexpression of wild-type heat shock protein 27 and a nonphosphorylatable heat shock protein 27 mutant protects against ischemia/reperfusion injury in a transgenic mouse model. Circulation 110:3544–3552

Hong Z, Zhang QY, Liu J, Wang ZQ, Zhang Y, Xiao Q, Lu J, Zhou HY, Chen SD (2009) Phosphoproteome study reveals Hsp27 as a novel signaling molecule involved in GDNF-induced neurite outgrowth. J Proteome Res 8:2768–2787

Horman S, Galand P, Mosselmans R, Legros N, Leclercq G, Mairesse N (1997) Changes in the phosphorylation status of the 27 kDa heat shock protein (HSP27) associated with the modulation of growth and/or differentiation in MCF-7 cells. Cell Prolif 30:21–35

Hsu FF, Lin TY, Chen JY, Shieh SY (2010) p53-mediated transactivation of LIMK2b links actin dynamics to cell cycle checkpoint control. Oncogene 29:2864–2876

Huang HD, Lee TY, Tseng SW, Horng JT (2005) KinasePhos: a web tool for identifying protein kinase-specific phosphorylation sites. Nucleic Acids Res 33:W226–W229

Huot J, Houle F, Spitz DR, Landry J (1996) HSP27 phosphorylation-mediated resistance against actin fragmentation and cell death induced by oxidative stress. Cancer Res 56:273–279

Jakob U, Gaestel M, Engel K, Buchner J (1993) Small heat shock proteins are molecular chaperones. J Biol Chem 268:1517–1520

Jonak C, Mildner M, Klosner G, Paulitschke V, Kunsfeld R, Pehamberger H, Tschachler E, Trautinger F (2011) The hsp27kD heat shock protein and p38-MAPK signaling are required for regular epidermal differentiation. J Dermatol Sci 61:32–37

Kampinga HH, Brunsting JF, Stege GJ, Konings AW, Landry J (1994) cells overexpressing Hsp27 show accelerated recovery from heat-induced nuclear protein aggregation. Biochem Biophys Res Commun 204:1170–1177

Kampinga HH, Hageman J, Vos MJ et al (2009) Guidelines for the nomenclature of the human heat shock proteins. Cell Stress Chaperones 14:105–111

Kang S, Elf S, Lythgoe K, Hitosugi T, Tauton J, Zhou W, Xiong L, Wang D, Muller S, Fan S, Sun SY, Marcus AI, Gu TL, Polakiewicz RD, Chen Z, Khuri FR, Shin DM, Chen, J (2010) P90 ribosomal S6 kinase 2 promotes invasion and metastasis of human head and neck squamous cell carcinoma cells. J Clin Invest 120:1165–117

Kappe G, Franck E, Verschuure P, Boelens WC, Leunissen JA, de Jong WW (2003) The human genome encodes 10 α-crystallin-related small heat shock proteins: HspB1–10. Cell Stress Chaperones 8:53–61

Kato K, Hasegawa K, Goto S, Inaguma Y (1994) Dissociation as a result of phosphorylation of an aggregated form of the small stress protein hsp27. J Biol Chem 269:11274–11278

Kato H, Takai, S, Matsushima-Nishiwaki R, Adachi S, Minamitani C, Otsuka T, Tokuda H, Akamatsu S, Doi T, Ogura S, Kozawa O (2008) HSP27 phosphorylation is correlated with ADP-induced platelet granule secretion. Arch Biochem Biophys 475:80–86

Kato H, Adachi S, Doi T, Matsushima-Nishiwaki R, Minamitani C, Akamatsu S, Enomoto Y, Tokuda H, Otsuka T, Iwama T, Kozawa O, Ogura S (2010) Mechanism of collagen-induced release of 5-HT, PDGF-AB and sCD40 L from human platelets: role of HSP27 phosphorylation via p44/p42 MAPK. Thromb Res 126:39–43

Kato K, Adachi S, Matsushima-Nishiwaki R, Minamitani C, Natsume H, Katagiri Y, Hirose Y, Mitzutani J, Tokuda H, Kozawa O, Otsuka T (2011) Regulation by heat shock protein 27 of osteocalcin synthesis in osteblasts. Endocrinol 152:1872–1882

Khalili AA, Kabapy NF, Deraz SF, Smith C (2011) Heat shock proteins in oncology: diagnostic biomarkers or therapeutic targets? Biochem Biophys Acta 1816:89–104

Kim MS, Kewalramani G, Puthanveetil P, Lee V, Kumar U, An D, Abramani A, Rodrigues B (2008a) Acute diabetes moderates trafficking of cardiac lipoprotein lipase through p38 mitogen-activated protein kinase-dependent actin cytoskeleton organization. Diabetes 57:64–76

Kim MS, Wang F, Puthanveetil P, Kewalramani G, Hosseini-Beheshti E, Ng N, Wang Y, Kumar U, Innis S, Proud CG, Abramani A, Rodrigues B (2008b) Protein kinase D is a key regulator of cardiomyocyte lipoprotein lipase secretion after diabetes. Circ Res 103:252–260

Knapinska AM, Gratacós FM, Krause CD, Hernandez K, Jensen AG, Bradley JJ, Wu X, Pestka S, Brewer G (2011) Chaperone Hsp27 modulates AUF1 proteolysis and AU-rich element-mediated mRNA degradation. Mol Cell Biol 31:1419–1431

Knauf U, Jakob U, Engel K, Buchner J, Gaestel M (1994) Stress- and mitogen-induced phosphorylation of the small heat shock protein Hsp25 by MAPKAP kinase 2 is not essential for chaperine properties and cellular thermoresistance. EMBO J 13:54–60

Kostenko S, Moens U (2009) Heat shock protein 27 phosphorylation: kinases, phosphatases, functions and pathology. Cell Mol Life Sci 66:3289–3307

Kostenko S, Johannessen M, Moens U (2009) PKA-induced F-actin rearrangement requires phosphorylation of Hsp27 by the MAPKAP kinase MK5. Cell Signal 21:712–718

Krueger-Naug AM, Plumier JC, Hopkins DA, Currie RW (2002) Hsp27 in the nervous system: expression in pathophysiology and in the aging brain. Prog Mol Subcell Biol 28:235–251

Laimer M, Kocher T, Chiocchetti A, Trost A, Lottspeich F, Richter K, Hintner H, Bauer JW, Onder K (2010) Proteomic profiling reveals a catalogue of new candidate proteins for human skin aging. Exp Dermatol 19:912–928

Lakshman M, Xu L, Ananthanarayanan V, Cooper J, Takimoto CH, Helenowski I, Pelling JC, Bergan RC (2008) Dietry genistein inhibits metastasis of human prostate cancer in mice. Cancer Res 68:2024–2032

Lambert H, Charette SJ, Bernier AF, Guimond A, Landry J (1999) Hsp27 multimerization mediated by phosphorylation-sensitive intermolecular interactions at the amino terminus. J Biol Chem 274:9378–9385

Lamont KR, Tindall DJ (2011) Minireview: alternative activation pathways for the androgen receptor in prostate cancer. Mol Endocrinol 25:897–907

Landry J, Huot J (1999) Regulation of actin dynamics by stress-activated protein kinase 2 (SAPK2)-dependent phosphorylation of heat-shock protein of 27 kDa (Hsp27). Biochem Soc Symp 64:79–89

Lanneau D, de Thonel A, Maurel S, Didelot C, Garrido C (2007) Apoptosis versus cell differentiation: role of heat shock proteins HSP90, HSP70 and HSP27. Prion 1:53–60

Lasa M, Mahtani KR, Finch A, Brewer G, Saklatvala J, Clark AR (2000) Regulation of cyclooxygenase 2 mRNA stability by the mitogen-activated protein kinase p38 signaling cascade. Mol Cell Biol 20:4265–4274

Lavoie JN, Hickey E, Weber LA, Landry J (1993) Modulation of actin microfilament dynamics and fluid phase pinocytosis by phosphorylation of heat shock protein 27. J Biol Chem 268:24210–24214

Lavoie JN, Lambert H, Hickey E, Weber LA, Landry J (1995) Modulation of cellular thermoresistance and actin filament stability accompanies phosphorylation-induced changes in the oligomeric structure of heat shock protein 27. Mol Cell Biol 15:505–516

Lee JW, Kwak HJ, Lee JJ, Kim YN, Lee JW, Park MJ, Jung SE, Hong SI, Lee JH, Lee JS (2008) HSP27 regulates cell adhesion and invasion via modulation of focal adhesion kinase and MMP-2 expression. Eur J Cell Biol 87:377–387

Lee HE, Berkowitz P, Jolly PS, Diaz LA, Chun MP, Rubenstein DS (2009) Biphasic activation of p38MAPK suggests that apoptosis is a downstream event in pemphigus acantholysis. J Biol Chem 284:12524–12532

Lee S, Kim JN, Lee HK, Yoon KS, Shin KD, Kwon BM, Han DC (2011) Biological evaluation of KRIBB3 analogs as a microtubule polymerization inhibitor. Bioorg Med Chem Lett 21:977–979

Leja-Szpak A, Jaworek J, Szklarczyk J, Konturek SJ, Pawlik WW (2007) Melatonin stimulates HSP27 phosphorylation in human pancreatic carcinoma cells (PANC-1). J Physiol Pharmacol 58(Suppl 3):177–188

Lelj-Garolla B, Mauk AG (2005) Self-association of a small heat shock protein. J Mol Biol 345:631–642

Levin EG, Santell L (1991) Phosphorylation of an Mr = 29,000 protein by IL-1 is susceptible to partial down-regulation after endothelial cell activation. J Immunol 146:3772–3778

Li L, Sevinsky JR, Rowland MD, Bundy JL, Stephenson JL, Sherry B (2010) Proteomic analysis reveals virus-specific Hsp25 modulation in cardiac myocytes. J Proteome Res 9:2460–2471

Linding R, Jensen LJ, Ostheimer GJ, van Vugt MA, Jørgensen C, Miron IM et al (2007) Systematic discovery of in vivo phosphorylation networks. Cell 129:1415–1426

Liu T, Guevara OE, Warburton RR, Hill NS, Gaestel M, Kayyali US (2009) Modulation of HSP27 hypoxia-induced endothelial permeability and related signalling pathways. J Cell Physiol 220:600–610

Loktionova SA, Kabakov AE (1998) Protein phosphatase inhibitors and heat preconditioning prevent Hsp27 dephosphorylation, F-actin disruption and deterioration of morphology in ATP-depleted endothelial cells. FEBS Lett 433:294–300

Lopes LB, Flynn C, Komalavilas P, Panitch A, Brophy CM, Seal BL (2009) Inhibition of Hsp27 phosphorylation by a cell-permeant MAPKAP kinase 2 inhibitor. Biochem Biophys Res Commun 382:535–539

Lu T, Pan Y, Kao SY, Li C, Kohane I, Chan J, Yankner BA (2004) Gene regulation and DNA damage in the ageing human brain. Nature 429:883–891

Mathew SS, Della Selva MP, Burch AD (2009) Modification and reorganization of the cytoprotective cellular chaperone Hsp27 during herpes simplex virus type 1 infection. J Virol 83:9304–9312

Matsushima-Nishiwaki R, Takai S, Adachi S, Minamitani C, Yasuda E, Noda T, Kato K, Toyoda, H, Kaneoka Y, Yamaguchi A, Kumada T, Kozawa O (2008) Phosphorylated heat shock protein 27 represses growth of hepatocellular carcinoma via inhibition of extracellular signal-regulated kinase. J Biol Chem 283:18852–18860

Matt P, Fu Z, Carrel T, Huso DL, Dirnhofer S, Lefkovits I, Zerkowski HR, Van Eyk JE (2007) Proteomic alterations in heat shock protein 27 and identification of phosphoproteins in ascending aortic aneurysm associated with bicuspid and tricuspid aortic valve. J Mol Cell Cardiol 43:792–801

Mehlen P, Hickey E, Weber LA, Arrigo AP (1997) Large unphosphorylated aggregates as the active form of hsp27 which controls intracellular reactive oxygen species and glutathione levels and generates a protection against TNF alpha in NIH-3t3-ras cells. Biochem Biophys Res Commun 241:187–192

Melle C, Ernst G, Escher N, Hartmann D, Schimmel B, Bleul A, Thieme H, Kaufmann R, Felix K, Friess HM, Settmacher U, Hommann M, Richter KK, Daffner W, Täubig H, Manger T, Claussen U, von Eggeling F (2007) Protein profiling of microdissected pancreas carcinoma and identification of HSP27 as a potential serum marker. Clin Chem 53:629–635

Michishita M, Satoh M, Yamaguchi M, Hirayoshi K, Okuma M, Nagata K (1991) Phosphorylation of the stress protein hsp27 is an early event in murine myelomonocytic leukemic cell differentiation induced by leukemia inhibitory factor/D-factor. Biochem Biophys Res Commun 176:979–984

Morris JK, Bomhoff GL, Stanford JA, Geiger PC (2010) Neurodegeneration in an animal model of Parkinson's disease is exacerbated by a high-fat diet. Am J Physiol Regul Integr Comp Physiol 299:R1082–R1090

Mosser DD, Morimoto RI (2004) Molecular chaperones and the stress of oncogenesis. Oncogene 23:2907–2918

Mounier N, Arrigo AP (2002) Actin cytoskeleton and small heat shock proteins: how do they interact? Cell Stress Chaperones 7:167–176

Nakatsue T, Katoh I, Nakamura S, Takahashi Y, Ikawa Y, Yoshinaka Y (1998) Acute infection of Sindbis virus induces phosphorylation and intracellular translocation of small heat shock protein HSP27 and activation of p38 MAP kinase signaling pathway. Biochem Biophys Res Commun 253:59–64

Nishio M, Endo T, Tsukada N, Ohata J, Kitada S, Reed JC, Zvaifler NJ, Kipps TJ (2005) Nurselike cells express BAFF and APRIL, which can promote survival of chronic lymphocytic leukemia cells via a paracrine pathway distinct from that of SFD-1alpha. Blood 106:1012–1020

Nomura N, Nomura M, Sugiyama K, Hamada JI (2007) Phorbol 12-myristate 13-acetate (PMA)-induced migration of glioblastoma cells is mediated via p38MAPK/Hsp27 pathway. Biochem Pharmacol 74:690–701

O'Callaghan-Sunol C, Gabai VL, Sherman MY (2007) Hsp27 modulates p53 signaling and suppresses cellular senescence. Cancer Res 67:11779–11788

O'Hayre M, Salange CL, Kipps TJ, Messmer D, Dorrestein PC, Handel TM (2010) Elucidating the CXCL12/CXCR4 signaling network in chronic lymphocytic leukemia through phosphoproteomics analysis. PLoS One 5:e11716

Olsen JV, Vermeulen M, Snatamaria A, Kumar C, Miller ML, Jensen LJ et al (2010) Quantitative phosphoproteomics reveals widespread full phosphorylation site occupancy during mitosis. Sci Signal 3:ra3

Orejuela D, Bergeron A, Morrow G, Tanguay, RM (2007) Small heat shock proteins in physiological and stress-related processes. In: Calderwood SK (ed) Cell stress proteins, vol 7. Springer, New York, pp 143–177

Panasenko OO, Kim MV, Marston SB, Gusev NB (2003) Interaction of the small heat shock protein with molecular mass 25 kDa (hsp25) with actin. Eur J Biochem 270:892–901

Parcellier A, Brunet M, Schmitt E, Col E, Didelot C, Hammann A, Nakayama K, Nakayama KI, Khochbin S, Solary E, Garrido C (2006) HSP27 favors ubiquitination and proteasomal degradation of p27Kip1 and helps S-phase re-entry in stressed cells. FASEB J 20:1179–1181

Park KJ, Gaynor RB, Kwak YT (2003) Heat shock protein 27 association with the IκB kinase complex regulates tumor necrosis factor α-induced NF-κB activation. J Biol Chem 278:35272–35278

Park HK, Park EC, Bae SW, Park MY, Kim SW, Yoo HS, Tudev M, Ko YH, Choi YH, Kim S, Kim DI, Kim YW, Lee BB, Yoon JB, Park JE (2006) Expression of heat shock protein 27 in human atherosclerotic plaques and increased plasma level of heat shock protein in patient with acute coronary syndrome. Circulation 114:886–893

Park JK, Ronkina N, Höft A, Prohl C, Menne J, Gaestel M, Haller H, Meier M (2008) Deletion of MK2 signalling *in vivo* inhibits small Hsp phosphorylation but not diabetic nephropathy. Nephrol Dial Transplant 23:1844–1853

Paul C, Simon S, Gilbert B, Virot S, Manero F, Arrigo AP (2010) Dynamic processes that reflect anti-apoptotic strategies set up by HspB1 (Hsp27). Exp Cell Res 316:1535–1552

Pichon S, Bryckaert M, Berrou E (2004) Control of actin dynamics by p38 MAP kinase- Hsp27 distribution in the lamellipodium of smooth muscle cells. J Cell Sci 117:2569–2577

Piotrowicz RS, Hickey E, Levin EG (1998) Heat shock protein 27 kDa expression and phosphorylation regulates endothelial cell migration. FASEB J 12:1481–1490

Pivovarova AV, Mikhailova VV, Chernik IS, Chebotareva NA, Levitsky DI, Gusev NB (2005) Effects of small heat shock proteins on the thermal denaturation and aggregation of F-actin. Biochem Biophys Res Commun 331:1548–1553

Pivovarova AV, Chebotareva NA, Chernik IS, Gusev NB, Levitsky DI (2007) Small heat shock protein Hsp27 prevents heat-induced aggregation of F-actin by forming soluble complexes with denatured actin. FEBS J 274:5937–5948

Robinson AA, Dunn MJ, McCormack A, dos Remendios C, Rose ML (2010) Protective effect of phosphorylated hsp27 in coronary artereis through actin stabilization. J Mol Cell Cardiol 49:370–379

Robitaille H, Simard-Bisson C, Larouche D, Tanguay RM, Blouin R, Germain L (2010) The small heat-shock protein Hsp27 undergoes ERK-dependent phosphorylation and redistribution to the cytoskeleton in response to dual zipper-bearing kinase expression. J Invest Dermatol 130:74–85

Rogalla T, Ehrnsperger M, Preville X, Kotlyarov A, Lutsch G, Ducasse C, Paul C, Wieske M, Arrigo AP, Buchner J, Gaestel M (1999) Regulation of Hsp27 oligomerization, chaperone function, and protective activity against oxidative stress/tumor necrosis factor alpha by phosphorylation. J Biol Chem 274:18947–18956

Rubin CI, Atweh GF (2004) The role of stathmin in the regulation of the cell cycle. J Cell Biochem 93:242–250

Rui Z, Jian-Guo J, Yuang-Peng T, Hai P, Bing-Gen R (2003) Use of serological proteomic methods to find biomarkers associated with breast cancer. Proteomics 3:433–439

Salinthone S, Ba M, Hanson L, Martin JL, Halayko AJ, Gerthoffer WT (2007) Overexpression of human Hsp27 inhibits serum-induced proliferation in airway smooth muscle monocytes and confers resistance to hydrogen peroxide cytotoxicity. Am J Physiol Lung Cell Mol Physiol 293:L1194–L1207

Schneider GB, Hamano H, Cooper LF (1998) In vivo evaluation of hsp27 as an inhibitor of actin polymerization: hsp27 limits actin stress fiber and focal adhesion formation after heat shock. J Cell Physiol 177:575–584

Shi B, Grahn JC, Reilly DA, Dizon TC, Isseroff RR (2008a) Responses of the 27-kDa heat shock protein to UVB irradiation in human epidermal melanocytes. Exp Dermatol 17:108–114

Shi GX, Jin L, Andres DA (2008b) Pituitary adenylate cyclase-activating polypeptide 38-mediated Rin activation requires Src and contributes to the regulation of HSP27 signaling during neuronal differentiation. Mol Cell Biol 28:4940–4051

Shin KD, Lee M-Y, Shin D-S, Lee S, Son K-H, Koh S, Paik Y-Y, Kwon B-M, Han DC (2005) Blocking tumor cell migration and invasion with biphenyl isoxazole derivative KRIBB3, a synthetic molecule that inhibits Hsp27 phosphorylation. J Biol Chem 280:41439–41448

Shiryaev A, Dimitriu G, Moens U (2011) Distinct roles of MK2 and MK5 in cAMP/PKA- and stress/p38MAPK-induced heat shock protein 27 phosphorylation. J Mol Signal 6:4

Singh D, McCann KL, Imani F (2007) MAPK and heat shock protein 27 activation are associated with respiratory syncytial virus induction of human bronchial epithelial monolayer disruption. Am J Physiol Lung Cell Mol Physiol 293:L436–L445

Smoyer WE, Gupta A, Mundel P, Ballew, JD, Welsh MJ (1996) Altered expression of glomerular heat shock protein 27 in experimental nephrotic syndrome. J Clin Invest 97:2697–2704

Spector NL, Mehlen P, Ryan C, Hardly L, Samson W, Levine H, Nadler LM, Fabre N, Arrigo AP (1994) Regulation of the 28 kDa heat shock protein by retinoic acid during differentiation of human leukemic HL-60 cells. FEBS Lett 337:184–188

Sun Y, MacRae TH (2005) Small heat shock proteins: molecular structure and chaperone function. Cell Mol Life Sci 62:2460–2476

Sun X, Welsh MJ, Benndorf R (2006) Conformational changes resulting from pseudophosphorylation of mammalian small heat shock proteins—a two-hybrid study. Cell Stress Chaperones 11:61–70

Tanabe K, Takai R, Kato K, Dohi S, Kozawa O (2008) Alpha2 adrenoreceptor agonist regulates protein kinase C-induced heat shock protein 27 phosphorylation in C6 glioma cells. J Neurochem 106:519–528

Tanabe K, Matsushima-Nishiwaki R, Dohi S, Kozawa O (2010) Phosphorylation status of heat shock protein 27 regulates the interleukin-1b-induced interleukin-6 synthesis in C6 glioma cells. Neuroscience 170:1028–1034

Tar K, Csortos C, Czikora I, Olah G, Ma SF, Wadgaonkar R, Gergely P, Garcia JG, Verin AD (2006) Role of protein phosphatase 2A in the regulation of endothelial cell cytoskeleton structure. J Cell Biochem 98:931–953

Thériault JR, Lambert H, Chávez-Zobel AT, Charest G, Lavigne P, Landry J (2004) Essential role of the NH2-terminal WD/EPF motif in the phosphorylation-activated protective function of mammalian Hsp27. J Biol Chem 279:23463–23471

Tremolada L, Magni F, Valsecchi C, Sarto C, Mocarelli P, Perego R, Cordani N, Favini P, Kienle MG, Sanchez JC, Hochstrasser D, Corthals GL (2005) Characterization of heat shock protein 27 phosphorylation sites in renal cell carcinoma. Proteomics 5:788–795

Trott D, McManus CA, Martin JL, Brennan B, Dunn MJ, Rose ML (2009) Effect of phosphorylated hsp27 on proliferation of human endothelial and smooth muscle cells. Proteomics 9:3383–3394

Tzivion G, Luo ZJ, Avruch J (2000) Calyculin A-induced vimentin phosphorylation sequesters 14-3-3 and displaces other 14-3-3 partners in vivo. J Biol Chem 275:29772–29778

Uddin S, Ah-Kang J, Ulaszek J, Mahmud D, Wickrema A (2004) Differentiation stage-specific activation of p38 mitogen-activated protein kinase isoforms in primary human erythroid cells. Proc Natl Acad Sci U S A 101:147–152

Uhlen M, Oksvold P, Fagerberg L et al (2010) Towards a knowledge-based human protein atlas. Nat Biotechnol 12:1248–1250

Venkatakrishnan CD, Dunsmore K, Wong H, Roy S, Sen SK, Wani A, Zweier JL, Ilangovan G (2008) HSP27 regulates p53 transcriptional activity in doxorubicin-treated fibroblasts and cardiac H9c2 cells: p21 upregulation and G2/M phase cell cycle arrest. Am J Physiol Heart Circ Physiol 294:H1736–H1744

Vidyasagar A, Reese S, Acun Z, Hullett D, Djamali A (2008) Hsp27 is involved in the pathogenesis of kidney tubulointerstitial fibrosis. Am J Physiol Renal Physiol 295:F707–F716

Villafan-Bernal JR, Sanchez-Enriquez S, Muoz-Valle JF (2011) Molecular modulation of osteocalcin and its relevante in diabetes. Int J Mol Med 28:283–293

Vos MJ, Hagemand J, Carra S, Kampinga HH (2008) Structural and functional diversities between members of the human HSPB, HSPH, HSPA, and DNAJ chaperone families. Biochemistry 47:7001–7011

Wang RE, Kao JL, Hilliard CA, Pandita RK, Roti Roti JL, Hunt CR, Taylor JS (2009) Inhibition of heat shock induction of heat shock protein 70 and enhancement of heat shock protein 27 phosphorylation by quercetin derivatives. J Med Chem 52:1912–1921

Webster M, Witkin KL, Cohen-Fix O (2009) Sizing up the nucleus: nuclear shape, size and nuclear-envelope assembly. J Cell Sci 122:1477–1486

Welle S, Brooks AI, Delehanty JM, Needler N, Bhatt K, Shah B, Thornton CA (2004) Skeletal muscle gene expression profiles in 20–29 year old and 65–71 year old women. Exp Gerontol 39:369–377

Wyttenbach A, Sauvageot O, Carmichael J, Diaz-Latoud C, Arrigo AP, Rubinsztein DC (2002) Heat shock protein 27 prevents cellular polyglutamine toxicity and suppresses the increase of reactive oxygen species caused by huntingtin. Hum Mol Gen 11:1137–1151

Wong JW, Shi B, Farboud B, McClaren M, Shibamoto T, Cross CE, Isseroff RR (2000) Ultraviolet B-mediated phosphorylation of the small heat shock protein HSP27 in human keratinocytes. J Invest Dermatol 115:427–434

Wong LL, Zhang D, Chang CF, Koay ES (2010) Silencing of the PP2A catalytic subunit causes HER-2/neu positive breast cancer cells to undergo apoptosis. Exp Cell Res 316:3387–3396

Xu L, Chen S, Bergan RC (2006) MAPKAPK2 and HSP27 are downstream effectors of p38 MAP kinase-mediated matrix metalloproteinase type 2 activation and cell invasion in human prostate cancer. Oncogene 18:2987–2998

Yamaguchi T, Arai H, Katayama N, Ishikawa T, Kikumoto K, Atomi Y (2007) Age-related increase of insoluble, phosphorylated small heat shock proteins in human skeletal muscle. J Gerontol A Biol Sci Med Sci 62:481–489

Yang F, Stenoien DL, Strittmatter EF, Wang J, Ding L, Lipton MS et al (2006) Phosphoproteome profiling of human skin fibroblast cells in response to low- and high-dose irradiation. J Proteome Res 5:1252–1260

Yasuda E, Kumada T, Takai S, Ishisaki A, Noda T, Matsushima-Nishiwaki R, Yoshimi N, Kato K, Toyoda H, Kaneoka Y, Yamaguchi A, Kozawa O (2005) Attenuated phosphorylation of heat shock protein 27 correlates with tumor progression in patients with hepatocellular carcinoma. Biochem Biophys Res Commun 337:337–342

Zhang D, Wong LL, Koay ESC (2007) Phosphorylation of Ser78 of Hsp27 correlates with HER-2/neu status and lymph node positivity in breast cancer. Mol Cancer 6:52

Zhu Y, O'Neill S, Saklatvala J, Tassi L, Mendelsohn ME (1994a) Phosphorylated HSP27 associates with the activation-dependent cytoskeleton in human platelets. Blood 84:3715–3723

Zhu Y, Tassi L, Lane W, Mendelsohn ME (1994b) Specific binding of the transglutaminase, platelet factor XIII, to HSP27. J Biol Chem 269:22379–22384

Zoubeidi A, Zardan A, Beraldi E, Fazli L, Sowery R, Rennie P, Nelson C, Gleave M (2007) Co-operative interactions between androgen receptor (AR) and heat-shock protein 27 facilitate AR transcriptional activity. Cancer Res 67:10455–10465

Chapter 5
Evidence on Cholesterol-Controlled Lipid Raft Interaction of the Small Heat Shock Protein HSPB11

Zsolt Török, Ana-Maria Pilbat, Imre Gombos, Enikö Hocsák, Balázs Sümegi, Ibolya Horváth and László Vígh

Abstract Small heat-shock proteins (sHSPs) are members of the family of molecular chaperones. Their major cellular function is considered to be the prevention of irreversible protein aggregation during stress conditions and subsequent promotion of the folding of partially denatured proteins. However, sHSPs may also be associated with biological membranes and participate in cellular "stress management" by acting as membrane-stabilizing factors. In spite of the great potential significance in the development of therapeutic strategies, the mechanisms of the membrane (and lipid) association of sHSPs are still unknown. A novel 16.2 kDa human sHSP, HSPB11, inhibits H_2O_2, taxol and etoposide-induced cell death through stabilization of the mitochondrial membrane system, the activation of HSP90, the stabilization of lipid rafts and activation of the PI-3-kinase—Akt cytoprotective pathway. We show here that HSPB11 binds to lipid membranes via a specific cholesterol-mediated interaction. The affinity of HSPB11 demonstrates a very distinct cholesterol-dependent binding to cholesterol/sphingomyelin Langmuir monolayers: If the cholesterol concentration increases above a certain level, HSPB11 binds to membranes much more efficiently. The possible roles of HSPB11 and other sHSPs in protection against stress-induced hydrophobic membrane defects are discussed.

Abbreviation

SM	egg sphingomyelin
Chol	cholesterol
DChol	dihydrocholesterol
POPC	1-palmitoyl-2-oleoyl-sn-glycero-3-phosphocholine

Z. Török (✉) · A.-M. Pilbat · I. Gombos · I. Horváth · L. Vígh
Institute of Biochemistry, Biological Research Centre, Hungarian Academy of Sciences, Temesvári krt 62, Szeged, H6726, Hungary

L. Vígh
e-mail: vigh@brc.hu

E. Hocsák · B. Sümegi
Department of Biochemistry and Medical Chemistry, University of Pécs, Pécs, Hungary

B. Henderson, A. G. Pockley (eds.), *Cellular Trafficking of Cell Stress Proteins in Health and Disease*, Heat Shock Proteins 6, DOI 10.1007/978-94-007-4740-1_5,

5.1 Introduction

Small heat-shock proteins (sHSPs) function as molecular chaperones which are inducible in response to stress (van Montfort et al. 2001), and which prevent the thermal and chemical-induced aggregation of a variety of targets both *in vitro* and *in vivo* (Basha et al. 2005; Franzmann et al. 2005; Haslbeck 2002; Stromer et al. 2003). sHSPs are made up of a highly conserved C-terminal domain, the "α-crystallin domain", and a variable N-terminal domain (Horwitz 2003; Narberhaus 2002). They are ATP-independent chaperones, which alone cannot regulate protein folding, but which interact with partially folded proteins and stabilize them to prevent their aggregation and precipitation during conditions of cellular stress. sHSPs can therefore create a reservoir of denatured proteins, which could be refolded in the presence of ATP-dependent chaperones (Franzmann et al. 2005; Haslbeck 2002; Stromer et al. 2003; Ehrnsperger et al. 1997; Lee et al. 1997). The monomeric masses of sHSPs's are between 12 and 43 kDa and they form oligomeric structures with mostly 12 or 24 subunits or even larger complexes (Horwitz 1992, 2003). The human sHSP family is comprised of 11 known members (Haslbeck et al. 2005; Kampinga et al. 2009), some of which can be ubiquitously expressed (HSPB1, HSPB5, HSPB6 and HSPB8), while the others are expressed in a tissue-specific manner (HSPB2, HSPB3, HSPB4, HSPB7, HSPB9, HSPB10 and HSPB11; Bellyei et al. 2007a; Pozsgai et al. 2007; Taylor and Benjamin 2005). sHSPs reside not only in the cytosol, but also in the cellular organelles. A subpopulation of sHSPs is present either on the surface or within the cellular membranes (Horvath et al. 2008; Nakamoto and Vigh 2007; Soti et al. 2005; Vigh and Maresca 2002; Vigh et al. 2005, 2007b). There are numerous indications of the important physiological roles of membrane-associated sHSPs. They can protect against stress conditions (heat, light and oxidative stress) in prokaryotes, unicellular eukaryotes and plants.

5.2 Membrane-Association of sHSPs: Overview

Several members of the sHSP family, present in mammalian cells, are also associated with membranes. The murine 25-kDa HSP was reported to be expressed on the surface of the plasma membrane (Bausero et al. 2004). HSPB1 (HSP27) is associated with various malignancies and is expressed at high levels in biopsies and in the serum of breast cancer patients (Bausero et al. 2006). It was recently found that the phosphorylation and cellular localization of HSPB1 were significantly altered following Doxorubicin treatment in breast cancer cells. In drug-resistant cells, phospho-HSPB1 was excluded from the nuclei and most of the cytoplasm and appeared to be associated with the cell membrane (Fujita et al. 2011). A comprehensive review of Hsp27 in health and disease is found in Chap. 4. HSPB2, expressed in heart and skeletal muscle, associates with the outer membrane of the mitochondria (Nakagawa et al. 2001). αB-Crystallin (HSPB5), the major protein component of the vertebrate lens, is thought to play a critical role in the maintenance of transparency through its ability to inhibit stress-induced protein aggregation. However, during aging an increase in the level of membrane binding of HSPB5 is an integral step in the pathogenesis of many forms of cataracts (Cobb and Petrash 2000, 2002). A missense mutation (R120G) of HSPB5 has been linked to a familial form of desmin-related myopathy (DRM). It is noteworthy that both HSPB5 and HSPB1 exhibit enhanced plasma membrane localization in the myotubes of dexamethasone-treated DRM patients (Nedellec et al. 2002). HSPB5 was found in detergent-resistant membrane microdomains and is secreted via exosomes from human retinal pigment epithelial cells. The secretion of HSPB5 was shown to be inhibited by the cholesterol-depleting drug, methyl β-cyclodextrin, suggesting that the physiological function of HSPB5 and the regulation of its export through exosomes may reside in its association with cholesterol of lipid rafts (Gangalum et al. 2010).

A significant proportion of human HSPB8 has been shown to be localized to the plasma membrane in a human neuroblastoma cell line. HSPB8 binds stably to lipid vesicles in a manner dependent on the nature of the lipid (Chowdary et al. 2007). It binds to vesicles consisting of lipids containing a phosphatidic acid, phosphatidylinositol or phosphatidylserine head-group (known to be present in the inner leaflet of the plasma membrane) more strongly as compared with lipid vesicles containing a phosphatidylcholine head-group alone. HSPB8 is involved in the regulation of cell proliferation, cardiac hypertrophy, apoptosis and carcinogenesis, and the expression of point mutants of HSPB8 correlates with the development of different neuromuscular diseases. HSPB11, the recently discovered human sHSP that features at the focus of this report, has been shown to be partially associated with and to stabilize mitochondrial membranes and rafts on the plasma membrane (Bellyei et al. 2007a, b). Other members of the chaperone family have been found to associate to a variable extent with detergent-resistant microdomains ("rafts"), and the association of the HSPs with these microdomains can be modulated by stress (Broquet et al. 2003). The membrane microdomain-associated HSPs can evidently participate in the orchestration and activity of distinct raft-associated signalling platforms (Wang et al. 2006).

By virtue of their specific membrane lipid interactions, sHSPs have been shown to modulate major features of the membrane lipid phase, such as the fluidity, permeability or non-bilayer propensity (Torok et al. 2001; Tsvetkova et al. 2002; Welker et al. 2010). The membrane binding of sHSPs through specific protein-lipid interactions may restrict the location of sHSPs to one or more membrane lipid domains (Balogi et al. 2005 and 2008).

It has been suggested that sHSPs can contribute to the development of various diseases, including cancer (Horwitz 2003; Pozsgai et al. 2007; Fujita et al. 2011; Launay et al 2010), and their inhibition or suppression can contribute to a more effective therapy, indicating the potential of the sHSPs as possible new drug targets (Zhu et al. 2010). In view of the widely documented presence of sHSPs in membranes, it has been suggested that a subset of sHSPs functions in cellular "stress management" by acting as membrane-stabilizing factors (see Horvath et al. 2008 and references therein).

In spite of the great potential in the development of therapeutic strategies, the mechanisms of the membrane (and lipid) association of sHSPs remain unknown. We show here that the novel human sHSP, HSPB11, binds to lipid membranes via a specific cholesterol-mediated interaction, which indicates a universal role of sHSPs in the structural homeostasis of biological membranes during stress.

5.3 HSPB11 is a Novel Molecular Chaperone

In previous studies, we have identified a novel, 16.2 kDa sHSP, HSPB11, which displays high homology to HSPB5 and whose expression is induced by heat stress. The recombinant protein prevents protein aggregation *in vitro*, and HSPB11-overexpressing *E. coli* cells become resistant to heat stress (Bellyei et al. 2007a, b). For these reasons, it is considered to be a new member of the sHSP family (HSPB11).

The level of HSPB11 expression is low in some cells, such as those of the 3T3 fibroblast line, but significantly higher in the others e.g. HepG2, HeLa, Panc-1 and WRL-68 cell lines. Moreover, a positive correlation has been found between the level of HSPB11 expression and the degree of anaplasia in different malignant tissue samples (Pozsgai et al. 2007). HSPB11 was detectable in the cytoplasm, in the mitochondria and in the nucleus, but in grade 4 tumours a very large increase in the quantity of HSPB11 was observed in the cytoplasmic fraction (Pozsgai et al. 2007; Bellyei et al. 2007b).

It has been clearly demonstrated that the sHSPs inhibit apoptosis (Zhu et al. 2010), and therefore we investigated the effects of HSPB11 on cell death processes; it proved that HSPB11 overexpression inhibits both H_2O_2 and taxol-induced cell death in NIH3T3 cells. The suppression of HSPB11 by siRNA sensitized HeLa cells to cell death (Bellyei et al. 2007a, b), indicating that HSPB11 behaves like other sHSPs.

5.3.1 HSPB11 Protects the Mitochondrial Membrane System

It is well known that the mitochondria play a major role in the cell death process, and that both oxidative stress and taxol treatment can lead to mitochondrial membrane permeabilization (Tapodi et al. 2005; Varbiro et al. 2001), resulting in collapse of the mitochondrial membrane potential and the release of proapoptotic proteins (Norberg et al. 2010). Investigation of the role of HSPB11 revealed that the overexpression of HSPB11 prevents the H_2O_2 or taxol-induced collapse of the mitochondrial membrane potential determined by JC-1, and decreases the release of AIF and endonuclease G (Bellyei et al. 2007a, b). This type of protective effect can be a consequence of different complex regulatory mechanisms, but it also raises the possibility that HSPB11 can interact directly with the mitochondrial membrane system. Through the use of recombinant HSPB11 and isolated mitochondria, it was demonstrated that

HSPB11 inhibits the calcium and phosphate-induced collapse of the mitochondrial membrane potential and mitochondrial permeability transition in a concentration-dependent way (Bellyei et al. 2007a).

5.3.2 Role of HSPB11 in Lipid Raft Formation

A search for the intracellular targets of HSPB11 indicated that immobilized recombinant HSPB11 protein binds specifically to HSP90 in HeLa and Panc-1 cell homogenates, suggesting that the protective effect of HSPB11 could be partially mediated by HSP90. To prove the physiological significance of this interaction, the HSP90 function was inhibited with geldanamycin (Bellyei et al. 2007a); this treatment attenuated the protective effect of HSPB11 in both the H_2O_2 and the taxol-induced cell death model, demonstrating that the HSPB11–HSP90 interaction plays a major role in the cytoprotective effect of HSPB11 (Bellyei et al. 2007a, b). Since it was clear that HSP90 is involved in the stabilization of lipid rafts and the microenvironment of lipid rafts can serve as a platform for the activation of cytoprotective kinase, Akt (protein kinase B; Chen et al. 2005; Shah et al. 2002; Sreedhar et al. 2003; Triantafilou et al. 2002; Waheed and Jones 2002), we investigated whether HSPB11 overexpression can influence the quantity of lipid rafts (the cholesterol and sphingolipid-enriched microdomain of the plasma membrane). It emerged that HSPB11 facilitated lipid raft formation, which was attenuated by geldanamycin, showing that, through the interaction with HSP90, HSPB11 facilitated lipid raft microdomain formation and protected the cells against H_2O_2 or taxol-induced cell death, and that the lipid raft-HSP90–HSPB11 complex can serve as a platform for the activation of kinases.

5.3.3 Regulation of Kinase Cascades by HSPB11

Previous data indicated that the activity of Akt on the lipid raft can be regulated in an HSP90-dependent manner (Adam et al. 2007; Cinar et al. 2007), and that inhibition of the HSP90 function also decreases the phosphorylation and activation of Akt (Neckers and Ivy 2003). Overexpression of HSPB11 indeed activated the PI-3-kinase Akt pathway and protected the cells against oxidative stress, and the inhibition of the PI-3-kinase-Akt pathway by Ly-294002 abolished Akt activation and protection against oxidative stress. In parallel with the effects on Akt, the HSPB11 overexpression activated ERK1/2, but inhibited p38 MAP kinase activation. These data indicate that one of the important cytoprotective mechanisms of HSPB11 relies on its interaction with HSP90, binding to and facilitating lipid raft formation, where Akt can be phosphorylated. Phosphorylated and activated Akt can protect mitochondrial membrane systems (Tapodi et al. 2005), thereby protecting cells against oxidative stress and taxol-induced cell death.

5.3.4 *Covalent Modification of HSPB11*

In some cell lines and in grade 1 and 2 tumours, a significant proportion of HSPB11 is located in the nucleus, whereas in grade 3 and 4 tumours, HSPB11 migrates in large quantities to the cytoplasm (Bellyei et al. 2007a; Pozsgai et al. 2007). There are therefore possibilities for the interactions of HSPB11 and proteins, leading to chemical modifications in both compartments. It has been shown that HSPB1 can be phosphorylated *in vivo* by MAPKAP kinase-2, p38 MAP kinase and Akt. The phosphorylation of HSPB5 on serine 59, or a protein mimicking it, can protect the mitochondrial membrane system (Whittaker et al. 2009) as an indicator that the chemical modification of sHSPs can significantly determine their biochemical role. We observed that HSPB11 can be serine-phosphorylated in HeLa cells, and that H_2O_2 increases the phosphorylation level. Since HSPB11 occurs in the nucleus, we studied ADP-ribosylation and acetylation. These data demonstrated PARP-1 poly-ADP-ribosylated HSPB11 and the acetylation of HSPB11 in HeLa cells (Bellyei et al. 2007b). It is likely, therefore, that HSPB11 can undergo different chemical modifications under stress conditions and these modifications may be of regulatory significance in the cytoprotective effects of this protein.

5.3.5 *HSPB11 Binds to a Lipid Monolayer in a Cholesterol-Dependent Manner*

To test whether HSPB11 binds directly to the lipid membrane, we employed Langmuir lipid monolayers. We measured the surface pressure at constant surface area after injecting HSPB11 into the subphase. In these experiments, dihydrocholesterol (DChol) was used rather than cholesterol (Chol) itself so as to minimize possible artifacts due to Chol oxidation (Radhakrishnan and McConnell 2000). DChol and Chol have very closely related chemical structures, and DChol is often used in place of Chol because of its better stability (Lancelot and Grauby-Heywang 2007). Control experiments suggested that there was no difference if DChol was used instead of Chol. The insertion of HSPB11 into a mixed monolayer containing phosphatidylcholin (POPC) and sphingomyelin (SM) in a ratio of 2/1 (POPC/SM = 2/1) proved to be negligible (Fig. 5.1). Addition of DChol to the lipid mix resulted in a gradual increase in surface pressure until a transition threshold concentration of about 50 % Dchol was reached; above this the affinity of HSPB11 for the monolayer increased significantly. Since rafts are already observed at 25 % Chol in this lipid mixture (Dietrich et al. 2001), this binding most probably occurs when free, chemically active Chol patches appear in the monolayer. It has been suggested that the condensed phospholipid-Chol complexes observed in lipid monolayers represent rafts in bilayers (Radhakrishnan and McConnell 2000). These complexes exert considerable effect on the relation between the membrane composition and the chemical activity of Chol. In monolayers, there is both experimental and theoretical evidence of a strong, almost step-like increase in the chemical activity of Chol when the Chol

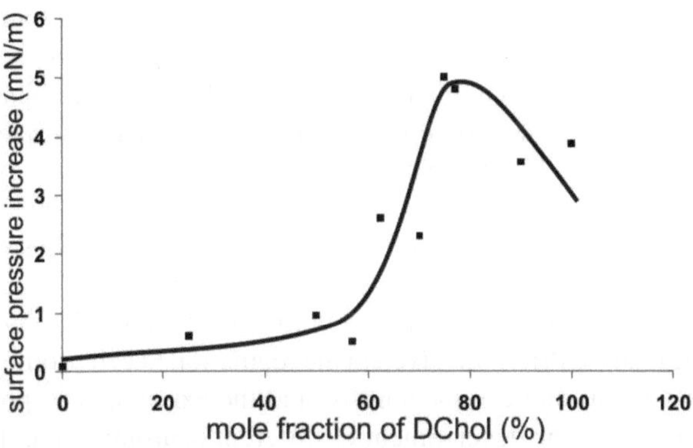

Fig. 5.1 Cholesterol-dependent interaction of HSPB11 with lipids. Monolayers of POPC/SM = 2/1/ with x mol% DChol were spread at an air-buffer (PBS, pH 7.2) interface from $CHCl_3$ solution to give an initial surface pressure of 22 mN/m. Purified recombinant HSPB11 (Bellyei et al. 2007a) was added to the subphase to give a final concentration of 6 μg/ml at constant surface area. The surface pressure was recorded 15 min after protein addition in a KSV3000 Langmuir-Blodget instrument as described previously (Torok et al. 1997)

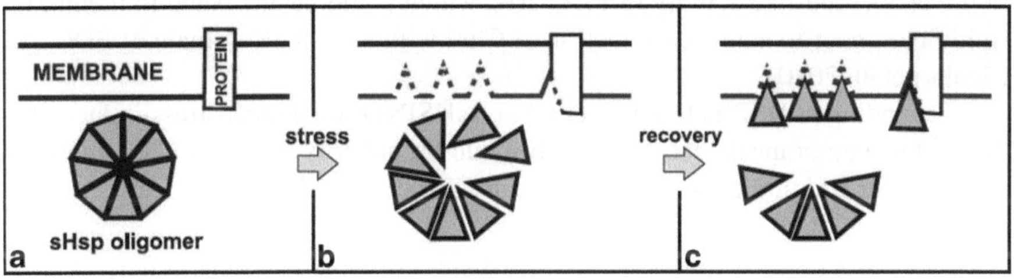

Fig. 5.2 Schematic representation of the possible role of sHSPs in the structural homeostasis of biological membranes during stress. Under physiological conditions, lipids and membrane proteins are in their native state in the heterogeneous fluid-mosaic bilayer (**a**). During stress, the native membrane structure could denature, resulting in unstable, highly hydrophobic membrane defects (**b**), which could be reversibly "covered" and protected by HSPB11 (and other sHSP relatives) subunits dissociated from larger oligomeric structures (**c**). An extensive membrane surface and also individual membrane-associated proteins could be shielded in this way during *stress*. Depending on the level of damage, in the *recovery* phase these defects could either undergo refolding or be eliminated by the cellular protein and membrane recycling apparatus

concentration increases beyond the complex stoichiometric composition (Radhakrishnan et al. 2000). Above this threshold concentration, Chol could form structurally unstable highly hydrophobic patches which could be protected by the binding of the amphitropic sHSP (Fig. 5.2).

Previous reports have described a similar lipid-dependent association of HSPB5 with lens membranes, which increases with age and cataract development. A higher binding capacity was associated with a higher relative amount of sphingolipid and lower relative amounts of phosphatidylethanolamine-related lipid and phosphatidylcholine. The binding capacity of HSPB5 to lens lipids, measured *in vitro*, increases with age and decreases in diabetic donors treated with insulin (Grami et al. 2005). An increased association of HSPB5 with membranes proved to be correlated to age-related lipid compositional changes (Tang et al. 1998). A significant fraction of human HSPB8 was shown to be localized to the plasma membrane in a human neuroblastoma cell line. HSPB8 binds stably to lipid vesicles, to extents depending on the nature of the lipid (Chowdary et al. 2007). HSPB8 binds more strongly to vesicles composed of lipids containing a phosphatidic acid, phosphatidylinositol or phosphatidylserine head-group (known to be present in the inner leaflet of the plasma membrane) than to lipid vesicles composed of a phosphatidylcholine head-group alone. Human HSPB8 is involved in the regulation of cell proliferation, cardiac hypertrophy, apoptosis and carcinogenesis, and the expression of point mutants of HSPB8 correlates with the development of different neuromuscular diseases.

A recent study revealed the correlation of membrane binding and hydrophobicity to the chaperone-like activity of PDC-109, the major protein of bovine seminal plasma, which exhibits significant similarities with sHSPs (Sankhala et al. 2011). The high-affinity binding of PDC-109 to membranes containing choline phospholipids (PCs) displays a higher chaperone-like activity than that of PDC-109 alone. Interestingly, the incorporation of Chol into the PC membrane resulted in a decrease in the chaperone-like activity of PDC-109, which could be attributed to membrane rigidification, or to a direct interaction of Chol with the CRAC domain of PDC-109 (Scolari et al. 2010).

The binding of HSPB11 and other related sHSPs could protect stress-induced hydrophobic membrane defects and may provide a regulatory factor for Chol transport, raft turnover and numerous other membrane-associated cellular functions.

5.4 Concluding Remarks

It is suggested that the associations between Hsps and membranes may constitute a general mechanism that preserves the membrane integrity during thermal fluctuations. The membrane association of Hsps could antagonize the heat-induced hyperfluidization of specific membrane domains, and thereby serve to preserve the structural and functional integrity of biomembranes. Moreover, a lipid-selective association of a subpopulation of Hsps with membranes, resulting in increased molecular order, may in turn lead to the downregulation of heat-shock gene expression (Torok et al. 1997; Vigh et al. 1998, 2002; Torok et al. 2001; Tsvetkova et al. 2002). Such "cross-talk" between the primary stress sensor in the membranes and the Hsps suggests a feedback mechanism in the regulation of heat-shock genes, and can explain the known temporality of the induction of stress responses (Vigh et al. 1998; 2007a, c).

Acknowledgments This work was supported in part by grants from the Hungarian National Scientific Research Foundation (OTKA K82097 and OTKA K84257) and the Hungarian National Development Agency (TAMOP-4.2.2/08/1-2008-0002, TAMOP-4.2.2/08/1-2008-0013 and TAMOP-4.2.2/08/1-2008-0014).

References

Adam RM, Mukhopadhyay NK, Kim J et al (2007) Cholesterol sensitivity of endogenous and myristoylated Akt. Cancer Res 67:6238–6246

Balogi Z, Torok Z, Balogh G et al (2005) “Heat shock lipid” in cyanobacteria during heat/light-acclimation. Arch Biochem Biophys 436:346–354

Balogi Z, Cheregi O, Giese KC et al (2008) A mutant small heat shock protein with increased thylakoid association provides an elevated resistance against UV-B damage in Synechocystis 6803. J Biol Chem 283:22983–22991

Basha E, Lee GJ, Demeler B, Vierling E (2005) Chaperone activity of cytosolic small heat shock proteins from wheat. Eur J Biochem 271:1426–1436

Bausero MA, Page DT, Osinaga E, Asea A (2004) Surface expression of HSP25 and HSP72 differentially regulates tumor growth and metastasis. Tumour Biol 25:243–251

Bausero MA, Bharti A, Page DT et al (2006) Silencing the hsp25 gene eliminates migration capability of the highly metastatic murine 4T1 breast adenocarcinoma cell. Tumour Biol 27:17–26

Bellyei S, Szigeti A, Boronkai A et al (2007a) Inhibition of cell death by a novel 16.2 kD heat shock protein predominantly via HSP90 mediated lipid rafts stabilization and Akt activation pathway. Apoptosis 12:97–112

Bellyei S, Szigeti A, Pozsgai E, Boronkai A, Gomori E, Hocsak E et al (2007b) Preventing apoptotic cell death by a novel small heat shock protein. Eur J Cell Biol 86:161–171

Broquet AH, Thomas G, Masliah J, Trugnan G, Bachelet M (2003) Expression of the molecular chaperone HSP70 in detergent-resistant microdomains correlates with its membrane delivery and release. J Biol Chem 278:21601–21606

Chen S, Bawa D, Besshoh S, Gurd JW, Brown IR (2005) Association of heat shock proteins and neuronal membrane components with lipid rafts from the rat brain. J Neurosci Res 81:522–529

Chowdary TK, Raman B, Ramakrishna T, Rao Ch M (2007) Interaction of mammalian HSP22 with lipid membranes. Biochem J 401:437–445

Cinar B, Mukhopadhyay NK, Meng G, Freeman MR (2007) Phosphoinositide 3-kinase-independent non-genomic signals transit from the androgen receptor to Akt1 in membrane raft microdomains. J Biol Chem 282:29584–29593

Cobb BA, Petrash JM (2000) Characterization of alpha-crystallin-plasma membrane binding. J Biol Chem 275:6664–6672

Cobb BA, Petrash JM (2002) Alpha-crystallin chaperone-like activity and membrane binding in age-related cataracts. Biochemistry 41:483–490

Dietrich C, Bagatolli LA, Volovyk ZN, Thompson NL, Levi M, Jacobson K, Gratton (2001) Lipid rafts reconstituted in model membranes. Biophys J 80:1417–1428

Ehrnsperger M, Gräber S, Gaestel M, Buchner J (1997) Binding of non-native protein to Hsp25 during heat shock creates a reservoir of folding intermediates for reactivation. EMBO J 16:221–229

Franzmann TM, Wühr M, Richter K, Walter S, Buchner J (2005) The activation mechanism of Hsp26 does not require dissociation of the oligomer. J Mol Biol 350:1083–1093

Fujita R, Ounzain S, Wang AC, Heads RJ, Budhram-Mahadeo VS (2011) HSP-27 induction requires POU4F2/Brn-3b TF in doxorubicin-treated breast cancer cells, whereas phosphorylation alters its cellular localisation following drug treatment. Cell Stress Chaperones 16:427–439

Gangalum RK, Atanasov IC, Zhou ZH, Bhat SP (2010) AlphaB-crystallin is found in detergent-resistant membrane microdomains and is secreted via exosomes from human retinal pigment epithelial cells. J Biol Chem 286:3261–3269

Grami V, Marrero Y, Huang L, Tang D, Yappert M, Borchman D (2005) α-Crystallin binding *in vitro* to lipids from clear human lenses. Exp Eye Res 81(2):138–146

Haslbeck M (2002) sHsps and their role in the chaperone network. Cell Mol Life Sci 59:1649–1657

Haslbeck M, Franzmann T, Weinfurtner D, Buchner J (2005) Some like it hot: The structure and function of small heat-shock proteins. Nat Struct Mol Biol 1:842–846

Horvath I, Multhoff G, Sonnleitner A, Vigh L (2008) Membrane-associated stress proteins: More than simply chaperones. Biochimica et Biophysica Acta – Biomembranes 1778:1653–1664
Horwitz J (1992) Alpha-crystallin can function as a molecular chaperone. Proc Natl Acad Sci USA 89:10449–10453
Horwitz J (2003) Alpha-crystallin. Exp Eye Res 76:145–153
Kampinga HH, Hageman J, Vos MJ, Kubota H, Tanguay RM, Bruford E et al. (2009) Guidelines for the nomenclature of the human heat shock proteins. Cell Stress Chaperones 1:105–111
Lancelot E, Grauby-Heywang C (2007) Comparison of the interaction of dihydrocholesterol and cholesterol with sphingolipid or phospholipid Langmuir monolayers. Colloids Surf B Biointerfaces 59:1–86
Launay N, Tarze A, Vicart P, Lilienbaum A (2010) Serine 59 phosphorylation of {alpha}B-crystallin down-regulates its anti-apoptotic function by binding and sequestering Bcl-2 in breast cancer cells. J Biol Chem 285:37324–37332
Lee JS, Satoh T, Shinoda H, Samejima T, Wu SH, Chiou SH (1997) Effect of heat-induced structural perturbation of secondary and tertiary structures on the chaperone activity of alpha-crystallin. Biochem Biophys Res Commun 237:277–282
Nakagawa M, Tsujimoto N, Nakagawa H, Iwaki T, Fukumaki Y, Iwaki A (2001) Association of HSPB2, a member of the small heat shock protein family, with mitochondria. Exp Cell Res 271:161–168
Nakamoto H, Vigh L (2007) The small heat shock proteins and their clients. Cell Mol Life Sci 64:294–306
Narberhaus F (2002) Alpha-crystallin-type heat shock proteins: Socializing minichaperones in the context of a multichaperone network. Microbiol Mol Biol Rev 66:64–93
Nedellec P, Edling Y, Perret E, Fardeau M, Vicart P (2002) Glucocorticoid treatment induces expression of small heat shock proteins in human satellite cell populations: Consequences for a desmin-related myopathy involving the R120G alpha B-crystallin mutation. Neuromuscul Disord 12:457–465
Neckers L, Ivy SP (2003) Heat shock protein 90. Curr Opin Oncol 15:419–442
Norberg E, Orrenius S, Zhivotovsky B (2010) Mitochondrial regulation of cell death: Processing of apoptosis-inducing factor (AIF). Biochem Biophys Res Commun 396:95–100
Pozsgai E, Gomori E, Szigeti A, Boronkai A, Gallyas F Jr, Sumegi B et al (2007) Correlation between the progressive cytoplasmic expression of a novel small heat shock protein (HSPB11) and malignancy in brain tumors. BMC Cancer 7:233
Radhakrishnan A, McConnell HM (2000) Chemical activity of cholesterol in membranes. Biochemistry 39:8119–8124
Radhakrishnan A, Anderson TG, McConnell HM (2000) Condensed complexes, rafts, and the chemical activity of cholesterol in membranes. Proc Natl Acad Sci USA 97:12422–12427
Sankhala RS, Damai RS, Swamy MJ (2011) Correlation of membrane binding and hydrophobicity to the chaperone-like activity of PDC-109, the major protein of bovine seminal plasma. PLoS ONE 6(3):e17330
Scolari S, Müller K, Bittman R, Herrmann A, Müller P (2010) Interaction of mammalian seminal plasma protein PDC-109 with cholesterol: Implications for a Putative CRAC Domain. Biochemistry 49:9027–9031
Shah M, Patel K, Fried VA et al (2002) Interactions of STAT3 with caveolin-1 and heat shock protein 90 in plasma membrane raft and cytosolic complexes. Preservation of cytokine signaling during fever. J Biol Chem 277:45662–45669
Soti C, Nagy E, Giricz Z, Vigh L, Csermely P, Ferdinandy P (2005) Heat shock proteins as emerging therapeutic targets. Br J Pharmacol 146:769–780
Sreedhar AS, Mihaly K, Pato B, Schnaider T, Steták A, Kis-Petik K et al (2003) HSP90 inhibition accelerates cell lysis. Anti-HSP90 ribozyme reveals a complex mechanism of HSP90 inhibitors involving both superoxide- and HSP90-dependent events. J Biol Chem 278:35231–35240
Stromer T, Ehrnsperger M, Gaestel M, Buchner J (2003) Analysis of the interaction of small heat shock proteins with unfolding proteins. J Biol Chem 278:18015–18021

Tang D, Borchman D, Yappert MC, Cenedella RJ (1998) Influence of cholesterol on the interaction of alpha-crystallin with phospholipids. Exp Eye Res 66:559–567

Tapodi A, Debreceni B, Hanto K, Bognar Z, Wittmann I, Gallyas F Jr, Varbiro G, Sumegi B (2005) Pivotal role of Akt activation in mitochondrial protection and cell survival by poly(ADP-ribose) polymerase-1 inhibition in oxidative stress. J Biol Chem 280:35767–35775

Taylor RP, Benjamin IJ (2005) Small heat shock proteins: A new classification scheme in mammals. J. Mol Cell Cardiol 38:433–444

Torok Z, Horváth I, Goloubinoff P, Kovács E, Glatz A, Balogh G, Vígh L (1997) Evidence for a lipochaperonin: association of active protein-folding GroESL oligomers with lipids can stabilize membranes under heat shock conditions. Proc Natl Acad Sci USA 94:2192–2197

Torok Z, Goloubinoff P, Horvath I, Tsvetkova NM, Glatz A, Balogh G et al (2001) Synechocystis HSP17 is an amphitropic protein that stabilizes heat-stressed membranes and binds denatured proteins for subsequent chaperone-mediated refolding. Proc Natl Acad Sci USA 98:3098–3103

Triantafilou M, Miyake K, Golenbock DT, Triantafilou K (2002) Mediators of innate immune recognition of bacteria concentrate in lipid raft and facilitate lipopolysaccharide-induced cell activation. J Cell Sci 115:2603–2611

Tsvetkova NM, Horvath I, Torok Z, Wolkers WF, Balogi Z, Shigapova N et al (2002) Small heat-shock proteins regulate membrane lipid polymorphism. Proc Natl Acad Sci USA 99:3504–13509

van Montfort RL, Basha E, Friedrich KL, Slingsby C, Vierling E (2001) Crystal structure and assembly of a eukaryotic small heat shock protein. Nat Struct Biol 1:1025–1030

Varbiro G, Veres B, Gallyas F Jr, Sumegi B (2001) Direct effect of Taxol on free radical formation and mitochondrial permeability transition. Free Radic Biol Med 31:548–558

Vigh L, Maresca B, Harwood JL (1998) Does the membrane's physical state control the expression of heat shock and other genes? Trends Biochem Sci 23:369–374

Vigh L, Maresca B (2002) Dual role of membranes in heat stress: As thermosensors modulate the expression of stress genes and, by interacting with stress proteins, re-organize their own lipid order and functionality. In: Cell and Molecular Responses to Stress, KB Storey and JM Storey (eds) (173–188) Elsevier, Amsterdam

Vigh L, Escriba PV, Sonnleitner A, Sonnleitner M, Piotto S, Maresca B et al (2005) The significance of lipid composition for membrane activity: New concepts and ways of assessing function. Prog Lipid Res 44:303–344

Vigh L, Torok Z, Balogh G, Glatz A, Piotto S, Horvath I (2007a). Membrane-regulated stress response: A theoretical and practical approach. Adv Exp Med Biol 594:114–131

Vigh L, Torok Z, Balogh G, Glatz A, Piotto S, Horvath I (2007b) Membrane-regulated stress response: a theoretical and practical approach. Adv Exp Med Biol 594:114–131

Vigh L, Horváth I, Maresca B, Harwood JL (2007c) Can the stress protein response be controlled by 'membrane-lipid therapy'? Trends Biochem Sci 32:357–363

Waheed AA, Jones TL (2002) HSP90 interactions and acylation target the G protein Galpha 12 but not Galpha 13 to lipid rafts. J Biol Chem 277:32409–32412

Wang R, Kovalchin JT, Muhlenkamp P, Chandawarkar RY (2006) Exogenous heat shock protein 70 binds macrophage lipid raft microdomain and stimulates phagocytosis, processing, and MHC-II presentation of antigens. Blood 107:1636–1642

Welker S, Rudolph B, Frenzel E, Hagn F, Liebisch G, Schmitz G et al (2010) HSP12 Is an intrinsically unstructured stress protein that folds upon membrane association and modulates membrane function. Molecular Cell 39:507–520

Whittaker R, Glassy MS, Gude N, Sussman MA, Gottlieb RA, Glembotski CC (2009) Kinetics of the translocation and phosphorylation of alphaB-crystallin in mouse heart mitochondria during ex vivo ischemia. Am J Physiol Heart Circ Physiol 296:H1633–1642

Zhu Z, Xu X, Yu Y, Graham M, Prince ME, Carey TE, Sun D (2010) Silencing heat shock protein 27 decreases metastatic behavior of human head and neck squamous cell cancer cells in vitro. Mol Pharm 7:1283–1290

[illegible]

Chapter 6
Hsp70 Chaperone Systems in Vesicular Trafficking

David A. Parfitt, Daniel C. Campbell and J. Paul Chapple

Abstract In humans there are 13 different Hsp70 proteins and 50 J proteins (also known as DnaJ or Hsp40 proteins; Kampinga et al., Cell Stress Chaperones 14:105–111, 2009). This diversity of the cochaperone J protein allows the recruitment of Hsp70 family members to multiple cellular locales and activities beyond their canonical protein folding functions (Cheetham and Caplan AJ, Cell Stress Chaperones 3:28–36, 1998; Kampinga and Craig, Nat Rev Mol Cell Biol 11:579–592, 2010). Significant amongst these roles is the participation of Hsp70 systems in intracellular traffic of membranes, including involvement in regulation of endocytosis and exocytosis mechanisms. In this chapter we describe the core components of Hsp70 chaperone systems and consider their multiple functions, from the perspective of J proteins acting as targeting factors for Hsp70 activity. We then focus on the role of Hsp70 chaperone systems in intracellular trafficking of vesicles and the fate of vesicular trafficked proteins.

6.1 The Hsp70 Chaperone System

Most organisms have more than one Hsp70 protein, with both constitutively expressed and stress inducible forms. For example, in humans, there are constitutively expressed Hsc70 (HSPA8) and the stress inducible Hsp70 (HSPA1A). In eukaryotic cells there are also subcellular compartment-specific Hsp70 family members e.g. the endoplasmic reticulum (ER) luminal Hsp70 family member BiP (also known as Grp78 or HSPA5). Some Hsp70 family members also have tissue specific expression profiles, these include the cytosolic HspA1L and HspA2, which are highly expressed in testis (Kampinga et al. 2009). Hsp70 proteins are conserved throughout evolution, with high levels of sequence identity between orthologs and paralogs. Client-protein recognition is typically promiscuous (Kampinga et al. 2009) and is determined by the sub-cellular localization of specific Hsp70s and their interactions with cochaperone partners. Moreover, cochaperone interaction has a role in determining the fate of the systems substrates, with cochaperones such as the E3 ligase C-terminus Hsp70

J. P. Chapple (✉) · D. A. Parfitt · D. C. Campbell
Centre for Endocrinology, William Harvey Research Institute, Barts and the London School of Medicine and Dentistry, Queen Mary University of London, London, UK
e-mail: j.p.chapple@qmul.ac.uk

B. Henderson, A. G. Pockley (eds.), *Cellular Trafficking of Cell Stress Proteins in Health and Disease*, Heat Shock Proteins 6, DOI 10.1007/978-94-007-4740-1_6,

interacting protein (CHIP) promoting ubiquitin proteasome system (UPS)-mediated degradation of Hsp70 clients (McDonough and Patterson 2003; Murata et al. 2003).

The function of Hsp70 family members is dependent on their ability to bind and release client proteins. Protein folding involves numerous cycles of client binding and release, coupled to conformational change of Hsp70s, driven by ATP hydrolysis and release. Hsp70s have two functional domains, a 40 kDa N-terminal nucleotide-binding domain (NBD) and a 25 kDa C-terminal substrate-binding domain (SBD) that recognizes exposed hydrophobic segments of clients. These domains are connected by an inter-domain linker, which mediates their allosteric coupling. In the mammalian system the SBD and NBD interact when Hsp70 is ADP bound (Swain et al. 2007). When Hsp70s are ATP bound they have a low affinity for client peptides, with fast kinetics of binding and release. In the ADP bound state client binding affinity is increased while kinetics of binding and release is reduced. Peptide binding by the SBD stimulates the ATPase activity of the NBD, helping promote client loading of Hsp70s.

Importantly, Hsp70s intrinsic ATPase activity is low and for efficient chaperone function a J protein is essential. Through a requisite histidine-proline-aspartic acid motif, in the loop between helix II and helix III of their ~70 amino acid J domain, J proteins are able to stimulate Hsp70s ATPase activity. The J domains interact with the NBD of Hsp70s, with the inter-domain linker also playing a role in this interaction (Jiang et al. 2007; Kumar et al. 2011; Swain et al. 2007). The mechanism by which J domains stimulate Hsp70s activity is only partially resolved. It is proposed that the J domain directs the inter-domain linker towards a hydrophobic patch on the NBD surface. This results in displacement of the SBD from the NBD, likely promoting client capture (Jiang et al. 2007).

Efficient chaperone activity of Hsp70s also requires the presence of nucleotide exchange factors (NEFs), which promote the exchange of ADP for ATP and thus stimulate client release. Four structurally different types of NEFs have been identified in eukaryotes. These are homologs of the *E. coli* NEF GrpE, Hsp70 binding protein 1 (HspB1), Bag (Bcl-2-associated athanogene) domain proteins, and Hsp110 family members (Cyr 2008).

The concentration of J proteins and other cochaperones, relative to Hsp70 partners, is central to how the chaperone system functions and likely plays a regulatory role. This could include increased stimulation of ATPase activity by J proteins having a negative effect on client binding and elevated NEF activity causing premature client release (Kampinga and Craig 2010).

6.2 Diversity of J Proteins

There are three major sub-groups of J proteins (Fig. 6.1). DNAJA or type I, which have an N-terminal J-domain followed by a glycine/phenylalanine (G/F) -rich region and a cysteine-rich region stabilised by two zinc ions; DNAJB or type II, which have

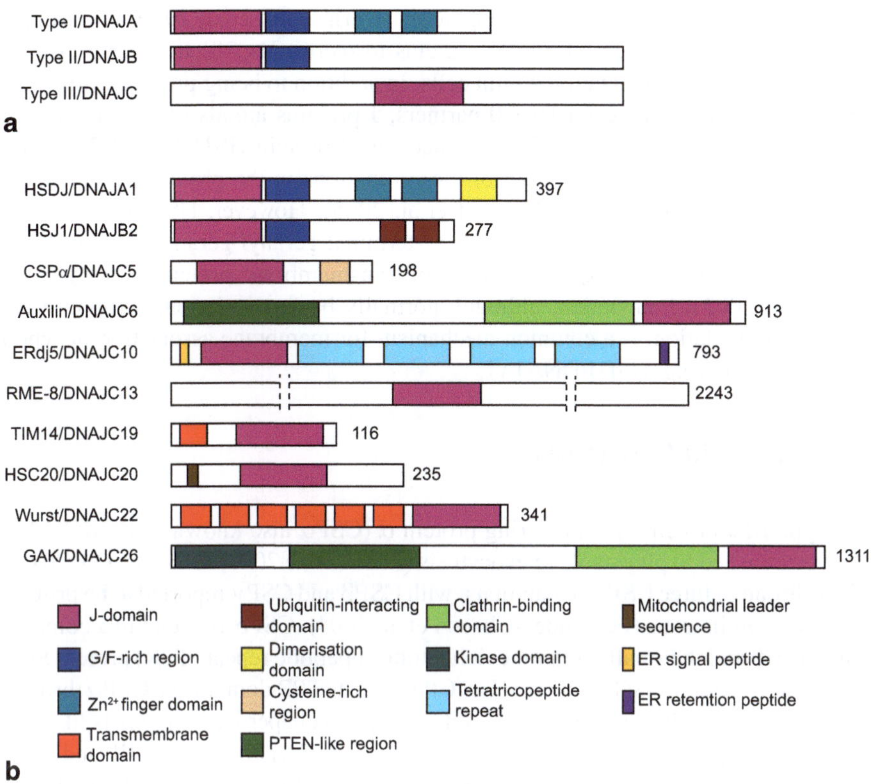

Fig. 6.1 Domain structure of the three classes of J proteins (**a**) and specific J proteins discussed in this chapter (**b**)

the N-terminal J domain and the G/F-rich region only; and DNAJC or type III, which have the J domain only, located anywhere within the protein (Cheetham and Caplan 1998; Kampinga et al. 2009). Type III J proteins represent the largest group of J domain proteins, however individually they are less likely to have such a diversity of clients as type I and II J proteins because they do not generally have a role in canonical protein folding. Therefore, many type III J proteins may be considered 'private' cochaperones for specific clients/complexes.

In the cytosol and other cellular compartments, individual Hsp70 proteins will interact with multiple J protein partners. Some chaperone/cochaperone pairings have a canonical function, with the J proteins inhibiting aggregation of clients and recruiting Hsp70 proteins to promote folding. However, many J proteins contain multiple domains not present in the archetypal J protein DnaJ. These recruit Hsp70s to specific clients where they perform diverse tasks, extending beyond maintenance of proteostasis. For example, the mitochondrial HSP70, Grp75/mortalin/HSPA9, works in conjunction with the J protein TIM14/DNAJC19 (TIM = translocase of the inner mitochondrial membrane) and the NEF protein SIL1 at the TIM23 complex to import proteins into the mitochondria, and is also recruited by HSC20/DNAJC20 to play a role in iron-sulphur cluster biogenesis (Uhrigshardt et al. 2010; Wiedemann et al. 2004). Part of the ability of J proteins to recruit Hsp70s to diverse tasks is dependent on their localization within cells. In addition to being present in different organelles, as is the case for Hsp70 partners, J proteins are also found associated with intracellular membranes. For instance, the J protein HSJ1/DNAJB2 has two isoforms, HSJ1a and HSJ1b. Both of these functions in the proteasome mediated degradation of Hsp70 clients (Westhoff et al. 2005). However, HSJ1b is localized to the cytoplasmic face of the ER via a C-terminal geranyl-geranyl anchor allowing it to participate in the quality control of transmembrane proteins (Chapple and Cheetham 2003). Prenylation, although normally by farnesylation, as opposed to gerenyl-gerenylation, is a common mechanism for membrane association of type I and II J proteins e.g. HSDJ/DNAJA1.

6.3 Hsc70 in Exocytosis

The type III J protein cysteine-string protein α (CSPα also known as DnaJC5) has been established to play a role in exocytosis (Evans et al. 2003; Johnson et al. 2010). CSPα is one of three CSPs in mammals, with CSPβ and CSPγ reported to be primarily expressed in testis (Fernandez-Chacon et al. 2004). CSPα is found in a complex with Hsc70 and the small glutamine rich tetratricopeptide repeat (TPR) protein SGT on secretory vesicles. SGT binds to Hsc70 through its TPR domain and CSPα through its N-terminus (Tobaben et al. 2001). Activity of this chaperone complex is likely to be regulated by further Hsc70 cochaperones including Hsp90, Hsp40, HIP and HOP, which have all been shown to associate with the CSPα:Hsc70:SGT complex (Johnson et al. 2010; Rosales-Hernandez et al. 2009). CSPα is targeted to membranes through a hydrophobic domain that contains the cysteine-string and is sorted dependent on palmitoylation (Greaves and Chamberlain 2006). CSP complexes are anchored to secretory vesicles, including synaptic vesicles, chromaffin granules, pancreatic zymogen granules and mucin granules (Braun and Scheller 1995; Chamberlain et al. 1996; Mastrogiacomo et al. 1994; Park et al. 2008). Loss of CSP function leads to severe neurodegeneration. In *Drosophila,* CSP deletion causes progressive neuronal loss and reduced lifespan (Zinsmaier et al. 1994), with CSP-null flies exhibiting impaired Ca^{2+}-dependent neurotransmission with abnormal presynaptic calcium levels and a reduction in synaptic boutons at neuromuscular junctions (Dawson-Scully et al. 2007; Johnson et al. 2010; Nie et al. 1999; Umbach et al. 1994). Hsc70 is also critical for neurotransmitter exocytosis in *Drosophila*, supporting that it functions with CSPα (Johnson et al. 2010). CSPα mice start to show signs of neurodegeneration 2–4 weeks after birth (Fernandez-Chacon et al. 2004). Moreover, mutations of human CSPα, which affect palmitoylation-dependent sorting, cause autosomal-dominant adult-onset neuronal ceroid lipofuscinosis, a disease characterized by accumulation of autofluorescent storage material in neuronal tissues and neurodegeneration (Noskova et al. 2011).

Evidence suggests that CSP plays multiple roles in synaptic transmission. It is thought that CSP is an organizer of protein-protein interactions at various stages of the secretory cycle, stabilising and/or refolding components of the regulated exocytotic machinery (Burgoyne and Morgan 2011; Evans et al. 2003). CSPα has been shown to interact with components of the core membrane fusion machinery, including the synaptic calcium sensor synaptotagmin and SNARE (N-ethylmaleimide sensitive factor attachment receptor) proteins, including syntaxin and synaptobrevin (Evans et al. 2003). CSPα was also shown to suppress a fly syntaxin overexpression phenotype (Nie et al. 1999; Wu et al. 1999). More recently CSPα has been shown to interact with SNAP-25 (Sharma et al. 2011). During exocytosis, at the neuronal synapse, syntaxin-1 forms a complex with synaptobrevin-2 and SNAP-25. This SNARE complex is required for membrane fusion and is dissociated by NSF (N-ethylmaleimide sensitive factor). Because of the rapid transmission of signals at nerve terminals, SNARE proteins rapidly go through many cycles of complex assembly and disassembly. In CSPα knockout mice, levels of SNAP25 and presynaptic SNARE complex assembly are markedly reduced (Chandra et al. 2005). Binding of a CSPα:Hsc70:SGT complex to monomeric SNAP-25 prevents its aggregation, and degradation by the UPS, with misfolded SNAP-25 inhibiting SNARE complex formation (Sharma et al. 2011). Normally J proteins bind substrates and recruit Hsp70, therefore it is unusual that the interaction between CSPα and SNAP-25 is indirect through Hsc70 (Burgoyne and Morgan 2011; Sharma et al. 2011). Another study has shown CSPα interacts indirectly with SNAP-25 and directly with the calcium sensor synaptotagmin 9 (Syt9), suggesting Syt9 can also acts as an intermediate between CSPα and SNAP-25 (Boal et al. 2011). The CSP:Hsc70:SGT complex represents a peripheral cell compartment chaperone system for SNARE proteins. However the presence of CSPs and SGT in non-neuronal cells indicates this system does not just function at nerve terminals remote from the cell body.

In neurons from CSPα knockout mice, cellular levels of Hsc70 are reduced and Hsc70 degradation is increased. This suggests that, despite the myriad cellular roles performed by Hsc70, the demand for its activity in pre-synaptic compartments is sufficient to translate into significant changes in its levels when the mechanism to recruit it to exocytic work (e.g. CSPα) is absent (Sharma et al. 2011).

6.4 Hsc70-Mediated Endocytosis

There are two main types of endocytosis, endocytosis that is clathrin independent and endocytosis that is mediated by clathrin, with the latter requiring Hsc70 action. Clathrin forms a triskelion structure with three clathrin 'legs' composed of a 190 kDa heavy chain and 30 kDa light chain each radiating from a central point (Kirchhausen et al. 1991; Lindner and Ungewickell 1991). Clathrin triskelions self-associate into a lattice-type structure, this acts to 'coat vesicles' (Fotin et al. 2004). In clathrin-mediated endocytosis, clathrin-coated pits form at the plasma membrane around the desired cargo before budding off from the membrane. In this process, clathrin does not directly interact with the membrane or cargo receptors at any time and relies on adaptor proteins such AP-2 and epsin to be recruited to the plasma membrane

(McMahon and Boucrot 2011). After the coated vesicle has formed, it needs to be uncoated before fusion with intracellular target membranes, such as endosomes. This disassembly of clathrin-coated vesicles is controlled by the recruitment of Hsc70 by the J protein auxilin (DnaJC6; Ahle and Ungewickell 1990; Chappell et al. 1986; Holstein et al. 1996).

Auxilin is a 100 kDa protein containing an N-terminal phosphoinositide phosphatase (PTEN)-like domain and C-terminal clathrin-binding and J-domains (Ungewickell et al. 1995). The auxilin PTEN-like domain has the phosphatase and C2 domains of PTEN. However, the crystal structure of the auxilin PTEN-like domain was recently elucidated and found to lack key residues needed to confer phosphatase activity, due to a change in one of its catalytic loops (Guan et al. 2010). In auxilin the PTEN-like domain is thought to be required for the burst of recruitment to a clathrin-coated pit, during the transition between membrane constriction and vesicle budding (Massol et al. 2006). This is based on evidence that auxilin, lacking its PTEN-like domain, can still mediate Hsc70 and ATP-dependent clathrin uncoating (Holstein et al. 1996), but has impaired recruitment dynamics (Massol et al. 2006). The C2 region was identified to be the region which binds the vesicle membrane, with mutation of key residues in this region destroying binding (Guan et al. 2010). The PTEN-like domain is presumed to project into the vesicle and contacts a phosphoinositide group after the clathrin-binding domain binds an internal face of the clathrin coat (Lee et al. 2006; Massol et al. 2006).

Mammalian species contain a second auxilin paralog called cyclin G-associated kinase (GAK/auxilin-2/DnaJC28), which has the same basic molecular structure to auxilin but with an additional N-terminal kinase domain (Greener et al. 2000; Kimura et al. 1997). The GAK kinase domain has been shown to phosphorylate the clathrin adaptor proteins AP-1 and AP-2 (Umeda et al. 2000). Furthermore, GAK is expressed ubiquitously whereas auxilin is neuron-specific.

Studies in multiple systems and organisms have investigated the depletion of auxilin and GAK. Disruption of the yeast auxilin-like gene, *SWA2*, results in increased accumulation of clathrin-coated vesicles, implying reduced auxilin-mediated uncoating (Gall et al. 2000). Reduced expression of auxilin in *Drosophila* and *C. elegans* disrupts clathrin-mediated endocytosis dynamics (Greener et al. 2001; Hagedorn et al. 2006). A recent study by Hirst et al. (2008) investigated targeted double knockdown of auxilin and GAK in HeLa cells. Auxilin depletion was shown to inhibit clathrin mediated endocytosis, reduce the number of clathrin coated pits at the plasma membrane, inhibit intracellular trafficking, increase association of clathrin and adaptor proteins with intracellular membranes, reduce clathrin trafficking, and increase accumulation of empty clathrin coats containing no cargo. These data suggest an additional role for auxilin and Hsc70 in prevention of formation of empty clathrin cages i.e. as a chaperone of clathrin before and after endocytosis (Hirst et al. 2008).

Knockout (KO) mice for both auxilin (Yim et al. 2010) and GAK (Lee et al. 2008) have been created. Interestingly, auxilin KO mice had higher GAK levels in the brain than controls at 3–5 weeks of age (80 % of auxilin KO mice had a 3-fold increase in GAK). However, increased levels of GAK cannot compensate fully for lack of auxilin in uncoating of clathrin coated vesicles, as auxilin KO mice showed increased amounts of clathrin coated vesicles and empty clathrin cages at synapses (Yim et al. 2010). Conventional GAK null mice were embryonically lethal, but conditional GAK knockout mouse embryonic fibroblasts showed markedly reduced endocytosis and mislocalisation of adaptor proteins AP-2, epsin, Eps15 and Eps15R at the plasma membrane (Lee et al. 2008).

Multiple recent studies have shed light on the molecular mechanism of clathrin uncoating (Bocking et al. 2011; Rothnie et al. 2011; Xing et al. 2010). Auxilin binds to clathrin at a ratio of one molecule per triskelion and initially recruits Hsc70-ATP at a ratio of one molecule per triskelion (Bocking et al. 2011). Interaction between the auxilin J-domain and Hsc70 causes hydrolysis of ATP to ADP, resulting in conformational change in the SBD of Hsc70, which attaches to clathrin at the peptide motif QLMLT on the clathrin heavy chain C-terminus (Rapoport et al. 2008; Xing et al. 2010). Auxilin dissociates from Hsc70-ADP and recruits another Hsc70-ATP, again causing hydrolysis and attachment of the second Hsc70 to clathrin (Rothnie et al. 2011). A third Hsc70-ATP is recruited by auxilin, ATP-hydrolysed and Hsc70 attached to clathrin. The conformational strain on the clathrin cage, with three Hsc70s bound, causes disassembly. Auxilin is released but Hsc70 remains bound to the triskelion, presumably until NEFs replace ADP for ATP (Bocking et al. 2011). Therefore clathrin uncoating is achieved by sequential binding of three Hsc70 molecules that put stress on the clathrin triskelions until the coat disassembles (Rothnie et al. 2011).

Other J proteins besides auxilin have a role in endocytosis. The *Drosophila* protein wurst was identified in a search for respiratory tube regulators in flies. It is a six-span transmembrane protein containing C-terminal clathrin-binding and J-domains (Behr et al. 2007) that are highly conserved down to placozoa (Wingen et al. 2009), including a human ortholog DnaJC22. *Drosophila* wurst mutants have increased respiratory tube length and diameter. This tracheal tube phenotype is caused by problems with extracellular matrix organisation ultimately compromising gas diffusion rates, resulting in larval death (Behr et al. 2007). Moreover, tracheal tube-specific targeted knockdown of wurst via RNAi in *Drosophila* demonstrated wurst is specifically required at embryonic stage 17 for endocytosis-mediated liquid clearance from respiratory tubes (Stumpges and Behr 2011). Wurst interacts with Hsc70 and clathrin and inserts into the plasma membrane (Stumpges and Behr 2011). In *Drosophila,* wurst is proposed to recruit Hsc70 and clathrin to the apical surface of the plasma membrane and thus coordinate the early steps in clathrin-mediated endocytosis. Furthermore, in immunofluorescence studies, wurst localisation partially overlaps with rab5, suggesting that wurst is localised in early endosomes (Behr et al. 2007).

Receptor mediated endocytosis-8 (RME-8, also known as DnaJC13) was originally identified in a screen for endocytic defects in *C. elegans* (Zhang et al. 2001). RME-8 is evolutionarily conserved, with orthologs in *Drosophila* and humans interacting with Hsc70 during endocytosis (Chang et al. 2004; Girard et al. 2005). Colocalisation studies for RME-8 showed an overlap between RME-8 and endosomes (Chang et al. 2004; Zhang et al. 2001); in particular, human RME-8 colocalised with EEA1 and endocytosed transferrin, indicating localisation on early and recycling endosomes (Girard et al. 2005) Furthermore, RME-8 did not colocalise with late endosomal markers LBPA and Rab7 (Fujibayashi et al. 2008). The J-domain of

RME-8, which binds Hsc70, is located in the middle of the protein, and the N-terminal region was found to be required for membrane binding (Fujibayashi et al. 2008). RME-8 knockdown has been shown to decrease endocytosis of epidermal growth factor (EGF; Girard et al. 2005). Also, EGF receptor levels are decreased after RME-8 depletion, most likely due to an increase of EGFR degradation (Girard and McPherson 2008). However, RME-8 knockdown does not affect cellular levels of receptors that are recycled to the plasma membrane after endocytosis, such as tranferrin receptor and insulin receptor (Girard et al. 2005). Interestingly, RME-8 knockdown also misdirected sorting of cation independent mannose-6-phosphate receptor by causing accumulation at the trans-Golgi network (TGN; Girard et al. 2005). Two recent studies have shown that RME-8 directly interacts with SNX-1, a component of the retromer complex (Popoff et al. 2009; Shi et al. 2009), which plays a central role in the retrieval of cargo proteins from the endosome to the *trans*-Golgi network. Targeted knockdown of *rme-8* or *snx-1* in *C. elegans* disrupts endosome to TGN transport of MIG-14, a retromer dependent cargo protein, and also causes clathrin accumulation on endosomes (Shi et al. 2009). Moreover, RME-8 depletion in HeLa cells causes a disruption of Shiga toxin subunit B (STxB) trafficking (Popoff et al. 2009). These data suggest a role of RME-8 in recruiting Hsc70 onto sites of vesicle formation, thereby regulating clathrin dynamics in retrograde sorting from early endosomes (Popoff et al. 2009).

6.5 Quality Control of Trafficked Proteins

The vesicular trafficking of ER-luminal and membrane proteins is regulated by molecular chaperones that function as quality control components in pathways that retain and degrade misfolded proteins. The luminal domains of partially folded ER protein interact with ER resident chaperones, such as calnexin, calreticulin and BiP. For proteins to traffic beyond the ER they must be released by these ER chaperones. For many proteins, recognition by cargo receptors, to allow concentration at sites of ER exit, is also required for trafficking (Braakman and Bulleid 2011). Therefore both ER chaperones and cargo receptors contribute to quality control. Moreover, it has been suggested that ER retention and export are competing processes (Brodsky and Skach 2011). Indeed terminally misfolded proteins normally remain associated with chaperones and are retained at the ER, prior to ER associated degradation (ERAD; Bernasconi and Molinari 2011; Vembar and Brodsky 2008). In ERAD, misfolded proteins undergo retrotranslocation from the ER and then cleavage by cytosolic proteases. ERAD is also dependent on the ER Hsp70 chaperone system, as the J protein ERdj5/DNAJC10 is a member of an ERAD complex with BiP. Through its thioredoxin domain ERdj5 reduces disulphide bonds in ERAD substrates. BiP then maintains these in a state compatible with transfer to the ERAD retrotranslocation channel (Ushioda et al. 2008). Interestingly there is evidence that not all misfolded ER proteins undergo ERAD. This includes the misfolded prion protein, which is not recognized by ER quality control systems and is instead trafficked to the Golgi and then sorted for lysosomal degradation (Ashok and Hegde 2009).

It is also known that some misfolded or partially assembled proteins exit the ER but are then retrieved from the early secretory system, rather than being transported to their final destination. This is mediated by recycling form post-ER to ER compartments by retrograde transport of COPI vesicles dependent on KDEL receptors (Capitani and Sallese 2009). Both BiP and ERdj5 contain a C-terminal KDEL retrieval motif allowing them to bind KDEL receptors in the cis-Golgi and trigger retrieval back to the ER (Hosoda et al. 2003; Yamamoto et al. 2001). In some instances, such as ER stress, BiP can be trafficked to the cell surface. It has been suggested this may be because in some conditions intracellular levels of KDEL proteins exceed the binding capacity of KDEL receptor, preventing efficient retrieval from the Golgi. Corroborating this concept, ER stress conditions in HeLa cells did not increase KDEL receptor levels, however other studies have shown BiP can traffic to the cell surface without ER stress suggesting mechanisms, such as direct blocking of the KDEL:KDEL receptor interaction, may play a role (Llewellyn et al. 1997; Ni et al. 2011; Zhang et al. 2010). BiP cochaperones also play a role in its trafficking, as the transmembrane J protein ERdj1/DNAJC1 has been shown to promote transport to the cell surface of BiP and to interact with it at this cellular locale (Misra et al. 2005). The role of client binding in cell surface expression of BiP, and whether folding intermediates of BiP bound client are trafficked to the cell surface with the chaperone is unclear.

Transmembrane proteins that bind BiP in the ER lumen can also have cytosolic domains that interact with Hsp70s and their cochaperone systems in the cytosol. For example, it has been reported that the testis enriched Hsc70t/HspA1L promotes expression and trafficking to the cell surface of odorant receptors (Neuhaus et al. 2006), while we have shown that another G-protein coupled receptor (GPCR), the obesity linked melanocortin-4 receptor (MC4R), interacts with Hsc70 and HSJ1. Moreover, Hsc70 was able to promote mobility at the ER and trafficking to the cell surface of misfolded MC4R mutants (Meimaridou et al. 2011). In addition to GPCRs, Hsc70 has been shown to bind directly to membrane channel proteins such as K^+ channels and cystic fibrosis transmembrane conductance regulator (CFTR), modulating their trafficking by mechanisms including degradation by ERAD. This is the case for immature CFTR that is targeted for proteasome degradation by CHIP, functioning with Hsc70, to promote its ubiquitination (Meacham et al. 2001). Additional Hsc70 cochaperones further regulate the degradation of CFTR, with HspBP1 and Bag-2 acting to inhibit CHIP activity (Alberti et al. 2004; Arndt et al. 2005).

Proteins that have reached the cell surface are further subject to quality control systems that utilise Hsp70 machineries (Apaja et al. 2010). Peripheral quality control occurs along late secretory and endocytic pathways, as well as at the plasma membrane, and is dependent on ubiquitination acting as a targeting signal for endocytosis and lysosomal degradation (Okiyoneda et al. 2011). A model has been suggested where Hsc70 is recruited to cytosolic domains of plasma membrane proteins by DNAJA1, CHIP then ubiquitinates these clients. The ubiquitinated proteins are endocytosed and then either sorted to the lysosome for degradation or recycled to the plasma membrane, if chaperone interaction has promoted their refolding. It has also been suggested that endocytosis may depend on ubiquitin binding endocytic adaptors (e.g. epsin1 and eps15/eps15R) promoting clathrin-mediated internalisation (Okiyoneda et al. 2011). It is fascinating that in separate quality control systems CHIP is able to both poly-ubiquitinate proteins for degradation and ubiquitinate epsin in the regulation of endocytosis. That epsin is not degraded by CHIP ubiquitination, may be because it is processed to a mono-ubiquitinated form by deubiquitinating enzymes (Timsit et al. 2005). It is known that ubiquitin tagging can target proteins to different degradation pathways (proteasomal, endosomal and autophagy) dependent on linkage type, i.e. which lysine residue the ubiquitin chain is linked through, as well as chain length (Clague and Urbe 2010). When protein degradation is impaired, for example by overloading of the UPS, misfolded proteins may be partitioned into a cellular compartment known as the aggresome. Hsp70 binds and directs the deposition of associated substrates into aggresomes, in a pathway regulated by CHIP and Bag-3 (Zhang and Qian 2011).

The degree to which the quality control pathways discussed above are used to regulate cell surface levels of specific channel and receptor proteins is unclear. One paradigm where this has been considered is regulation of ErbB growth factor receptor tyrosine kinase levels (Carraway 2010). CHIP mediates degradation, ErbB2 and it is suggested quantity control mechanisms play central roles in suppressing receptor overexpression in normal cells. This is important as ErbB overexpression promotes progression of ErbB-dependent tumors (Carraway 2010). Exocytosis and endocytosis play essential roles in regulating levels of cell surface receptors, such as GPCRs, with many receptors being internalised and recycled or targeted for degradation after ligand binding as part of the control of signal termination, propagation and resensitisation. If there is any connectivity between chaperone mediated quality control systems and chaperone mediated vesicular trafficking systems remains to be elucidated.

6.6 Conclusions

J proteins direct Hsp70s power to do conformational work to multiple cellular roles. Particularly important J protein:Hsp70 functions are the maintenance of cellular proteostasis and regulation of vesicular trafficking (Fig. 6.2). Moreover, it appears that vesicular transport processes such as exocytosis represent a significant cellular commitment for Hsp70 action, particularly in neurons.

The importance of maintaining cellular proteostasis is clear and it is established that some cells types, such as neurons, are highly sensitive to disruption of protein folding and degradation pathways. Moreover, regulation of folding and degradation of individual proteins represents a way to control their trafficking. The retention of misfolded transmembrane proteins in the ER is a common disease mechanism for pathologically diverse conditions including autosomal dominant retinitis pigmentosa, cystic fibrosis and monogenic obesity. It is therefore attractive to develop strategies to promote escape of disease proteins that harbor misfolding mutations from ER quality control systems, providing rescue of trafficking to the cell surface

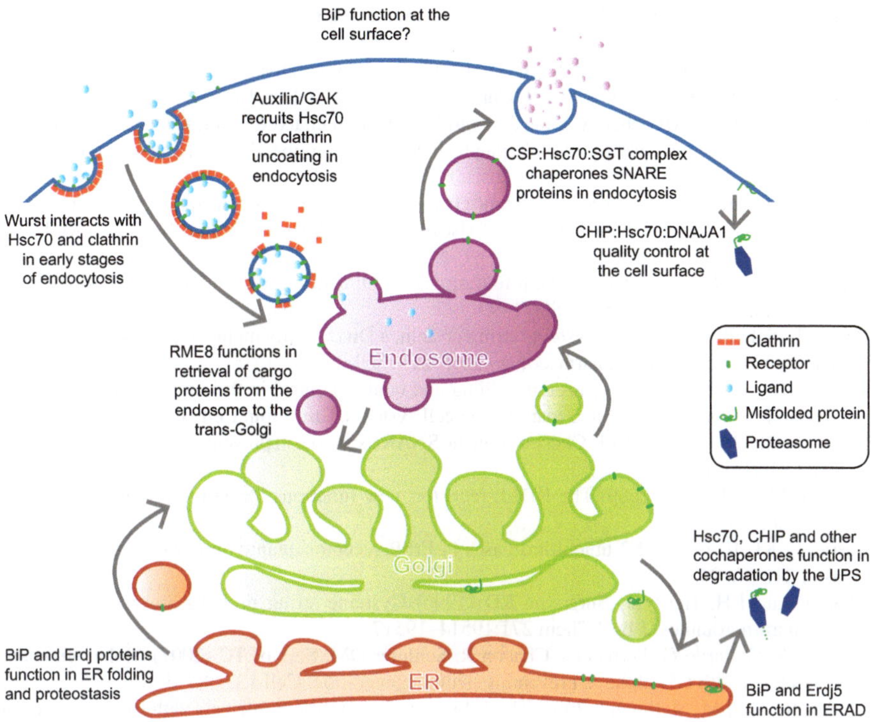

Fig. 6.2 Role of Hsp70 chaperone systems in vesicular transport and quality control of membrane proteins

has the potential to restore a degree of function. Modulation of endogenous Hsp70 chaperone systems represents a potential strategy to achieve such rescue, potentially in combination with other protein stabilising strategies such as the use of chemical chaperones.

References

Ahle S, Ungewickell E (1990) Auxilin, a newly identified clathrin-associated protein in coated vesicles from bovine brain. J Cell Biol 111:19–29

Alberti S, Bohse K, Arndt V, Schmitz A, Hohfeld J (2004) The cochaperone HspBP1 inhibits the CHIP ubiquitin ligase and stimulates the maturation of the cystic fibrosis transmembrane conductance regulator. Mol Biol Cell 15:4003–4010

Apaja PM, Xu H, Lukacs GL (2010) Quality control for unfolded proteins at the plasma membrane. J Cell Biol 191:553–570

Arndt V, Daniel C, Nastainczyk W, Alberti S, Hohfeld J (2005) BAG-2 acts as an inhibitor of the chaperone-associated ubiquitin ligase CHIP. Mol Biol Cell 16:5891–5900

Ashok A, Hegde RS (2009) Selective processing and metabolism of disease-causing mutant prion proteins. PLoS Pathog 5:e1000479

Behr M, Wingen C, Wolf C, Schuh R, Hoch M (2007) Wurst is essential for airway clearance and respiratory-tube size control. Nat Cell Biol 9:847–853
Bernasconi R, Molinari M (2011) ERAD and ERAD tuning: Disposal of cargo and of ERAD regulators from the mammalian ER. Curr Opin Cell Biol 23:176–183
Boal F, Laguerre M, Milochau A, Lang J, Scotti PA (2011) A charged prominence in the linker domain of the cysteine-string protein Cspalpha mediates its regulated interaction with the calcium sensor synaptotagmin 9 during exocytosis. FASEB J 25:132–143
Bocking T, Aguet F, Harrison SC, Kirchhausen T (2011) Single-molecule analysis of a molecular disassemblase reveals the mechanism of Hsc70-driven clathrin uncoating. Nat Struct Mol Biol 18:295–301
Braakman I, Bulleid NJ (2011) Protein folding and modification in the mammalian endoplasmic reticulum. Annu Rev Biochem 80:71–99
Braun JE, Scheller RH (1995) Cysteine string protein, a DnaJ family member, is present on diverse secretory vesicles. Neuropharmacology 34:1361–1369
Brodsky JL, Skach WR (2011) Protein folding and quality control in the endoplasmic reticulum: Recent lessons from yeast and mammalian cell systems. Curr Opin Cell Biol 23:464–475
Burgoyne RD, Morgan A (2011) Chaperoning the SNAREs: A role in preventing neurodegeneration? Nat Cell Biol 13:8–9
Capitani M, Sallese M (2009) The KDEL receptor: new functions for an old protein. FEBS Lett 583:3863–3871
Carraway KL III (2010) E3 ubiquitin ligases in ErbB receptor quantity control. Semin Cell Dev Biol 21:936–943
Chamberlain LH, Henry J, Burgoyne RD (1996) Cysteine string proteins are associated with chromaffin granules. J Biol Chem 271:19514–19517
Chandra S, Gallardo G, Fernandez-Chacon R, Schluter OM, Sudhof TC (2005) Alpha-synuclein cooperates with CSPalpha in preventing neurodegeneration. Cell 123:383–396
Chang HC, Hull M, Mellman I (2004) The J-domain protein Rme-8 interacts with Hsc70 to control clathrin-dependent endocytosis in Drosophila. J Cell Biol 164:1055–1064
Chappell TG, Welch WJ, Schlossman DM, Palter KB, Schlesinger MJ, Rothman JE (1986) Uncoating ATPase is a member of the 70 kilodalton family of stress proteins. Cell 45:3–13
Chapple JP, Cheetham ME (2003) The chaperone environment at the cytoplasmic face of the endoplasmic reticulum can modulate rhodopsin processing and inclusion formation. J Biol Chem 278:19087–19094
Cheetham ME, Caplan AJ (1998) Structure, function and evolution of DnaJ: conservation and adaptation of chaperone function. Cell Stress Chaperones 3:28–36
Clague MJ, Urbe S (2010) Ubiquitin: same molecule, different degradation pathways. Cell 143:682–685
Cyr DM (2008) Swapping nucleotides, tuning Hsp70. Cell 133:945–947
Dawson-Scully K, Lin Y, Imad M, Zhang J, Marin L, Horne JA, Meinertzhagen IA, Karunanithi S, Zinsmaier KE, Atwood HL (2007) Morphological and functional effects of altered cysteine string protein at the Drosophila larval neuromuscular junction. Synapse 61:1–16
Evans GJ, Morgan A, Burgoyne RD (2003) Tying everything together: the multiple roles of cysteine string protein (CSP) in regulated exocytosis. Traffic 4:653–659
Fernandez-Chacon R, Wolfel M, Nishimune H, Tabares L, Schmitz F, Castellano-Munoz M, Rosenmund C, Montesinos ML, Sanes JR, Schneggenburger R, Sudhof TC (2004) The synaptic vesicle protein CSP alpha prevents presynaptic degeneration. Neuron 42:237–251
Fotin A, Cheng Y, Grigorieff N, Walz T, Harrison SC, Kirchhausen T (2004) Structure of an auxilin-bound clathrin coat and its implications for the mechanism of uncoating. Nature 432:649–653
Fujibayashi A, Taguchi T, Misaki R, Ohtani M, Dohmae N, Takio K, Yamada M, Gu J, Yamakami M, Fukuda M, Waguri S, Uchiyama Y, Yoshimori T, Sekiguchi K (2008) Human RME-8 is involved in membrane trafficking through early endosomes. Cell Struct Funct 33:35–50
Gall WE, Higginbotham MA, Chen C, Ingram MF, Cyr DM, Graham TR (2000) The auxilin-like phosphoprotein Swa2p is required for clathrin function in yeast. Curr Biol 10:1349–1358

Girard M, McPherson PS (2008) RME-8 regulates trafficking of the epidermal growth factor receptor. FEBS Lett 582:961–966

Girard M, Poupon V, Blondeau F, McPherson PS (2005) The DnaJ-domain protein RME-8 functions in endosomal trafficking. J Biol Chem 280:40135–40143

Greaves J, Chamberlain LH (2006) Dual role of the cysteine-string domain in membrane binding and palmitoylation-dependent sorting of the molecular chaperone cysteine-string protein. Mol Biol Cell 17:4748–4759

Greener T, Zhao X, Nojima H, Eisenberg E, Greene LE (2000) Role of cyclin G-associated kinase in uncoating clathrin-coated vesicles from non-neuronal cells. J Biol Chem 275:1365–1370

Greener T, Grant B, Zhang Y, Wu X, Greene LE, Hirsh D, Eisenberg E (2001) Caenorhabditis elegans auxilin: a J-domain protein essential for clathrin-mediated endocytosis in vivo. Nat Cell Biol 3:215–219

Guan R, Dai H, Harrison SC, Kirchhausen T (2010) Structure of the PTEN-like region of auxilin, a detector of clathrin-coated vesicle budding. Structure 18:1191–1198

Hagedorn EJ, Bayraktar JL, Kandachar VR, Bai T, Englert DM, Chang HC (2006) Drosophila melanogaster auxilin regulates the internalization of Delta to control activity of the Notch signaling pathway. J Cell Biol 173:443–452

Hirst J, Sahlender DA, Li S, Lubben NB, Borner GH, Robinson MS (2008) Auxilin depletion causes self-assembly of clathrin into membraneless cages in vivo. Traffic 9:1354–1371

Holstein SE, Ungewickell H, Ungewickell E (1996) Mechanism of clathrin basket dissociation: separate functions of protein domains of the DnaJ homologue auxilin. J Cell Biol 135:925–937

Hosoda A, Kimata Y, Tsuru A, Kohno K (2003) JPDI, a novel endoplasmic reticulum-resident protein containing both a BiP-interacting J-domain and thioredoxin-like motifs. J Biol Chem 278:2669–2676

Jiang J, Maes EG, Taylor AB, Wang L, Hinck AP, Lafer EM, Sousa R (2007) Structural basis of J cochaperone binding and regulation of Hsp70. Mol Cell 28:422–433

Johnson JN, Ahrendt E, Braun JE (2010) CSPalpha: the neuroprotective J protein. Biochem Cell Biol 88:157–165

Kampinga HH, Craig EA (2010) The HSP70 chaperone machinery: J proteins as drivers of functional specificity. Nat Rev Mol Cell Biol 11:579–592

Kampinga HH, Hageman J, Vos MJ, Kubota H, Tanguay RM, Bruford EA, Cheetham ME, Chen B, Hightower LE (2009) Guidelines for the nomenclature of the human heat shock proteins. Cell Stress Chaperones 14:105–111

Kimura SH, Tsuruga H, Yabuta N, Endo Y, Nojima H (1997) Structure, expression, and chromosomal localization of human GAK. Genomics 44:179–187

Kirchhausen T, Davis AC, Frucht S, Greco BO, Payne GS, Tubb B (1991) AP17 and AP19, the mammalian small chains of the clathrin-associated protein complexes show homology to Yap17p, their putative homolog in yeast. J Biol Chem 266:11153–11157

Kumar DP, Vorvis C, Sarbeng EB, Cabra Ledesma VC, Willis JE, Liu Q (2011) The four hydrophobic residues on the hsp70 inter-domain linker have two distinct roles. J Mol Biol 411:1099–1113

Lee DW, Wu X, Eisenberg E, Greene LE (2006) Recruitment dynamics of GAK and auxilin to clathrin-coated pits during endocytosis. J Cell Sci 119:3502–3512

Lee DW, Zhao X, Yim YI, Eisenberg E, Greene LE (2008) Essential role of cyclin-G-associated kinase (Auxilin-2) in developing and mature mice. Mol Biol Cell 19:2766–2776

Lindner R, Ungewickell E (1991) Light-chain-independent binding of adaptors, AP180, and auxilin to clathrin. Biochemistry 30:9097–9101

Llewellyn DH, Roderick HL, Rose S (1997) KDEL receptor expression is not coordinatedly up-regulated with ER stress-induced reticuloplasmin expression in HeLa cells. Biochem Biophys Res Commun 240:36–40

Massol RH, Boll W, Griffin AM, Kirchhausen T (2006) A burst of auxilin recruitment determines the onset of clathrin-coated vesicle uncoating. Proc Natl Acad Sci U S A 103:10265–10270

Mastrogiacomo A, Parsons SM, Zampighi GA, Jenden DJ, Umbach JA, Gundersen CB (1994) Cysteine string proteins: a potential link between synaptic vesicles and presynaptic Ca^{2+} channels. Science 263:981–982

McDonough H, Patterson C (2003) CHIP: a link between the chaperone and proteasome systems. Cell Stress Chaperones 8:303–308

McMahon HT, Boucrot E (2011) Molecular mechanism and physiological functions of clathrin-mediated endocytosis. Nat Rev Mol Cell Biol 12:517–533

Meacham GC, Patterson C, Zhang W, Younger JM, Cyr DM (2001) The Hsc70 co-chaperone CHIP targets immature CFTR for proteasomal degradation. Nat Cell Biol 3:100–105

Meimaridou E, Gooljar SB, Ramnarace N, Anthonypillai L, Clark AJ, Chapple JP (2011) The cytosolic chaperone Hsc70 promotes traffic to the cell surface of intracellular retained melanocortin-4 receptor mutants. Mol Endocrinol 25:1650–1660

Misra UK, Gonzalez-Gronow M, Gawdi G, Pizzo SV (2005) The role of MTJ-1 in cell surface translocation of GRP78, a receptor for alpha 2-macroglobulin-dependent signaling. J Immunol 174:2092–2097

Murata S, Chiba T, Tanaka K (2003) CHIP: a quality-control E3 ligase collaborating with molecular chaperones. Int J Biochem Cell Biol 35:572–578

Neuhaus EM, Mashukova A, Zhang W, Barbour J, Hatt H (2006) A specific heat shock protein enhances the expression of mammalian olfactory receptor proteins. Chem Senses 31:445–452

Ni M, Zhang Y, Lee AS (2011) Beyond the endoplasmic reticulum: atypical GRP78 in cell viability, signalling and therapeutic targeting. Biochem J 434:181–188

Nie Z, Ranjan R, Wenniger JJ, Hong SN, Bronk P, Zinsmaier KE (1999) Overexpression of cysteine-string proteins in Drosophila reveals interactions with syntaxin. J Neurosci 19:10270–10279

Noskova L, Stranecky V, Hartmannova H, Pristoupilova A, Baresova V, Ivanek R, Hulkova H, Jahnova H, Van Der Zee J, Staropoli JF, Sims KB, Tyynela J, Van Broeckhoven C, Nijssen PC, Mole SE, Elleder M, Kmoch S (2011) Mutations in DNAJC5, encoding cysteine-string protein alpha, cause autosomal-dominant adult-onset neuronal ceroid lipofuscinosis. Am J Hum Genet 89:241–252

Okiyoneda T, Apaja PM, Lukacs GL (2011) Protein quality control at the plasma membrane. Curr Opin Cell Biol 23:483–491

Park J, Fang S, Crews AL, Lin KW, Adler KB (2008) MARCKS regulation of mucin secretion by airway epithelium in vitro: interaction with chaperones. Am J Respir Cell Mol Biol 39:68–76

Popoff V, Mardones GA, Bai SK, Chambon V, Tenza D, Burgos PV, Shi A, Benaroch P, Urbe S, Lamaze C, Grant BD, Raposo G, Johannes L (2009) Analysis of articulation between clathrin and retromer in retrograde sorting on early endosomes. Traffic 10:1868–1880

Rapoport I, Boll W, Yu A, Bocking T, Kirchhausen T (2008) A motif in the clathrin heavy chain required for the Hsc70/auxilin uncoating reaction. Mol Biol Cell 19:405–413

Rosales-Hernandez A, Beck KE, Zhao X, Braun AP, Braun JE (2009) RDJ2 (DNAJA2) chaperones neural G protein signaling pathways. Cell Stress Chaperones 14:71–82

Rothnie A, Clarke AR, Kuzmic P, Cameron A, Smith CJ (2011) A sequential mechanism for clathrin cage disassembly by 70-kDa heat-shock cognate protein (Hsc70) and auxilin. Proc Natl Acad Sci U S A 108:6927–6932

Sharma M, Burre J, Sudhof TC (2011) CSPalpha promotes SNARE-complex assembly by chaperoning SNAP-25 during synaptic activity. Nat Cell Biol 13:30–39

Shi A, Sun L, Banerjee R, Tobin M, Zhang Y, Grant BD (2009) Regulation of endosomal clathrin and retromer-mediated endosome to Golgi retrograde transport by the J-domain protein RME-8. EMBO J 28:3290–3302

Stumpges B, Behr M (2011) Time-specific regulation of airway clearance by the Drosophila J-domain transmembrane protein Wurst. FEBS Lett 585:3316–3321

Swain JF, Dinler G, Sivendran R, Montgomery DL, Stotz M, Gierasch LM (2007) Hsp70 chaperone ligands control domain association via an allosteric mechanism mediated by the interdomain linker. Mol Cell 26:27–39

Timsit YE, Miller SL, Mohney RP, O'Bryan JP (2005) The U-box ligase carboxyl-terminus of Hsc 70-interacting protein ubiquitylates Epsin. Biochem Biophys Res Commun 328:550–559

Tobaben S, Thakur P, Fernandez-Chacon R, Sudhof TC, Rettig J, Stahl B (2001) A trimeric protein complex functions as a synaptic chaperone machine. Neuron 31:987–999

Uhrigshardt H, Singh A, Kovtunovych G, Ghosh M, Rouault TA (2010) Characterization of the human HSC20, an unusual DnaJ type III protein, involved in iron-sulfur cluster biogenesis. Hum Mol Genet 19:3816–3834
Umbach JA, Zinsmaier KE, Eberle KK, Buchner E, Benzer S, Gundersen CB (1994) Presynaptic dysfunction in Drosophila csp mutants. Neuron 13:899–907
Umeda A, Meyerholz A, Ungewickell E (2000) Identification of the universal cofactor (auxilin 2) in clathrin coat dissociation. Eur J Cell Biol 79:336–342
Ungewickell E, Ungewickell H, Holstein SE, Lindner R, Prasad K, Barouch W, Martin B, Greene LE, Eisenberg E (1995) Role of auxilin in uncoating clathrin-coated vesicles. Nature 378:632–635
Ushioda R, Hoseki J, Araki K, Jansen G, Thomas DY, Nagata K (2008) ERdj5 is required as a disulfide reductase for degradation of misfolded proteins in the ER. Science 321:569–572
Vembar SS, Brodsky JL (2008) One step at a time: endoplasmic reticulum-associated degradation. Nat Rev Mol Cell Biol 9:944–957
Westhoff B, Chapple JP, Van Der Spuy J, Hohfeld J, Cheetham ME (2005) HSJ1 is a neuronal shuttling factor for the sorting of chaperone clients to the proteasome. Curr Biol 15:1058–1064
Wiedemann N, Frazier AE, Pfanner N (2004) The protein import machinery of mitochondria. J Biol Chem 279:14473–14476
Wingen C, Aschenbrenner AC, Stumpges B, Hoch M, Behr M (2009) The Wurst protein: a novel endocytosis regulator involved in airway clearance and respiratory tube size control. Cell Adh Migr 3:14–18
Wu MN, Fergestad T, Lloyd TE, He Y, Broadie K, Bellen HJ (1999) Syntaxin 1 A interacts with multiple exocytic proteins to regulate neurotransmitter release in vivo. Neuron 23:593–605
Xing Y, Bocking T, Wolf M, Grigorieff N, Kirchhausen T, Harrison SC (2010) Structure of clathrin coat with bound Hsc70 and auxilin: mechanism of Hsc70-facilitated disassembly. EMBO J 29:655–665
Yamamoto K, Fujii R, Toyofuku Y, Saito T, Koseki H, Hsu VW, Aoe T (2001) The KDEL receptor mediates a retrieval mechanism that contributes to quality control at the endoplasmic reticulum. EMBO J 20:3082–3091
Yim YI, Sun T, Wu LG, Raimondi A, De Camilli P, Eisenberg E, Greene LE (2010) Endocytosis and clathrin-uncoating defects at synapses of auxilin knockout mice. Proc Natl Acad Sci U S A 107:4412–4417
Zhang X, Qian SB (2011) Chaperone-mediated hierarchical control in targeting misfolded proteins to aggresomes. Mol Biol Cell 22:3277–3288
Zhang Y, Grant B, Hirsh D (2001) RME-8, a conserved J-domain protein, is required for endocytosis in Caenorhabditis elegans. Mol Biol Cell 12:2011–2021
Zhang Y, Liu R, Ni M, Gill P, Lee AS (2010) Cell surface relocalization of the endoplasmic reticulum chaperone and unfolded protein response regulator GRP78/BiP. J Biol Chem 285:15065–15075
Zinsmaier KE, Eberle KK, Buchner E, Walter N, Benzer S (1994) Paralysis and early death in cysteine string protein mutants of Drosophila. Science 263:977–980

Chapter 7
Pathways of Hsp70 Release: Lessons from Cytokine Secretion

Ayesha Murshid and Stuart K. Calderwood

Abstract Heat shock proteins (HSP) have an essential role in the cytoplasm of all cells in mediating protein quality control. However, in addition to these molecular chaperone properties, HSP play additional extracellular roles as mediators of inflammation and immunity. Because of their lack of a signal sequence and exclusion from the classical secretion pathways, it was initially assumed that extracellular HSPs resulted from cell necrosis and release of cell contents. However, non-canonical protein secretion pathways have been described, and have been studied intensively for the cytokine interleukin-1β (IL-1β). At least four mechanisms have been described, including: (1) IL-1β entry into secretary lysosomes and secretion into the medium along with lysosomal enzymes; (2) shedding of microvesicles from the membrane that contain IL-1β; (3) release of IL-1β directly through the membrane bound to a conjectured secreted-protein-transporter and; (4) formation of multivesicular bodies containing IL-1β and MHC class II within recycling endosomes and release of IL-1β entrapped in exosomes enclosed within the vesicles. Each of these mechanisms may be operative in HSP secretion and there is particularly strong evidence for mechanisms (1) and (4), in which either free Hsp70 is released through the endolysosome pathway or HSPs are secreted in association with exosomes. HSPs released through these mechanisms have intercellular signalling properties and can regulate phagocytosis, T and NK cell activation and release of cytokines in the acute inflammatory response. We discuss triggering mechanisms for HSP release, the pathways involved in HSP traverse of lipid membranes and the physiological consequences of HSP secretion.

7.1 Introduction

Heat shock proteins (HSP) are essential intracellular chaperones required to maintain the cellular proteome in a functional and folded conformation (Calderwood et al. 2009). However, it has slowly become apparent that HSPs can also be found in the extracellular spaces and in the circulation (Hightower and Guidon 1989; Pockley

S. K. Calderwood (✉) · A. Murshid
Molecular and Cellular Radiation Oncology, Beth Israel Deaconess Medical Center, Harvard Medical School, 99 Brookline Avenue, Boston, MA 02215, USA
e-mail: scalderw@bidmc.harvard.edu

B. Henderson, A. G. Pockley (eds.), *Cellular Trafficking of Cell Stress Proteins in Health and Disease*, Heat Shock Proteins 6, DOI 10.1007/978-94-007-4740-1_7,

2002; Pockley et al. 1998; Tytell et al. 1986). The subject of extracellular chaperones is also discussed in Chaps. 3 and 18. The existence of such 'displaced' HSP molecules might be regarded as evidence of: (i) cell necrosis and release from necrotic bodies; (ii) active secretion from viable cells, or; (iii) a consequence of both processes (Mambula and Calderwood 2006a, b; Mambula et al. 2007). It seems unlikely that most extracellular HSPs could function significantly as molecular chaperones when secreted, due to their requirement for ATP and co-chaperones, and indeed an additional class of extracellular chaperones exists in the form of plasma proteins such as clusterin (Wilson et al. 2008). However, extracellular HSP have been shown to possess potent immunobiological properties and can induce both immune stimulation and immune suppression (Srivastava 2005; Srivastava and Maki 1991; van Eden et al. 2005; Wieten et al. 2009). Therefore insights into HSP release mechanisms would seem potentially valuable in unravelling their immune functions. In addition, the mechanisms and processes involved in HSP release are not uniformly agreed on and further inspection of the published data may illuminate some of the controversial issues (De Maio 2010).

Most proteins targeted for release from cells are secreted by the canonical pathway, in which they are inserted co-translationally into the endoplasmic reticulum, progress through the Golgi apparatus and are then released into extracellular spaces (Halban and Irminger 1994; Stanley and Lacy 2010). Proteins released through this pathway encode a hydrophobic N-terminal signal sequence that is inserted into the ER membrane during translation and then is cleaved before secretion (Halban and Irminger 1994; Stanley and Lacy 2010). However, not all secretion pathways employ this route and non-canonical routes exist for release of proteins devoid of signal sequences (such as HSPs) to be secreted. These non-classical pathways have been carefully explored in study of cytokines such as interleukin-1α (IL-1α), interleukin-1β (IL-1β) fibroblast growth factor (FGF), interleukin-15 (IL-15) and interleukin-18 (IL-18) and the mechanisms deduced may serve as useful paradigms for understanding HSP secretion (Andrei et al. 1999; Duitman et al. 2008, 2010; MacKenzie et al. 2001; Prudovsky et al. 2003).

7.2 HSP Secretion: Lessons from Cytokines

IL-1β is an essential cytokine that mediates the acute phase response and fever and is rapidly synthesized on free ribosomes in response to inflammatory signals in the cytoplasm, prior to secretion (Dinarello 2009, 2010). Secretion of the leaderless IL-1β polypeptide in monocytes, macrophages and dendritic cells appears to require at least two triggering signals at the cell surface, including: (1) transcriptional activation by agents such as bacterial endotoxins and; (2) stimulation of release by extracellular ATP (Wewers 2004). There are at least four known pathways for IL-1β release that are each supported by convincing data and illustrated in Fig. 7.1. These include:

1. IL-1β entry into secretary lysosomes and secretion into the medium along with lysosomal enzymes and markers (Andrei et al. 1999, 2004).

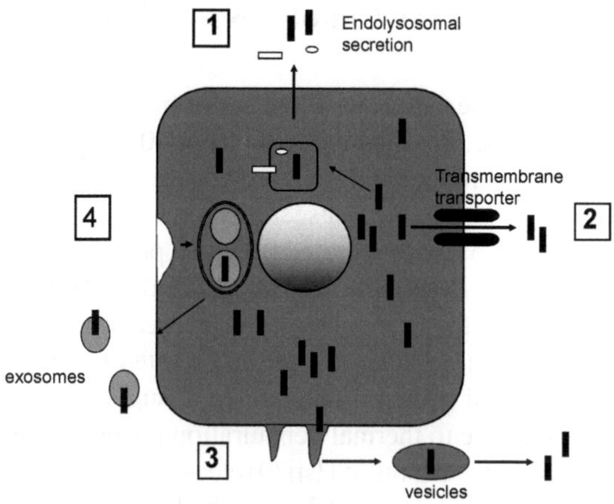

Fig. 7.1 Four pathways of non-canonical protein secretion. Proteins such as HSP (■) or IL-1β could be secreted through: (*1*) The endolysosomal pathway. HSP enter the lysosome after stress or other signals. Lysosomes also contain membrane LAMP1 (□) and intralysosomal cathepsin D (○). When the secretory lysosome fuses with the plasma membrane, free HSP is released along with LAMP1 and cathepsin D. (*2*) Secreted HSP could also be transported across the membrane directly by an, as-yet-uncharacterized transporter as wit IL-1β. (*3*) The HSP may also enter membrane blebs formed after treatment with agents such as ATP. These membrane protuberances may be pinched off and released into the extracellular space as vesicles containing HSP. (*4*) HSP may become packaged in MVB. These are formed after endocytosis from the plasma membrane, when endosomes are formed and then vesicles (exosomes) subsequently bud from the interior of the endosome to form enclosed exosomes. Hsp70 has been found protruding from the exosomal surface. When MVB fuse with the plasma membrane, exosomes are released

2. Shedding of microvesicles from the membrane that form on stimulation with ATP and contain IL-1β (Pizzirani et al. 2007).
3. Release of IL-1β directly through the membrane bound to a conjectured secreted-protein-transporter that may, or may not be the ATP-binding-cassette-1 transporter ABC-A1 (Brough and Rothwell 2007).
4. Formation of multivesicular bodies containing IL-1β and MHC class II within recycling endosomes and release of IL-1β entrapped in exosomes enclosed within the vesicles (Qu et al. 2007).

In comparing secretion pathways of HSP with mechanisms related to IL-1β, it is as well to consider also differences between the two processes. IL-1β is expressed at minimal levels in unstimulated monocytes and macrophages and on stimulation synthesized as a larger polypeptide (pro-interleukin-1β) that must be processed by the enzyme caspase-1 before secretion (Ogura et al. 2006; Perregaux and Gabel 1994; Wewers 2004). Many tissue culture cells contain abundant levels of constitutively expressed Hsp70 and there is currently no evidence for a role of proteolytic processing in secretion.

7.3 Hsp70 Secretion via the Endolysosome Pathway

We have carefully examined the pathways of release of free Hsp70 from cells. As we did not know the nature of the proximal inducer of Hsp70 release, we initially used heat shock as a trigger for secretion, based on previous studies which indicated that mild heat shock can lead to non-canonical secretion of FGF and IL-1α (Prudovsky et al. 2003). We found that Hsp70 was secreted from two prostate cell lines, PC-3 and LnCap at both 40 °C and 43 °C (Mambula and Calderwood 2006b). Release occurred from viable cells only during the heat shock itself, ceased when cells recovered at 37 °C and appeared to be a direct response to the elevated temperature. Exposure to more toxic forms of thermal stress inhibited release, suggesting a secretion pathway with components sensitive to thermal denaturation (Mambula and Calderwood 2006a). On comparing the mechanisms of Hsp70 release to those utilized in secretion of IL-1β, we found that mechanisms 1 and 2, described above appeared to be operative in Hsp70 secretion (Mambula and Calderwood 2006b) (Fig. 7.1). Hsp70 release was reduced by the lysosomotropic agents: methylamine and NH_4Cl. In addition, Hsp70 became localized to the lysosomal fraction marked by cathepsin D during the heat shock and its release was accompanied by the trafficking of lysosomal surface marker, LAMP1, to the cell surface of heated cells (Mambula and Calderwood 2006b). A mechanism involving stress-mediated entry of Hsp70 into lysosomes, fusion of these organelles to the plasma membrane and release of contents to the extracellular milieu is indicated. Further studies also suggested that Hsp70 release may require ATP (as with IL-1β) and elevated levels of Mg^{++}, a known inhibitory condition for purinergic receptors inhibited secretion (Mambula and Calderwood 2006b). Results consistent with these findings were observed in human subjects *in vivo* in which elevated levels of HSP72 were strongly correlated with increases in external ATP (Ogawa et al. 2011).There was some support for secretion mechanism (3) in Hsp70 release in that inhibitors of the transporter protein ABC-B1 (glibemclemide and DIDS) inhibited thermally-induced Hsp70 secretion. It is possible that ABC-B1 could be involved in either entry of the Hsp70 into secretory lysosomes (1) or in direct secretion across the plasma membrane (3). In these studies we did not find evidence for Hsp70 in membrane-bounded microvesicles or vesicular bodies, as ultracentrifugation of cell medium failed to sediment detectable amounts of Hsp70. We have more recently examined these mechanisms in macrophage cell lines stimulated by microorganisms and PAMPs derived from *E. coli* (S.S. Mambula, A. Murshid & S.K. Calderwood, submitted). Stimulation by exposure to *E. coli* caused Hsp70 release from Raw 264.7 mouse macrophages coordinately with IL-1β and the lysosomal marker LAMP1. In addition, lysosomotropic agents inhibited bacterial induction of Hsp70 release suggesting a similar mechanism, involving the endolysosome pathway for chaperone release after *E. coli* exposure, as was seen with thermal stress. Free Hsp70 may be released through the endolysosomal pathway following exposure to microorganisms and can have a number of effects including bystander activation of phagocytosis (S.S. Mambula, A. Murshid & S.K. Calderwood, submitted) and cytokine induction in monocytes, macrophages and DC (Asea et al. 2000,

2002). The cell surface role of Hsp70 in recognising the Gram-negative component LPS is described by Triantafilou in Chap. 9.

7.4 HSP Release in Vesicles

A number of studies indicate that HSPs can be released when entrapped in lipid vesicles (mechanisms 2, 4) described above (Chen et al. 2006; Clayton et al. 2005; Elsner et al. 2007; Gupta and Knowlton 2007; Lancaster and Febbraio 2005; Mathew et al. 1995; Taylor et al. 2007; Thery et al. 1999; Xie et al. 2010). These studies indicate that such HSPs are present in exosomes derived from a wide array of cell types, including: reticulocytes, peripheral blood mononuclear cells (PBMCs), B cells, dendritic cells, hepatocytes, and a range of cancer cells (mesothelioma, colon cancer, K562, mammary carcinoma—reviewed in De Maio 2010 (see Fig. 7.1)). This process involves the internalization of regions of the plasma membrane into endosomes that recycle to the cytoplasm, with specific sorting of proteins in the endosomes into internal vesicles pinched off from the endosomal membrane, the exosomes (Chaput et al. 2006). These structures have been described as multivesicular bodies (MVB) and are involved in secretion of both IL-β and Hsp70 (Chalmin et al. 2010; Qu et al. 2007). MVB next fuse with the membrane leading to release of the exosomes into the extracellular spaces (Chalmin et al. 2010). The key question regarding exosomal release, as with each of the pathways of non-canonical secretion, is—how do proteins without signal sequences become inserted into/cross lipid membranes? Even more intriguingly, Hsp70 has been found in the exposed surface of the exosomes and may serve a recognition/signalling role in targeting the exosomes (Chalmin et al. 2010; Vega et al. 2008). These questions have been addressed in detail by A. De Maio and co-workers (Vega et al. 2008). These workers have shown heat shock-induced secretion of exosomal-like structures that they refer to as ECM (extracellular membrane), with Hsp70 exposed in the outer leaflet of the membrane. They have proposed a mechanism whereby Hsp70 can bind specifically to phosphatidylserine (PS) and then become an integral membrane protein associated with this lipid and promote ion channel forming activities (Arispe et al. 2004; Schilling et al. 2009). The mechanisms involved in this process are not clear although *flipase* molecules known to be present in the plasma membrane and that transport phosphatidylserined (PS) from the inner leaflet into the outer leaflet of the membrane could potentially be involved (Vega et al. 2008). Hsp70 could thus be transported as a passenger when Hsp70-bound PS molecules are flipped from the inner to the outer leaflet of the plasma membrane. An alternative mechanism for non-canonical secretion of FGF during heat shock was proposed by Prudovsky and coworkers and involves conversion of the secreted protein into a partially unfolded "molten globule" form that could then cross the membrane (Prudovsky et al. 2003). Exosomes have been shown to have a specialized lipid content compared to the bulk plasma membrane, with elevated cholesterol and sphingolipids characteristic of detergent resistant microdomains or lipid rafts (De Maio 2010 see also Chap. 6). Hsp70 has been detected previously within the lipid raft fraction (Triantafilou et al. 2002—see also Chap. 9). Exosomes were also previously shown to be enriched in GPI-anchored proteins such as CD55 and CD59 that partition into lipid raft domains as well as tetraspanins, proteins that are also found in these domains (Chaput et al. 2006). Exosomes also have distinct protein contents depending on their cell of origin (Fevrier and Raposo 2004; Stoorvogel et al. 2002).

The physiological consequences of exosomal secretion also differ according to the cell of origin. For instance Hsp70-containing exosomes are found in dendritic cells (DC), are enriched in major histocompatability class II complex molecules and appear to play a key role in cross priming and activating T lymphocytes (Chaput et al. 2006). Bone marrow-derived DC pulsed with acid-eluted peptides were shown to secrete immunogenic exosomes, mediate CTL responses in mice and lead to retardation of tumor growth (Zitvogel et al. 1998). Interestingly, the exosomes did not interact directly with T cells but required the mediation of the DC for cross priming and T cell activation. However, Hsp70-containing exosomes derived from tumor cells may have the opposite consequence in terms of tumor immunity depending on the cell type that they interact with. Tumor-derived exosomes from EL4 thymoma, TS/A mammary carcinoma and CT26 colon carcinoma cells were shown to be immunosuppressive on encountering and binding to myeloid-derived suppressor cells (MDSC) (Chalmin et al. 2010). The Hsp70 moiety found on the external surface of such tumor-derived exosomes was bound to Toll-Like Receptor 2 (TLR-2) and gave rise to signaling that led to activation of STAT3 in MDSC and down-regulation of tumor immunity *in vivo* (Chalmin et al. 2010). These investigators found that the Na^+/H^+ export channel inhibitor, amiloride, could inhibit exosomal release from tumor cells *in vivo* and reverse immune suppression (Chalmin et al. 2010). By contrast, when Hsp70 surface-positive exosomes, from tumor cells, interact with natural killer (NK) cells their migratory and cytolytic activities were shown to be activated, indicating immune stimulation (Gastpar et al. 2005). These different studies indicate the various properties of Hsp70-containing exosomes that depend on the nature of the cell of origin and the target cell. The exosomal response to stress has been termed the *stress-observation system* (SOS) and it has been proposed that export of Hsp70 in exosomes from stressed cells may be a form of intercellular communication that informs macrophages of cell stress and may arm such cells for innate immune responses (De Maio 2010). This group showed that treatment of HepG2 cells with heat shock at 43 °C leads to Hsp70 expression on the surface of the cells within lipid raft domains, leading to release of exosomes containing surface-orientated hsp70. Such vesicles were highly enriched in Hsp70 and when incubated with macrophages induced abundant secretion of tumor necrosis factor-α (Vega et al. 2008). The pro-inflammatory effectiveness of stress-induced exosomes depended on ongoing transcription and treatment of the Hep-G2 cells with actinomycin D led to inhibition of Hsp70 synthesis that correlated with loss of tumor necrosis factor-α inducing activity by exosomes. Overall therefore, secretion of Hsp70-containing exosomes is a widely-observed phenomenon with powerful immune and inflammatory consequences and can have a myriad of effects depending of the nature of the exosome-donor and acceptor cells.

7.5 Other Potential Mechanisms of HSP Secretion?

Another form of secretion utilized by cytokines is the IL-15 pathway (Duitman et al. 2008, 2010). IL-15 has the unusual property of being secreted while bound to its receptor, IL-15Rα. Such IL-15 ligand-receptor complexes are required for the IL-15 to passage through the ER and Golgi networks and be transported to the cell surface (Duitman et al. 2008, 2010). The presence of such complexes on the surface of fibroblasts, DC and monocytes then leads to expansion of NK and $CD8^+$ memory cell populations (Bergamaschi et al. 2008). One possible pathway for HSP release could therefore involve its coupling to known HSP receptors such as SRECI, LOX-1 and FEEL-1/stabilin-1 and transport of the molecular chaperones to the cell surface through the canonical secretion pathway (Calderwood et al. 2007). Again the question would arise—how could cytoplasmic HSP enter the ER or Golgi and bind to HSP receptors?

7.6 Triggers for HSP Release?

The proximal signal(s) leading to HSP sorting to the secretory compartments and release from cells are not known. Earlier studies showed that secretion of the DNA-binding protein, *high mobility group B1* (HMGB1), and the homeodomain protein *engrailed-2*, through the non-canonical pathway involves posttranslational modification (PTM). The triggering PTMs for these proteins are respectively, hyperacetylation and dephosphorylation (Bonaldi et al. 2003; Maizel et al. 2002; Mambula and Calderwood 2006a; Wisniewski et al. 1999). Hyperacetylation of HMGB1 may diminish its DNA binding affinity and permit nuclear exit prior to secretion. In the case of engrailed-2, phosphorylation within its homeodomain by casein kinase II deters secretion and this effect is reversed by dephosphorylation (Maizel et al. 2002). Interestingly, secretion of HMGB1 is attenuated during the heat shock response or when Hsp70 is overexpressed due to a block in nuclear export (Shi et al. 2006). There may, thus, exist cross-talk between proteins such as HMGB1 and Hsp70 secreted through the non-canonical pathways. It may also be significant that hyperacetylation of Hsp90α, although associated with lack of chaperone function, increases its binding to extracellular MMP-2 and enhances tumor invasion properties (Yang et al. 2008). The precise effects of these PTMs are not clear although they may involve loss of normal intracellular protein function, such as binding to DNA or chaperoning intracellular proteins, and may thus liberate the proteins to enter secretion pathways.

7.7 Conclusions

Non-canonical secretion of cytokines thus seems an enlightening paradigm for study of allied processes in molecular chaperone release. Like IL-1αβ secretion, release of Hsp70 appears to involve multiple independent pathways each supported by impressive data, suggesting their independent involvement in delivering Hsp70 to the outside. Hsp70 for instance may enter the extracellular spaces either as free protein or packaged in exosomes and exert regulatory influences in its milieu. It is however not clear by which mechanisms Hsp70 crosses lipid membranes. Our studies on Hsp70 release during heat shock indicate that the chaperone is only secreted during the time of heating, suggesting that a triggering biophysical response to warming such as partial protein unfolding may occur (Mambula and Calderwood 2006b). This would be consistent with the idea of secretion of FGF and IL-1β during heat shock through a partially unfolded "molten globule" state that can cross membranes (Prudovsky et al. 2003). It may be significant that when cytoplasmic Hsp70 enters the lysosome in the process of chaperone-mediated autophagy or when it is transported into mitochondria, the protein is first unfolded in order to cross channels in the lipid membrane prior to refolding inside the organelle (Dice 2007; Pfanner and Truscott 2002). A similar unfolding process might trigger Hsp70 entry into secretory lysosomes or exosomal membranes in stress (De Maio 2010; Mambula and Calderwood 2006b; Mambula et al. 2007). The mechanisms through which HSP release is triggered under more physiological conditions are not so clear, although initiating steps could involve PTM of the HSPs, as with secretion of HMGB1 and engrailed-2.

We have concentrated in this review largely on HSP secretion through more physiological secretion mechanisms. However it is apparent that HSPs are also released when cells undergo necrosis (Mambula and Calderwood 2006a). Under these conditions Hsp70 for instance causes marked inflammatory effects that can be manipulated to mediate tumor rejection (Daniels et al. 2004; Kottke et al. 2007). Thus release of HSPs both from viable cells and from cells undergoing necrosis can have profound effects of immune and inflammatory responses.

Acknowledgments We thank the department of Radiation Oncology, Beth Israel Deaconess Medical Center for support and encouragement. SKC is supported by NIH research grants R01CA047407; R01CA094397 and R01CA119045.

References

Andrei C, Dazzi C, Lotti L, Torrisi MR, Chimini G, Rubartelli A (1999) The secretory route of the leaderless protein interleukin 1beta involves exocytosis of endolysosome-related vesicles. Mol Biol Cell 10:1463–1475

Andrei C, Margiocco P, Poggi A, Lotti LV, Torrisi MR, Rubartelli A (2004) Phospholipases C and A2 control lysosome-mediated IL-1 beta secretion: Implications for inflammatory processes. Proc Natl Acad Sci U S A 101:9745–9750

Arispe N, Doh M, Simakova O, Kurganov B, De Maio A (2004) Hsc70 and Hsp70 interact with phosphatidylserine on the surface of PC12 cells resulting in a decrease of viability. FASEB J 18:1636–1645

Asea A, Kraeft SK, Kurt-Jones EA, Stevenson MA, Chen LB, Finberg RW, Koo GC, Calderwood SK (2000) HSP70 stimulates cytokine production through a CD14-dependant pathway, demonstrating its dual role as a chaperone and cytokine. Nat Med 6:435–442

Asea A, Rehli M, Kabingu E, Boch JA, Bare O, Auron PE, Stevenson MA, Calderwood SK (2002) Novel signal transduction pathway utilized by extracellular HSP70: Role of toll-like receptor (TLR) 2 and TLR4. J Biol Chem 277:15028–15034

Bergamaschi C, Rosati M, Jalah R, Valentin A, Kulkarni V, Alicea C, Zhang GM, Patel V, Felber BK, Pavlakis GN (2008) Intracellular interaction of interleukin-15 with its receptor alpha during production leads to mutual stabilization and increased bioactivity. J Biol Chem 283:4189–4199

Bonaldi T, Talamo F, Scaffidi P, Ferrera D, Porto A, Bachi A, Rubartelli A, Agresti A, Bianchi ME (2003) Monocytic cells hyperacetylate chromatin protein HMGB1 to redirect it towards secretion. EMBO J 22:5551–5560

Brough D, Rothwell NJ (2007) Caspase-1-dependent processing of pro-interleukin-1beta is cytosolic and precedes cell death. J Cell Sci 120:772–781

Calderwood SK, Murshid A, Prince T (2009) The shock of aging: Molecular chaperones and the heat shock response in longevity and aging – A mini-review. Gerontology 55:55–58

Calderwood SK, Theriault J, Gray PJ, Gong J (2007) Cell surface receptors for molecular chaperones. Methods 43:199–206

Chalmin F, Ladoire S, Mignot G, Vincent J, Bruchard M, Remy-Martin JP, Boireau W, Rouleau A, Simon B, Lanneau D De, Thonel A, Multhoff G, Hamman A, Martin F, Chauffert B, Solary E, Zitvogel L, Garrido C, Ryffel B, Borg C, Apetoh L, Rebe C and Ghiringhelli F (2010) Membrane-associated Hsp72 from tumor-derived exosomes mediates STAT3-dependent immunosuppressive function of mouse and human myeloid-derived suppressor cells. J Clin Invest 120:457–471

Chaput N, Flament C, Viaud S, Taieb J, Roux S, Spatz A, Andre F, LePecq JB, Boussac M, Garin J, Amigorena S, Thery C, Zitvogel L (2006) Dendritic cell derived-exosomes: Biology and clinical implementations. J Leukoc Biol 80:471–478

Chen W, Wang J, Shao C, Liu S, Yu Y, Wang Q, Cao X (2006) Efficient induction of antitumor T cell immunity by exosomes derived from heat-shocked lymphoma cells. Eur J Immunol 36:1598–15607

Clayton A, Turkes A, Navabi H, Mason MD, Tabi Z (2005) Induction of heat shock proteins in B-cell exosomes. J Cell Sci 118:3631–3638

Daniels GA, Sanchez-Perez L, Diaz RM, Kottke T, Thompson J, Lai M, Gough M, Karim M, Bushell A, Chong H, Melcher A, Harrington K, Vile RG (2004) A simple method to cure established tumors by inflammatory killing of normal cells. Nat Biotechnol 22:1125–1132

De Maio A (2010) Extracellular heat shock proteins, cellular export vesicles, and the stress observation system: A form of communication during injury, infection, and cell damage. It is never known how far a controversial finding will go! Dedicated to Ferruccio Ritossa. Cell Stress Chaperones 16:235–249

Dice JF (2007) Chaperone-mediated autophagy. Autophagy 3:295–299

Dinarello CA (2009) Interleukin-1beta and the autoinflammatory diseases. N Engl J Med 360:2467–2470

Dinarello CA (2010) IL-1: Discoveries, controversies and future directions. Eur J Immunol 40:599–606

Duitman EH, Orinska Z, Bulanova E, Paus R, Bulfone-Paus S (2008) How a cytokine is chaperoned through the secretory pathway by complexing with its own receptor: Lessons from interleukin-15 (IL-15)/IL-15 receptor alpha. Mol Cell Biol 28:4851–4861

Duitman EH, Orinska Z, Bulfone-Paus S (2010) Mechanisms of cytokine secretion: A portfolio of distinct pathways allows flexibility in cytokine activity. Eur J Cell Biol 90:476–483

Elsner L, Muppala V, Gehrmann M, Lozano J, Malzahn D, Bickeboller H, Brunner E, Zientkowska M, Herrmann T, Walter L, Alves F, Multhoff G, Dressel R (2007) The heat shock protein HSP70 promotes mouse NK cell activity against tumors that express inducible NKG2D ligands. J Immunol 179:5523–5533

Fevrier B, Raposo G (2004) Exosomes: Endosomal-derived vesicles shipping extracellular messages. Curr Opin Cell Biol 16:415–421

Gastpar R, Gehrmann M, Bausero MA, Asea A, Gross C, Schroeder JA, Multhoff G (2005) Heat shock protein 70 surface-positive tumor exosomes stimulate migratory and cytolytic activity of natural killer cells. Cancer Res 65:5238–5247

Gupta S, Knowlton AA (2007) HSP60 trafficking in adult cardiac myocytes: Role of the exosomal pathway. Am J Physiol Heart Circ Physiol 292:H3052–3056

Halban PA, Irminger JC (1994) Sorting and processing of secretory proteins. Biochem J 299(1):1–18
Hightower LE, Guidon PT Jr (1989) Selective release from cultured mammalian cells of heat-shock (stress) proteins that resemble glia-axon transfer proteins. J Cell Physiol 138:257–266
Kottke T, Sanchez-Perez L, Diaz RM, Thompson J, Chong H, Harrington K, Calderwood SK, Pulido J, Georgopoulos N, Selby P, Melcher A, Vile R (2007) Induction of hsp70-mediated Th17 autoimmunity can be exploited as immunotherapy for metastatic prostate cancer. Cancer Res 67:11970–11979
Lancaster GI, Febbraio MA (2005) Exosome-dependent trafficking of HSP70: A novel secretory pathway for cellular stress proteins. J Biol Chem 280:23349–23355
MacKenzie A, Wilson HL, Kiss-Toth E, Dower SK, North RA, Surprenant A (2001) Rapid secretion of interleukin-1beta by microvesicle shedding. Immunity 15:825–835
Maizel A, Tassetto M, Filhol O, Cochet C, Prochiantz A, Joliot A (2002) Engrailed homeoprotein secretion is a regulated process. Development 129:3545–3553
Mambula SS, Calderwood SK (2006a) Heat induced release of Hsp70 from prostate carcinoma cells involves both active secretion and passive release from necrotic cells. Int J Hyperthermia 22:575–585
Mambula SS, Calderwood SK (2006b) Heat shock protein 70 is secreted from tumor cells by a nonclassical pathway involving lysosomal endosomes. J Immunol 177:7849–7857
Mambula SS, Stevenson MA, Ogawa K, Calderwood SK (2007) Mechanisms for Hsp70 secretion: crossing membranes without a leader. Methods 43:168–175
Mathew A, Bell A, Johnstone RM (1995) Hsp-70 is closely associated with the transferrin receptor in exosomes from maturing reticulocytes. Biochem J 308(3):823–830
Ogawa K, Seta R, Shimizu T, Shinkai S, Calderwood SK, Nakazato K, Takahashi K (2011) Plasma adenosine triphosphate and heat shock protein 72 concentrations after aerobic and eccentric exercise. Exerc Immunol Rev 17:136–149
Ogura Y, Sutterwala FS, Flavell RA (2006) The inflammasome: First line of the immune response to cell stress. Cell 126:659–662
Perregaux D, Gabel CA (1994) Interleukin-1 beta maturation and release in response to ATP and nigericin. Evidence that potassium depletion mediated by these agents is a necessary and common feature of their activity. J Biol Chem 269:15195–15203
Pfanner N, Truscott KN (2002) Powering mitochondrial protein import. Nat Struct Biol 9:234–236
Pizzirani C, Ferrari D, Chiozzi P, Adinolfi E, Sandona D, Savaglio E, Di Virgilio F (2007) Stimulation of P2 receptors causes release of IL-1beta-loaded microvesicles from human dendritic cells. Blood 109:3856–3864
Pockley AG (2002) Heat shock proteins, inflammation, and cardiovascular disease. Circulation 105:1012–1017
Pockley AG, Shepherd J, Corton JM (1998) Detection of heat shock protein 70 (Hsp70) and anti-Hsp70 antibodies in the serum of normal individuals. Immunol Invest 27:367–377
Prudovsky I, Mandinova A, Soldi R, Bagala C, Graziani I, Landriscina M, Tarantini F, Duarte M, Bellum S, Doherty H, Maciag T (2003) The non-classical export routes: FGF1 and IL-1alpha point the way. J Cell Sci 116:4871–4881
Qu Y, Franchi L, Nunez G, Dubyak GR (2007) Nonclassical IL-1 beta secretion stimulated by P2X7 receptors is dependent on inflammasome activation and correlated with exosome release in murine macrophages. J Immunol 179:1913–1925
Schilling D, Gehrmann M, Steinem C, De Maio A, Pockley AG, Abend M, Molls M, Multhoff G (2009) Binding of heat shock protein 70 to extracellular phosphatidylserine promotes killing of normoxic and hypoxic tumor cells. FASEB J 23:2467–2477
Shi Y, Tu Z, Tang D, Zhang H, Liu M, Wang K, Calderwood S K, Xiao X (2006) The inhibition of LPS-induced production of inflammatory cytokines by HSP70 involves inactivation of the NF-kappaB pathway but not the MAPK pathways. Shock 26:277–284
Srivastava PK (2005) Immunotherapy for human cancer using heat shock protein-peptide complexes. Curr Oncol Rep 7:104–108
Srivastava PK, Maki RG (1991) Stress-induced proteins in immune response to cancer. Curr Top Microbiol Immunol 167:109–123

Stanley AC, Lacy P (2010) Pathways for cytokine secretion. Physiology (Bethesda) 25:218–229

Stoorvogel W, Kleijmeer MJ, Geuze HJ, Raposo G (2002) The biogenesis and functions of exosomes. Traffic 3:321–330

Taylor AR, Robinson MB, Gifondorwa DJ, Tytell M, Milligan CE (2007) Regulation of heat shock protein 70 release in astrocytes: role of signaling kinases. Dev Neurobiol 67:1815–1829

Thery C, Regnault A, Garin J, Wolfers J, Zitvogel L, Ricciardi-Castagnoli P, Raposo G, Amigorena S (1999) Molecular characterization of dendritic cell-derived exosomes. Selective accumulation of the heat shock protein hsc73. J Cell Biol 147:599–610

Triantafilou M, Miyake K, Golenbock DT, Triantafilou K (2002) Mediators of innate immune recognition of bacteria concentrate in lipid rafts and facilitate lipopolysaccharide-induced cell activation. J Cell Sci 115:2603–2611

Tytell M, Greenberg SG, Lasek RJ (1986) Heat shock-like protein is transferred from glia to axon. Brain Res 363:161–164

van Eden W, Van Der Zee R, Prakken B (2005) Heat-shock proteins induce T-cell regulation of chronic inflammation. Nat Rev Immunol 5:318–330

Vega VL, Rodriguez-Silva M, Frey T, Gehrmann M, Diaz JC, Steinem C, Multhoff G, Arispe N, De Maio A (2008) Hsp70 translocates into the plasma membrane after stress and is released into the extracellular environment in a membrane-associated form that activates macrophages. J Immunol 180:4299–4307

Wewers MD (2004) IL-1beta: An endosomal exit. Proc Natl Acad Sci USA 101:10241–10242

Wieten L, Berlo SE, Ten Brink CB, van Kooten PJ, Singh M, Van Der Zee R, Glant TT, Broere F, van Eden W (2009) IL-10 is critically involved in mycobacterial HSP70 induced suppression of proteoglycan-induced arthritis. PLoS One 4:e4186

Wilson MR, Yerbury JJ, Poon S (2008) Potential roles of abundant extracellular chaperones in the control of amyloid formation and toxicity. Mol Biosyst 4:42–52

Wisniewski JR, Szewczuk Z, Petry I, Schwanbeck R, Renner U (1999) Constitutive phosphorylation of the acidic tails of the high mobility group 1 proteins by casein kinase II alters their conformation, stability, and DNA binding specificity. J Biol Chem 274:20116–20122

Xie Y, Bai O, Zhang H, Yuan J, Zong S, Chibbar R, Slattery K, Qureshi M, Wei Y, Deng Y, Xiang J (2010) Membrane-bound HSP70-engineered myeloma cell-derived exosomes stimulate more efficient CD8 (+) CTL- and NK-mediated antitumour immunity than exosomes released from heat-shocked tumour cells expressing cytoplasmic HSP70. J Cell Mol Med 14:2655–2666

Yang Y, Rao R, Shen J, Tang Y, Fiskus W, Nechtman J, Atadja P, Bhalla K (2008) Role of acetylation and extracellular location of heat shock protein 90 alpha in tumor cell invasion. Cancer Res 68:4833–4842

Zitvogel L, Regnault A, Lozier A, Wolfers J, Flament C, Tenza D, Ricciardi-Castagnoli P, Raposo G, Amigorena S (1998) Eradication of established murine tumors using a novel cell-free vaccine: Dendritic cell-derived exosomes. Nat Med 4:594–600

Chapter 8
Nucleolin: A Novel Intracellular Transporter of HSPA1A

Punit Kaur, Appukuttan R. Pradeep and Alexzander Asea

Abstract The author's previous studies demonstrated that thermal stress induces the release of HSPA1A from cells by a mechanism independent of the classical protein transport pathway. However, the exact mechanism by which HSPA1A, a leaderless protein, gains access to the extracellular milieu remains unknown. Our study further demonstrated that the plasma membrane of cells functions as a reservoir for HSPA1A and thermal stress induces the redistribution of plasma membrane bound HSPA1A into subcellular cytosolic components. From there, the nucleolin-mediated transport system carries the HSPA1A to the plasma membrane for its relocalization and final release into the extracellular milieu. In this chapter, we focus on the role of nucleolin as a transporter of intracellular HSPA1A.

Abbreviation

CAP2	Adenylyl cyclase-associated protein 2
EGFR	Epidermal growth factor receptor
ER	Endoplasmic reticulum
gp96	Glucose regulated protein 96
HSP	Heat shock protein family
hsp	Heat shock protein gene
HSPA1A	Seventy-two kilo-Dalton heat shock protein
MHC	Major histocompatibility complex
MVE	Multivesicular endosomes
NK	Natural killer
TF	Tissue factor
TLR	Toll-like receptors

A. Asea (✉) · A. R. Pradeep · P. Kaur
Department of Microbiology, Biochemistry and Immunology (MBI),
Morehouse School of Medicine, 720 Westview Ave., SW, Atlanta,
GA 30310, USA
e-mail: aasea@msm.edu

B. Henderson, A. G. Pockley (eds.), *Cellular Trafficking of Cell Stress Proteins in Health and Disease*, Heat Shock Proteins 6, DOI 10.1007/978-94-007-4740-1_8,

8.1 Introduction

The major problem of understanding HSPA1A trafficking and release from cells is that the genes encoding HSPA1A do not encode ER signal sequence. Therefore, HSPA1A cannot effectively utilize this classical protein secretion pathway to gain access to the extracellular milieu. This is not, however, unique for HSPA1A. A non-classical protein trafficking and section pathway is also utilized by other proteins, including: interleukin-1β (IL-1β), IL-1α, IL-18, IL-33 and IL-1 receptor antagonist IL1-Ra (Ferrari et al. 2006, see also Chap. 7). These proteins do not encode a leader sequence; however, they are still able to accomplish their primary function as secreted proteins to interact with receptors on adjacent or distant target cells (Ferrari et al. 2006). Studies of IL-1β secretion indicate three potential mechanisms by which this cytokine is secreted, including; (a) lysis of IL-1β-containing secretory cells and release of contents; (b) cell surface blebbing and release of IL-1β in microvesicles that lyse in the extracellular fluid and; (c) entry of IL-1β into intracellular structures known as endolysosomes that transport the cytokine to the cell surface where they then fuse with the plasma membrane and release the contents upon fusion (Wewers 2004). We, and others, have demonstrated that each of these mechanisms may be utilized for HSPA1A release (Bausero et al. 2005; Gastpar et al. 2005; Mambula and Calderwood 2006a, b; Mambula et al. 2007, see Murshi and Calderwood, Chap. 7). For therapeutic gain, the release of HSPA1A from cells by necrotic cell death rather than apoptotic cell death is suggested to enhance immunogenicity (Srivastava 2003; Daniels et al. 2004). Mild levels of heat shock within the fever range lead to HSPA1A release while more severe conditions inhibit release, consistent with the inactivation of a protein-based secretion mechanism (Mambula and Calderwood 2006a, b; Mambula et al. 2007). Severe heat shock (45–55 °C) induces delayed necrosis and a gradual release of HSPA1A as necrosis develops (Mambula and Calderwood 2006a, b; Mambula et al. 2007).

Nucleolin is a highly conserved 100–110 kDa protein, known to be one of the most abundant nonribosomal proteins in the nucleolus (for review see Tajrishi et al. 2011), and has been shown to account for approximately 10 % of the protein content within the nucleolus (Andersen et al. 2002; Andersen et al. 2005; Leung et al. 2006; Lam et al. 2007). The primary function ascribed to nucleolin is in regulating rRNA synthesis and ribosome biogenesis, and has been implicated in many aspects of cell biology that include functions such as gene silencing, senescence and cell cycle regulation (Olson et al. 2000; Tajrishi et al. 2011). Nucleolin has also been shown to increase mRNA stability in mitosis (Wang et al. 2011). Nucleolin has also been shown to shuttle between the nucleus, cytoplasm and cell surface, and is a receptor for DNA nanoparticles at the cell surface (Chen et al. 2008). In this chapter we focus on the role of nucleolin as a novel transporter of intracellular HSPA1A.

8.2 HSPA1A: A Leaderless Protein

The 72 kDa heat shock protein (HSPA1A) belongs to a family of highly conserved proteins involved in protein folding, stabilization and cytoplasmic transportation. The biological significance of plasma membrane-bound HSPA1A is demonstrated by its ability to act as a target structure for natural killer (NK) cell-mediated cytotoxicity (Multhoff et al. 1997) and elicit anti-tumor immunity (Chen et al. 2002). Different classes of HSP on tumor cells differentially activate monocytes and granulocytes to generate pro-inflammatory cytokines (Negulyaev et al. 1996; Asea et al. 2000), transfer antigen to APC to stimulate tumor specific T cells (Albert et al. 1998; Srivastava et al. 1998), enhance the ability of tumors to process and present MHC-class I antigens directly to T cells (Wells et al. 1998) or induce tumor cell lysis mediated by non-MHC restricted NK cells (Mosser et al. 1997; Gross et al. 2003). However, the over-expression of plasma membrane-bound HSPA1A induced negative feedback to prevent activation of HSF and subsequent HSPA1A production (Balakrishnan and De Maio 2006).

HSPA1A expression is induced by various stressors, including heat shock, radiation, heavy metals and microbial infections (Noonan et al. 2007). The HSPA1A is not only expressed within the cytoplasm, but also in the plasma membrane of cells, from where it is actively released into the extracellular milieu to bind to and be internalized by professional antigen presenting cells (Tytell et al. 1986; Mambula and Calderwood 2006, more details in Chap. 1). The extracellular release of HSPA1A was initially reported in neuronal cells by Hightower and Guidon (Hightower and Guidon 1989, see Chap. 1). Recently, the scavenger receptor, LOX-1 (Wang et al. 2004; Theriault et al. 2005; Theriault et al. 2006; Murshid et al. 2011), Toll-like receptors (TLR) (Asea et al. 2002; Vabulas et al. 2002a, b; Takeda et al. 2003; Wang et al. 2006), CD14 (Kol et al. 1999; Asea et al. 2000), CD91 (Basu et al. 2001) and CD40 (Wang et al. 2001) have been demonstrated to effectively bind extracellular HSPA1A. For more details on HSPA1A receptor binding, refer to Chap. 14. Internalization of extracellular HSPA1A has been suggested to require endocytosis through binding with lipid components present in the plasma membrane (Wang et al. 2006). Although the chaperokine activity of extracellular HSPA1A is now well accepted (Asea 2008, see Chap. 13), the exact mechanism by which HSPA1A passes through the plasma membrane and thus gains access to the extracellular milieu is less well understood. Recently it has been shown that plasma membrane HSPA1A enhances the plasma membrane density of CD56/CD94 and initiates the cytolytic activity of NK cells (Gross et al. 2008). In addition, plasma membrane HSPA1A has been shown to bind with nucleotide binding and substrate binding domains at distinct sites that are differentially localized on the membrane of macrophage cell lines (Zitzler et al. 2008). The selective depletion of HSPA1A was found to enhance malignant cell immunogenicity in rat colon cells (Gurbuxani et al. 2001), and its internalization was facilitated by its binding with lipid rafts present in the cell membrane (Arispe et al. 2002; Bausero et al. 2005; Wang et al. 2006). We recently demonstrated that the plasma membrane of cells functions as a reservoir for HSPA1A and thermal stress

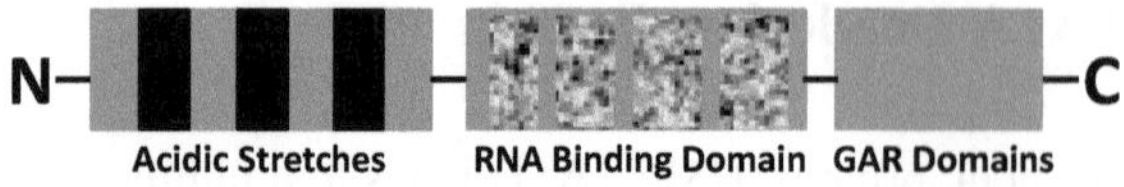

Fig. 8.1 Schematic representation of the primary sequence of nucleolin in mammals. There are three structural and multifunctional domains including acidic stretches in the N-terminus, four RNA-binding domains in the center and a GAR domain at the C-terminus

induces the redistribution of plasma membrane bound HSPA1A into subcellular cytosolic components, and from there the nucleolin-mediated transport system carries the HSPA1A to the plasma membrane for its relocalization and final release into the extracellular milieu (Pradeep et al. 2012).

8.3 Nucleolin: A Protein Transporter

The primary sequence of nucleolin has been determined (for review see Tajrishi et al. 2011). The protein sequence of nucleolin in mammals consists of three structural and multifunctional domains, including: (1) a N-terminal domain containing several acidic stretches; (2) four RNA-binding domains known as RNA recognition motifs (RRM) in the center, and; (3) a glycine/arginine-rich domain or GAR domain at the C-terminus (Fig. 8.1). Although nucleolin lacks a transmembrane domain or signal sequence, it is nevertheless found on the surface of various cell types. Several studies establish nucleolin's involvement in protein shuttling between cytoplasm and nucleus (Borer et al. 1989; Yu et al. 1998) and between the plasma membrane and nucleus (Schmidt-Zachmann and Nigg 1993; Said et al. 2002; Shibata et al. 2002; Christian et al. 2003). Plasma membrane nucleolin is a signaling receptor for P-selectin, in human colon carcinoma (Reyes-Reyes and Akiyama 2008). Earlier observations on the simultaneous increase in HSPA1A and nucleolin mRNAs in rat hepatocytes (Ohmori et al. 1990) and binding of plasma membrane nucleolin with the anti-HIV cytokine, midkine (Said et al. 2002) support the notion that nucleolin is an important transporter of intracellular proteins. Our lab (Asea et al. 2002) and others (Wang et al. 2006) have previously demonstrated that HSPA1A recruits pattern recognition molecules such as TLRs to move across the membrane over lipid rafts. TLR-mediated HSPA1A action requires intermediate molecules which have yet to be identified. Suggestively, nucleolin may be capable of direct interaction with TLR which may be a "missing link adapter" that exists in the nucleolin-TLR-HSPA1A association.

We recently demonstrated that plasma membrane-bound HSPA1A colocalizes with six proteins including Hsp90, CAP2, Gp96, TLR2, TLR4 and nucleolin colocalizes with HSPA1A on the plasma membrane (Pradeep et al. 2012; Fig. 8.2). Of the HSPA1A-interacting partners, we (Asea et al. 2002) and others (Vabulas et al. 2002a, b) have previously demonstrated that TLR2 and TLR4 are surface receptors for exogenous HSPA1A. Hsp90, CAP2 and gp96 are well known partners of HSPA1A (Kim et al. 2009).

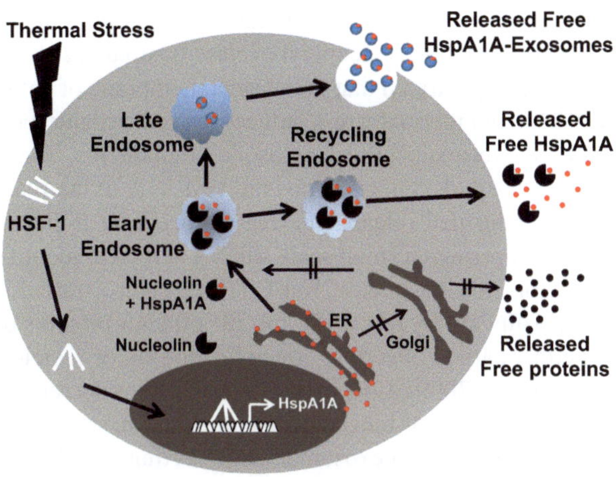

Fig. 8.2 Schematic representation of a hypothetical model by which nucleolin transports HSPA1A in cells. Thermal stress (*lightning bolt*) stimulates the trimerization of HSF-1 (*rods*) which translocate to the nucleus and activate the synthesis of HSPA1A (*circles*). From the ER (*organelle*), HSPA1A does not traffic through the cells *via* the classic protein transport pathway which requires transit through the Golgi apparatus. Our working hypothesis is that HSPA1A traffics *via* two distinct pathways assisted by nucleolin: (1) from the ER to the early endosomes (*EE*) then to recycling endosomes (*RE*) and is released as free HSPA1A chaperoning TAA (HSPA1A-TAA; *circles*), which HSPA1A has bound from while trafficking within the cell and from the plasma membrane as it exits the cell and/or (2) from the ER to EE to late endosomes (*LE*) where it combines with multivesicular bodies (*MVB*) and fuses with Ly or the lipid rafts on the plasma membrane before it exits the cell within exosomes (HSPA1A-exosomes)

Gp96 is a glycoprotein of the endoplasmic reticulum (ER) and an ER paralog of Hsp90 that is involved in antigen processing as an intermediate peptide carrier. Gp96 has previously been identified as a tumor-specific antigen that regulates antigen presenting cells (Schild and Rammensee 2000). Interaction of gp96 with HSPA1A on the plasma membrane reflects the peptide-binding activity of gp96 to serve as a cross-priming antigen to initiate T lymphocyte responses (Srivastava 1993; Srivastava et al. 1998). Binding of gp96 with associated proteins induces immune responses (Srivastava et al. 1986; Zheng et al. 2001; Chen et al. 2002) and protects cells from complete degradation peptides (Demine and Walden 2005). Cytosolic Hsp90 regulates numerous important cellular activities and trafficking of a large number of signal transducing proteins (Richter and Buchner 2001; Pratt and Toft 2003). The finding of a differential interaction of the two isoforms of Hsp90 has recently been reported with another chaperone GCUNC45 (Chadli et al. 2008), which has been suggested to be due to the greater presence of a monomeric form of Hsp90β in the cell lysates (Minami et al. 1991). Adenylyl cyclase-associated protein 2 (CAP2) is a multifunctional protein in which the N-terminal region with amino acids 1–168 is required for heat shock sensitivity (Shima et al. 1997). Adenylyl cyclase forms a complex with CAP (Field et al. 1990) which was associated with RAS2val-19-dependent heat shock sensitivity (Gerst et al. 1991; Wang et al. 1993). CAP may not be directly involved in physical association with Ras protein, but instead acts through alternated confirmation of adenylyl cyclase (Osipiuk et al. 1999).

Our study further demonstrated that the plasma membrane of cells functions as a reservoir for HSPA1A and thermal stress induces the redistribution of plasma membrane bound HSPA1A into subcellular cytosolic components, and from there the nucleolin-mediated transport system carries the HSPA1A to the plasma membrane for its relocalization and final release into the extracellular milieu (Pradeep et al. 2012). Ostrowski et al. present a small scale shRNA screen to identify possible Rab GTPases important for exosome release. Rab GTPases control trafficking in the endocytic and secretory pathways by recruiting specific effector proteins onto membrane surfaces to drive either cargo collection, organelle motility or vesicle docking at target membranes (Stenmark 2009). Loss of these Rab proteins decreased the amount of exosomes recovered but not their biochemical composition, suggesting that Rab27A and Rab27B do not participate in exosomal cargo sorting. Rab27A may be needed for MVE docking and fusion with the plasma membrane, whereas Rab27B may link MVEs to an outward directed motor protein. Rab27 effectors contain C2A/C2B domains containing synaptotagmin-related proteins, which suggests calcium regulation is involved in some forms of exosome release (Simons and Raposo 2009). The role of synaptotagmin in HSPA1A trafficking is also discussed in Chap. 6. Moreover, even relatively efficient depletion of Rab27A did not fully block exosome release. It is possible that the shRNA depletion was still not sufficient or that another Rab could take the place of Rab27A. Because neither exogenous Rab27A nor Rab27B colocalized to a large extent with CD63-containing MVEs and lysosomes, only a small subset of MVEs are probably capable of plasma membrane fusion.

Our hypothetical model suggests that nucleolin transports HSPA1A *via* two distinct pathways: (1) from the ER to the early endosomes (EE) then to recycling endosomes (RE) and then released as free HSPA1A chaperoning TAA (HSPA1A-TAA), which HSPA1A has bound from while trafficking within the cell and from the plasma membrane as it exits the cell and/or (2) from the ER to EE to late endosomes (LE) where it combines with multivesicular bodies (MVB) and fuses with Ly or the lipid rafts on the plasma membrane before it exits the cell within exosomes (HSPA1A-exosomes). Further work to definitively support this hypothesis is currently ongoing in our laboratory.

Acknowledgments We thank Princess Bempong, Shahrum Lillard and Viraj Mehta for expert technical assistance, and all the faculty and staff members of the Division of Investigative Pathology for helpful discussions. This work was supported in part by a Research Advancement Award from Scott & White Memorial Hospital and Clinic (to P.K.), the Department of Biotechnology (DBT) Government of India Overseas Associateship Award (to A.R.P.), the US National Institutes of Health grant RO1CA91889, Scott & White Hospital and Clinic, the Texas A&M Health Science Center College of Medicine, the Central Texas Veterans Health Administration and an Endowment from the Cain Foundation (to A.A.). The funders had no role in study design, data collection and analysis, decision to publish or preparation of the chapter.

References

Albert ML, Pearce SF, Francisco LM, Sauter B, Roy P, Silverstein RL, Bhardwaj N (1998) Immature dendritic cells phagocytose apoptotic cells via alphavbeta5 and CD36, and cross-present antigens to cytotoxic T lymphocytes. J Exp Med 188:1359–1368

Andersen JS, Lyon CE, Fox AH, Leung AK, Lam YW, Steen H, Mann M, Lamond AI (2002) Directed proteomic analysis of the human nucleolus. Curr Biol 12:1–11

Andersen JS, Lam YW, Leung AK, Ong SE, Lyon CE, Lamond AI, Mann M (2005) Nucleolar proteome dynamics. Nature 433:77–83

Arispe N, Doh M, De Maio A (2002) Lipid interaction differentiates the constitutive and stress-induced heat shock proteins Hsc70 and Hsp70. Cell Stress Chaperones 7:330–338

Asea A (2008) Hsp70: a chaperokine. Novartis Found Symp 291:173–179 (discussion 179–183, 221–174)

Asea A, Kraeft SK, Kurt-Jones EA, Stevenson MA, Chen LB, Finberg RW, Koo GC, Calderwood SK (2000) HSP70 stimulates cytokine production through a CD14-dependant pathway, demonstrating its dual role as a chaperone and cytokine. Nat Med 6:435–442

Asea A, Rehli M, Kabingu E, Boch JA, Bare O, Auron PE, Stevenson MA, Calderwood SK (2002) Novel signal transduction pathway utilized by extracellular HSP70: role of toll-like receptor (TLR) 2 and TLR4. J Biol Chem 277:15028–15034

Balakrishnan K, De Maio A (2006) Heat shock protein 70 binds its own messenger ribonucleic acid as part of a gene expression self-limiting mechanism. Cell Stress Chaperones 11:44–50

Basu S, Binder RJ, Ramalingam T, Srivastava PK (2001) CD91 is a common receptor for heat shock proteins gp96, hsp90, hsp70 and calreticulin. Immunity 14:303–313

Bausero MA, Gastpar R, Multhoff G, Asea A (2005) Alternative mechanism by which IFN-gamma enhances tumor recognition: active release of heat shock protein 72. J Immunol 175:2900–2912

Borer RA, Lehner CF, Eppenberger HM, Nigg EA (1989) Major nucleolar proteins shuttle between nucleus and cytoplasm. Cell 56:379–390

Chadli A, Felts SJ, Toft DO (2008) GCUNC45 is the first Hsp90 co-chaperone to show alpha/beta isoform specificity. J Biol Chem 283:9509–9512

Chen X, Tao Q, Yu H, Zhang L, Cao X (2002) Tumor cell membrane-bound heat shock protein 70 elicits antitumor immunity. Immunol Lett 84:81–87

Chen X, Kube DM, Cooper MJ, Davis PB (2008) Cell surface nucleolin serves as receptor for DNA nanoparticles composed of pegylated polylysine and DNA. Mol Ther 16:333–342

Christian S, Pilch J, Akerman ME, Porkka K, Laakkonen P, Ruoslahti E (2003) Nucleolin expressed at the cell surface is a marker of endothelial cells in angiogenic blood vessels. J Cell Biol 163:871–878

Daniels GA, Sanchez-Perez L, Diaz RM, Kottke T, Thompson J, Lai M, Gough M, Karim M, Bushell A, Chong H, Melcher A, Harrington K, Vile RG (2004) A simple method to cure established tumors by inflammatory killing of normal cells. Nat Biotechnol 22:1125–1132

Demine R, Walden P (2005) Testing the role of gp96 as peptide chaperone in antigen processing. J Biol Chem 280:17573–17578

Ferrari D, Pizzirani C, Adinolfi E, Lemoli RM, Curti A, Idzko M, Panther E, Di Virgilio F (2006) The P2X7 receptor: a key player in IL-1 processing and release. J Immunol 176:3877–3883

Field J, Vojtek A, Ballester R, Bolger G, Colicelli J, Ferguson K, Gerst J, Kataoka T, Michaeli T, Powers S, et al (1990) Cloning and characterization of CAP, the S. cerevisiae gene encoding the 70 kd adenylyl cyclase-associated protein. Cell 61:319–327

Gastpar R, Gehrmann M, Bausero MA, Asea A, Gross C, Schroeder JA, Multhoff G (2005) Heat shock protein 70 surface-positive tumor exosomes stimulate migratory and cytolytic activity of natural killer cells. Cancer Res 65:5238–5247

Gerst JE, Ferguson K, Vojtek A, Wigler M, Field J (1991) CAP is a bifunctional component of the Saccharomyces cerevisiae adenylyl cyclase complex. Mol Cell Biol 11:1248–1257

Gross C, Hansch D, Gastpar R, Multhoff G (2003) Interaction of heat shock protein 70 peptide with NK cells involves the NK receptor CD94. Biol Chem 384:267–279

Gross C, Holler E, Stangl S, Dickinson A, Pockley AG, Asea AA, Mallappa N, Multhoff G (2008) An Hsp70 peptide initiates NK cell killing of leukemic blasts after stem cell transplantation. Leuk Res 32:527–534
Gurbuxani S, Bruey JM, Fromentin A, Larmonier N, Parcellier A, Jaattela M, Martin F, Solary E, Garrido C (2001) Selective depletion of inducible HSP70 enhances immunogenicity of rat colon cancer cells. Oncogene 20:7478–7485
Hightower LE, Guidon PT, Jr (1989) Selective release from cultured mammalian cells of heat-shock (stress) proteins that resemble glia-axon transfer proteins. J Cell Physiol 138:257–266
Kim SR, Imoto S, Nakajima T, Ando K, Mita K, Taniguchi M, Sasase N, Matsuoka T, Kudo M, Hayashi Y (2009) Well-differentiated hepatocellular carcinoma smaller than 15 mm in diameter totally eradicated with percutaneous ethanol injection instead of radiofrequency ablation. Hepatol Int 3:411–415
Kol A, Bourcier T, Lichtman AH, Libby P (1999) Chlamydial and human heat shock protein 60s activate human vascular endothelium, smooth muscle cells, and macrophages. J Clin Invest 103:571–577
Lam YW, Lamond AI, Mann M, Andersen JS (2007) Analysis of nucleolar protein dynamics reveals the nuclear degradation of ribosomal proteins. Curr Biol 17:749–760
Leung AK, Trinkle-Mulcahy L, Lam YW, Andersen JS, Mann M, Lamond AI (2006) NOPdb: Nucleolar Proteome Database. Nucleic Acids Res 34:D218–220
Mambula SS, Calderwood SK (2006a) Heat induced release of Hsp70 from prostate carcinoma cells involves both active secretion and passive release from necrotic cells. Int J Hyperthermia 22:575–585
Mambula SS, Calderwood SK (2006b) Heat shock protein 70 is secreted from tumor cells by a nonclassical pathway involving lysosomal endosomes. J Immunol 177:7849–7857
Mambula SS, Stevenson MA, Ogawa K, Calderwood SK (2007) Mechanisms for Hsp70 secretion: crossing membranes without a leader. Methods 43:168–175
Minami Y, Kawasaki H, Miyata Y, Suzuki K, Yahara I (1991) Analysis of native forms and isoform compositions of the mouse 90-kDa heat shock protein, HSP90. J Biol Chem 266:10099–10103
Mosser DD, Caron AW, Bourget L, Denis-Larose C, Massie B (1997) Role of the human heat shock protein hsp70 in protection against stress- induced apoptosis. Mol Cell Biol 17:5317–5327
Multhoff G, Botzler C, Jennen L, Schmidt J, Ellwart J, Issels R (1997) Heat shock protein 72 on tumor cells: a recognition structure for natural killer cells. J Immunol 158:4341–4350
Murshid A, Theriault J, Gong J, Calderwood SK (2011) Investigating receptors for extracellular heat shock proteins. Methods Mol Biol 787:289–302
Negulyaev YA, Vedernikova EA, Kinev AV, Voronin AP (1996) Exogenous heat shock protein hsp70 activates potassium channels in U937 cells. Biochim Biophys Acta 1282:156–162
Noonan EJ, Place RF, Giardina C, Hightower LE (2007) Hsp70B' regulation and function. Cell Stress Chaperones 12:393–402
Ohmori H, Murakami T, Furutani A, Higashi K, Hirano H, Gotoh S, Kuroiwa A, Masui A, Nakamura T, Amalric F (1990) Simultaneous activation of heat shock protein (hsp 70) and nucleolin genes during in vivo and in vitro prereplicative stages of rat hepatocytes. Exp Cell Res 189:227–232
Olson MO, Dundr M, Szebeni A (2000) The nucleolus: an old factory with unexpected capabilities. Trends Cell Biol 10:189–196
Osipiuk J, Walsh MA, Freeman BC, Morimoto RI, Joachimiak A (1999) Structure of a new crystal form of human Hsp70 ATPase domain. Acta Crystallogr D Biol Crystallogr 55:1105–1107
Pradeep ANR, Nagaraja GM, Kaur P, Asea EE, Zheng H, Gizachew D, Shapiro LA, Asea A (2012) Nucleolin transports Hsp72 (HSPA1A) to the plasma membrane preparatory to its release into the microenvironment. PLoS ONE (in press)
Pratt WB, Toft DO (2003) Regulation of signaling protein function and trafficking by the hsp90/hsp70-based chaperone machinery. Exp Biol Med (Maywood) 228:111–133
Reyes-Reyes EM, Akiyama SK (2008) Cell-surface nucleolin is a signal transducing P-selectin binding protein for human colon carcinoma cells. Exp Cell Res 314:2212–2223
Richter K, Buchner J (2001) Hsp90: chaperoning signal transduction. J Cell Physiol 188:281–290

Said EA, Krust B, Nisole S, Svab J, Briand JP, Hovanessian AG (2002) The anti-HIV cytokine midkine binds the cell surface-expressed nucleolin as a low affinity receptor. J Biol Chem 277:37492–37502

Schild H, Rammensee HG (2000) gp96 – The immune system's swiss army knife. Nat. Immunol 1:100–101

Schmidt-Zachmann MS, Nigg EA (1993) Protein localization to the nucleolus: a search for targeting domains in nucleolin. J Cell Sci 105:799–806

Shibata Y, Muramatsu T, Hirai M, Inui T, Kimura T, Saito H, McCormick LM, Bu G, Kadomatsu K (2002) Nuclear targeting by the growth factor midkine. Mol Cell Biol 22:6788–6796

Shima F, Yamawaki-Kataoka Y, Yanagihara C, Tamada M, Okada T, Kariya K, Kataoka T (1997) Effect of association with adenylyl cyclase-associated protein on the interaction of yeast adenylyl cyclase with Ras protein. Mol Cell Biol 17:1057–1064

Simons M, Raposo G (2009) Exosomes–vesicular carriers for intercellular communication. Curr Opin Cell Biol 21:575–581

Srivastava PK (1993) Peptide-binding heat shock proteins in the endoplasmic reticulum: role in immune response to cancer and in antigen presentation. Adv Cancer Res 62:153–177

Srivastava PK (2003) Hypothesis: controlled necrosis as a tool for immunotherapy of human cancer. Cancer Immun 3:4

Srivastava PK, DeLeo AB, Old LJ (1986) Tumor rejection antigens of chemically induced sarcomas of inbred mice. Proc Natl Acad Sci U S A 83:3407–3411

Srivastava PK, Menoret A, Basu S, Binder RJ, McQuade KL (1998) Heat shock proteins come of age: primitive functions acquire new roles in an adaptive world. Immunity 8:657–665

Stenmark H (2009) Rab GTPases as coordinators of vesicle traffic. Nat Rev Mol Cell Biol 10:513–525

Tajrishi MM, Tuteja R, Tuteja N (2011) Nucleolin: The most abundant multifunctional phosphoprotein of nucleolus. Commun Integr Biol 4:267–275

Takeda K, Kaisho T, Akira S (2003) Toll-like receptors. Annu Rev Immunol 21:335–376

Theriault JR, Adachi H, Calderwood SK (2006) Role of scavenger receptors in the binding and internalization of heat shock protein 70. J Immunol 177:8604–8611

Theriault JR, Mambula SS, Sawamura T, Stevenson MA, Calderwood SK (2005) Extracellular HSP70 binding to surface receptors present on antigen presenting cells and endothelial/epithelial cells. FEBS Lett 579:1951–1960

Tytell M, Greenberg SG, Lasek RJ (1986) Heat shock-like protein is transferred from glia to axon. Brain Res 363:161–164

Vabulas RM, Ahmad-Nejad P, Ghose S, Kirschning CJ, Issels RD, Wagner H (2002a) HSP70 as endogenous stimulus of the Toll/interleukin-1 receptor signal pathway. J Biol Chem 277:15107–15112

Vabulas RM, Wagner H, Schild H (2002b) Heat shock proteins as ligands of toll-like receptors. Curr Top Microbiol Immunol 270:169–184

Wang J, Suzuki N, Nishida Y, Kataoka T (1993) Analysis of the function of the 70-kilodalton cyclase-associated protein (CAP) by using mutants of yeast adenylyl cyclase defective in CAP binding. Mol Cell Biol 13:4087–4097

Wang Y, Kelly CG, Karttunen JT, Whittall T, Lehner PJ, Duncan L, MacAry P, Younson JS, Singh M, Oehlmann W, Cheng G, Bergmeier L, Lehner T (2001) CD40 is a cellular receptor mediating mycobacterial heat shock protein 70 stimulation of CC-chemokines. Immunity 15:971–983

Wang Y, Theriault JR, He H, Gong J, Calderwood SK (2004) Expression of a dominant negative heat shock factor-1 construct inhibits aneuploidy in prostate carcinoma cells. J Biol Chem 279:32651–32659

Wang R, Town T, Gokarn V, Flavell RA, Chandawarkar RY (2006) HSP70 enhances macrophage phagocytosis by interaction with lipid raft-associated TLR-7 and upregulating p38 MAPK and PI3K pathways. J Surg Res 136:58–69

Wang SA, Li HY, Hsu TI, Chen SH, Wu CJ, Chang WC, Hung JJ (2011) Heat shock protein 90 stabilizes nucleolin to increase mRNA stability in mitosis. J Biol Chem 286:43816–43829

Wells AD, Rai SK, Salvato MS, Band H, Malkovsky M (1998) Hsp72-mediated augmentation of MHC class I surface expression and endogenous antigen presentation. Int Immunol 10:609–617
Wewers MD (2004) IL-1beta: an endosomal exit. Proc Natl Acad Sci U S A 101:10241–10242
Yu D, Schwartz MZ, Petryshyn R (1998) Effect of laminin on the nuclear localization of nucleolin in rat intestinal epithelial IEC-6 cells. Biochem Biophys Res Commun 247:186–192
Zheng H, Dai J, Stoilova D, Li Z (2001) Cell surface targeting of heat shock protein gp96 induces dendritic cell maturation and antitumor immunity. J Immunol 167:6731–6735
Zitzler S, Hellwig A, Hartl FU, Wieland F, Diestelkotter-Bachert P (2008) Distinct binding sites for the ATPase and substrate-binding domain of human Hsp70 on the cell surface of antigen presenting cells. Mol Immunol 45:3974–3983

Chapter 9
The Hsp90-Based Protein Trafficking System and Linkage to Protein Quality Control

William B. Pratt, Yoshihiro Morishima, Andrew P. Lieberman and Yoichi Osawa

Abstract Through their interaction with ligand-binding folding clefts in natively folded proteins, Hsp90 and Hsp70 affect the trafficking and turnover of several hundred signalling proteins. These chaperones each contain C-terminal acceptor sites that bind proteins containing tetratricopeptide repeat (TPR) domains. The interaction of Hsp90 with TPR domain immunophilins, such as FKBP52, accounts for Hsp90-based trafficking, and the binding of the ubiquitin E3 ligase CHIP via a TPR domain to Hsp70 mediates proteasomal degradation. The TPR domain immunophilins link via their peptidylprolyl isomerase domains to the dynamitin component of the dynein motor system for retrograde trafficking. Hsp90 stabilizes an open state of ligand binding clefts, and when Hsp90 can no longer interact because cleft opening proceeds to nascent unfolding with exposure of hydrophobic protein interior, Hsp70/CHIP-dependent ubiquitination occurs. In this way the Hsp90/Hsp70-based chaperone machinery plays a key role in the triage of damaged and aberrant proteins for degradation via the ubiquitin-proteasome pathway. We use the example of spinal and bulbar muscular atrophy, an androgen-dependent neurodegenerative disorder caused by an aberrant androgen receptor with an expanded polyglutamine repeat, to illustrate how trafficking and protein quality control are mechanistically and spatially integrated functions of the Hsp90/Hsp70-based chaperone machinery.

9.1 Introduction

The ubiquitous and abundant protein chaperone Hsp90 regulates the function, turnover and trafficking of a wide variety of signaling proteins (reviewed in Pratt and Toft 2003). Since 1994, when it became clear that certain benzoquinone antibiotics, such as geldanamycin, were specific Hsp90 inhibitors (Whitesell et al. 1994), Hsp90 has been found to participate in virtually all forms of protein trafficking. For example, Hsp90 is involved at all stages of vesicular protein transport, including

Y. Osawa (✉) · W. B. Pratt · Y. Morishima
Department of Pharmacology, University of Michigan Medical School,
1301 MSRB III, Ann Arbor, MI 48109-0630, USA
e-mail: osawa@umich.edu

A. P. Lieberman
Department of Pathology, University of Michigan Medical School,
1301 MSRB III, Ann Arbor, MI 48109-0630, USA

B. Henderson, A. G. Pockley (eds.), *Cellular Trafficking of Cell Stress Proteins in Health and Disease*, Heat Shock Proteins 6, DOI 10.1007/978-94-007-4740-1_9,

traverse of the endoplasmic reticulum (Xu et al. 2002), Rab1-dependent ER to Golgi trafficking (Chen and Balch 2006), trafficking between Golgi stacks and from Golgi to the plasma membrane (Lotz et al. 2008). A requirement for Hsp90 in ER to plasma membrane transport has been shown in plants (Chen et al. 2010), suggesting a conserved, general role for Hsp90 in vesicular transport. At synapses, Hsp90 plays a role in Rab3A-dependent synaptic vesicle fusion (Sakisaka et al. 2002), and, in addition to neurotransmitter release, Hsp90 was found to mediate the continuous synaptic cycling of AMPA-type glutamine receptors (Gerges et al. 2004). Thus, Hsp90 is involved in different stages of vesicular trafficking, but multiple different mechanisms have been proposed such that no uniform model of Hsp90 action in vesicular trafficking has evolved.

In contrast to the case in vesicular trafficking, the role of Hsp90 in the transport of protein solutes is better defined. Most of our understanding is based on the study of the movement of transcription factors, in particular steroid receptors. Because the glucocorticoid receptor (GR) is a transcription factor that moves through the cytoplasm to its sites of action within the nucleus in ligand-dependent manner, it has proven to be a facile model for studying cytoplasmic-nuclear trafficking. In this chapter, we will focus on the dynein -dependent retrograde trafficking of steroid receptors and the role of Hsp90 and Hsp90-binding immunophilins in linking the receptor cargo to the dynein motor complex. This trafficking system has been the subject of two previous detailed reviews (Pratt et al. 1999, 2004), and our goal here is to present an overview of the Hsp90-dependent trafficking model.

The chapter is written as a perspective, and we feel free to speculate on how Hsp90-dependent movement through the cytoplasm is related to such issues as the delivery of preproteins for import into organelles, formation of protein aggregates, and protein degradation via the ubiquitin-proteasome pathway. At the end of the chapter, we discuss how drugs that target the Hsp90/Hsp70-dependent chaperone machinery may be useful in the treatment of Kennedy disease (spinal and bulbar muscular atrophy, SBMA), an androgen-dependent, progressive neuromuscular disease caused by androgen receptor with an expanded polyglutamine tract (Lieberman and Pratt 2011).

9.1.1 Steroid Receptor Trafficking

9.1.1.1 Historical Perspective

Unliganded steroid receptors are present in cells in heterocomplexes containing Hsp90, Hsp70, and Hsp90-binding immunophilins (reviewed in Pratt and Toft 1997) that constantly shuttle into and out of the nucleus (reviewed in DeFranco et al. 1995). In hormone-free cells, estrogen and progesterone receptors are localized predominantly in the nucleus, whereas glucocorticoid and mineralocorticoid (MR) receptors are located predominantly in the cytoplasm (Pratt et al. 1999). Movement of a transcription factor from the cytoplasm to the nucleus can be considered as a four-step process: (1) movement through the cytoplasm to the nuclear pores;

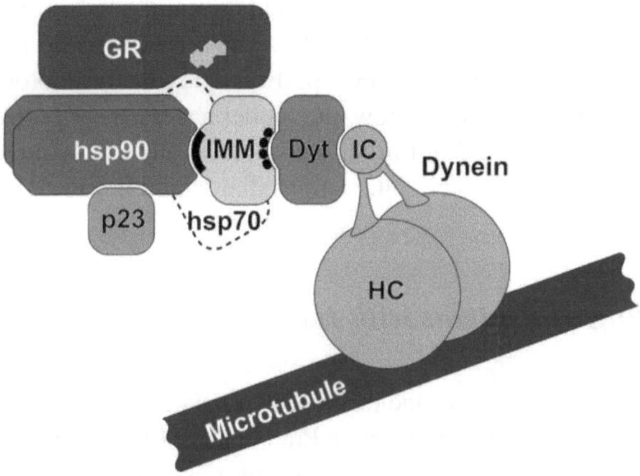

Fig. 9.1 TPR domain immunophilins, such as FKBP52, link the GR•Hsp90 heterocomplex to the dynamitin (*Dyt*) component of the dynein/dynactin motor complex. The entire movement complex, including tubulin, can be immunoprecipitated from paclitaxel-stabilized cytosol (Harrell et al. 2004; Galigniana et al. 2010). Dynein is a large multisubunit complex (~ 1.2 MDa) comprised of two heavy chains (*HC*) that have the processive motor activity, three intermediate chains (*IC*), and some light chains that are not shown. Also not shown is the dynein-associated dynactin complex, of which dynamitin is the component that connects the motor to cargo. The immunophilin (IMM) links to the GR-bound Hsp90 via its TPR domain (*solid black crescent*) and it links to dynamitin via its PPIase domain (*dotted crescent*). p23 binds to Hsp90 to stabilize its ATP-dependent conformation. p23 is present in unliganded GR trafficking complexes, but it is probably not present during rapid trafficking of the ligand transformed receptor

(2) importin-dependent transfer across the nuclear pore complex; (3) movement within the nucleus to loci for transcriptional activation, and; (4) subsequent dynamic exchange of the GR between chromatin and the nucleoplasm. In this chapter we will focus on retrograde movement through the cytoplasm, and the reader is referred to a review for steps 2–4 of the trafficking journey (Pratt et al. 2004). Suffice it to say that steroid receptors traverse the cytoplasm and the nuclear pore complex as the receptor•Hsp90 heterocomplex (Savory et al. 1999; Galigniana et al. 1999; Galigniana et al. 2010) and their mobility within the nucleus requires Hsp90 (Elbi et al. 2004).

The model for retrograde GR movement by dynein motors along microtubules that is illustrated in Fig. 9.1 was first proposed in 1993 (Pratt 1993). At that time, it was known that retrograde movement of vesicles in neurites and cell bodies was dynein-dependent and occurred on microtubular tracts (reviewed in Hirokawa 1998), and we reasoned that protein solutes that were not associated with vesicles might utilize a similar movement system. It was also known that some Hsp90 localized to microtubules (Sanchez et al. 1988) and that steroid receptor heterocomplexes contained FKBP52 (52 kDa FK506 binding protein), an Hsp90-binding immunophilin (reviewed in Pratt et al. 1999). Although it localizes predominantly to the nucleus, the portion of FKBP52 that is cytoplasmic localizes to microtubules (Ruff et al. 1992; Czar et al. 1994; Perrot-Applanat et al. 1995). Importantly, immunoadsorption of

FKBP52 from brain cytosol was found to yield co-immunoadsorption of the heavy and light chains of dynein (Czar et al. 1994). Thus, we proposed that one function of FKBP52 was to link the receptor•Hsp90 complex to the motor for retrograde movement. As we will describe below, the model has been supported by studies of cytoplasmic-nuclear GR and MR trafficking in intact cells where movement has been disrupted by blocking the protein interactions diagramed in Fig. 9.1.

9.1.2 Cycling of Proteins with Hsp90

The Hsp90/Hsp70-based chaperone machinery regulates signaling proteins by modulating ligand binding clefts (reviewed in Pratt et al. 2008, 2010), and these proteins constantly undergo cycles of Hsp90 heterocomplex assembly and disassembly in the cytoplasm and nucleoplasm (Pratt and Toft 2003). Two types of cycling with Hsp90 occur. The classical Hsp90 'client' proteins, such as steroid receptors and many protein kinases, form Hsp90 heterocomplexes that are stable enough to be isolated and analyzed biochemically. We call this 'stable cycling' with Hsp90, and the turnover of these proteins is stringently regulated by the chaperone (Pratt et al. 2008). Formation of heterocomplexes with Hsp90 inhibits client protein turnover, and treatment with an Hsp90 inhibitor, such as geldanamycin, uniformly triggers client protein degradation (Isaacs et al. 2003). Other signaling proteins, such as the nitric oxide synthase (NOS) enzymes, form Hsp90 heterocomplexes that rapidly disassemble such that no (or only trace amount of) Hsp90 heterocomplexes are recovered from cell lysates. We call this 'dynamic cycling', and the turnover of these proteins is not as affected by Hsp90 inhibitors as the classical client proteins (Pratt et al. 2008). Degradation of both types of Hsp90-regulated signaling proteins occurs via the ubiquitin-proteasome pathway, which in this case is initiated by Hsp70-dependent E3 ubiquitin ligases, such as CHIP (C-terminus of Hsp70-interacting protein) (Cyr et al. 2002) and parkin (Morishima et al. 2008).

Ligand binding clefts are hydrophobic clefts that must be open to allow access of ligands, such as steroids or ATP, to their binding sites within the protein's interior. In the absence of the chaperone machinery, ligand binding clefts are dynamic, shifting to varying extents between closed and open states. When the cleft opens, hydrophobic residues of the protein's interior are exposed to solvent, and continued opening may progress to protein unfolding. Therefore, the extent to which the ligand binding cleft is open determines ligand access and thus protein function, but clefts are inherent sites of conformational instability. Binding of Hsp90 stabilizes the open state of the cleft preventing further unfolding and Hsp70-dependent ubiquitination (Peng et al. 2009).

Stable cycling of Hsp90 with the GR requires ATP, Hsp70 and p23, and both receptor-bound Hsp70 and Hsp90 must pass through at least one complete ATPase cycle (Pratt and Toft 2003). As diagrammed in Fig. 9.2, formation of the GR•Hsp90 heterocomplex favors the open state of the cleft (Fig. 9.2b), and this accounts for the high affinity ligand binding activity of the receptor. On the basis of biochemical

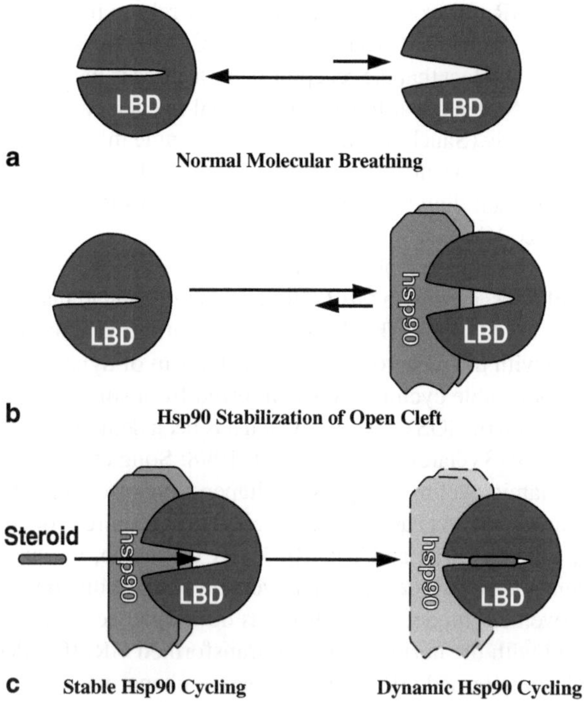

Fig. 9.2 States of the steroid binding cleft in the GR ligand binding domain (*LBD*) and Hsp90 cycling. **a** In the absence of Hsp90, the cleft in the GR LBD is predominantly closed, opening only very transiently during the course of normal molecular breathing; thus, high concentrations of steroid are required to initiate the hormone effect. **b** When stable complexes are assembled with Hsp90, nearly all the ligand binding clefts are open at any time, and low concentrations of steroid are now sufficient for binding. **c** Steroid binding within the cleft promotes a temperature-dependent collapse of the cleft around the ligand to the closed state which cycles dynamically with Hsp90. Thus, receptors that have bound steroid under physiological conditions in the cell are not recovered in association with Hsp90

observations, it was thought for many years that ligand-dependent transformation of the GR resulted in release from Hsp90, leaving the chaperone-free receptor to move to the nucleus. However, it is now clear that GR translocation occurs in association with Hsp90 (Savory et al. 1999; Galigniana et al. 1998, 1999), and it is thought that ligand-dependent transformation converts the receptor from a state with a metastable cleft that forms stable complexes with Hsp90 to a state with a more closed cleft that cycles dynamically with Hsp90 (Fig. 9.2c). The proposal that dynamic GR•Hsp90 heterocomplex assembly is required for rapid, retrograde receptor movement is supported by a variety of observations that have been previously reviewed (Pratt et al. 1999; DeFranco et al. 1998).

As mentioned above, unliganded steroid receptors that undergo stable cycling with Hsp90 continuously shuttle into and out of the nucleus (DeFranco et al. 1995). What determines the steady-state localization of the receptor is unclear. The unliganded mouse GR, for example, is localized entirely to the cytoplasm in mouse L cells, whereas the mouse GR overexpressed in CHO cells is localized entirely to the nucleus, despite the fact that the receptor in both cases is in a heterocomplex with Hsp90 and that the NL1 nuclear localization signal in each case reacts equivalently with anti-NL1 antibody (Sanchez et al. 1990). We assume that this shuttling of stable Hsp90 cycling receptors occurs along microtubules as illustrated in Fig. 9.1 because the entire complex, including tubulin, can be immunoadsorbed with the unliganded GR (Harrell et al. 2004).

The rapid ($t_{1/2} \sim 4.5$ min) cytoplasmic-nuclear GR translocation that is triggered by steroid and readily observed with GFP-GR is inhibited by geldanamycin (Czar et al. 1997; Galigniana et al. 1998). This Hsp90-dependent movement involves dynamic cycling of Hsp90 with the receptor, and the mechanism of dynamic cycling appears to be different from stable cycling. This is inferred from studies of dynamic Hsp90 binding and activation of NOS enzymes, which has been demonstrated in the absence of ATP, Hsp70 and p23 (Garcia-Cardena et al. 1998; Song et al. 2001). Thus, in the case of NOS, dynamic cycling is a passive chaperoning effect that does not require Hsp90 to pass through its ATPase cycle, a cycle that is required for activation of GR steroid binding activity (Pratt and Toft 2003). By extension, we suspect that rapid, ligand-dependent and Hsp90-dependent movement of GR along microtubules is also passive. The movement model of Fig. 9.1 was developed according to proteins co-immunoadsorbed with the hormone-free, untransformed GR. If, indeed, movement of the liganded, transformed GR reflects a passive chaperone interaction of Hsp90 with the receptor, then the Hsp90 co-chaperone p23 (required for stable heterocomplex assembly) would not be present during rapid trafficking after ligand-dependent activation of the receptor.

In those cases where the site of Hsp90 interaction with a substrate has been determined, it interacts with the domain containing the major ligand binding cleft and not with other regions of the protein. For example, Hsp90 interacts with the ligand binding domain of steroid receptors (Pratt and Toft 1997), with the kinase domain (which contains the ATP binding clefts) of protein kinase clients such as v-Raf (Stancato et al. 1993) and ErbB-2 (Xu et al. 2001; Citri et al. 2002), and with the oxygenase domain (contains the heme/substrate binding cleft) of NOS (Fontana et al. 2002; Xu et al. 2007). Mutational studies of the GR suggest that there is a very focal site of Hsp90 interaction that lies on the surface of the ligand binding domain at the opening of the hydrophobic steroid binding cleft (Xu et al. 1998; Giannoukos et al. 1999). Regions where hydrophobic clefts merge with the protein surface are a general topologic feature of all proteins in native conformation, and we have suggested that these openings are the general feature recognized by Hsp90 (Pratt and Toft 2003). When both stable and dynamic cycling are considered, Hsp90 could potentially interact with all proteins. This ability to interact with a general topological feature of all proteins would allow the movement system illustrated in Fig. 9.1 to play a general role in the trafficking of all of the protein solutes that shuttle around in the cell.

9.1.3 The Hsp90 Binding Immunophilins

In addition to Hsp90, steroid receptor heterocomplexes contain one of several tetratricopeptide repeat (TPR) domain immunophilins. TPR domains are degenerative sequences of 34 amino acids (Sikorski et al. 1990) that are required for the binding of these proteins to Hsp90 (Radanyi et al. 1994; Hoffman and Handschumacher 1995; Ratajczak and Carrello 1996). The TPR acceptor site is located at the C-terminus of Hsp90 (Chen et al. 1998; Young et al. 1998; Cerrello et al. 1999; Russell et al. 1999), and the core of the TPR binding site on Hsp90 is the MEEVD sequence (Scheufler et al. 2000; Brinker et al. 2002). Hsp90 functions biologically as a homodimer, the TPR binding site lies within the dimerization domain, and there is one TPR binding site per dimer (Young et al. 1998; Silverstein et al. 1999). Thus, the stoichiometry of the steroid receptor heterocomplex shown in Fig. 9.1 is 1 receptor: 2 Hsp90: 1 immunophilin (Pratt and Toft 1997). TPR domain immunophilins are widely distributed among animal and plant cells, and TPR domain binding to Hsp90 is conserved (Owens-Grillo et al. 1996b; Reddy et al. 1998; Harrell et al. 2002). This suggests that immunophilin binding to Hsp90 is essential both for the TPR domain immunophilins and for a major function of Hsp90.

The common feature of the immunophilins is that they bind immunosuppressant drugs, such as FK506, rapamycin and cyclosporine A. All members of the immunophilin family have peptidylprolyl isomerase (PPIase) activity, and there are two main classes: the FKBPs bind FK506 and rapamycin, and the cyclophilins (CyPs) bind cyclosporine A. The immunosuppressant drugs occupy the PPIase site on the immunophilin, blocking its ability to direct *cis-trans* isomerization of peptidylprolyl bonds. Three Hsp90 binding, TPR domain immunophilins have been found in steroid receptor heterocomplexes – FKBP52, FKBP51 and CyP40. The heterocomplexes may also contain protein phosphatase 5 (PP5), a protein-serine phosphatase with three TPRs and a PPIase homology domain with weak FK506 binding activity but no isomerase activity (Silverstein et al. 1997) (see Pratt et al. 2004 for domain structures of the Hsp90 binding immunophilins). Inasmuch as these TPR proteins exchange for binding to Hsp90, a single receptor•Hsp90 heterocomplex can theoretically be associated over time with more than one immunophilin. However, at any instant, the immunophilins exist in separate GR•Hsp90 heterocomplexes (Owens-Grillo et al. 1995; Renoir et al. 1995). The relative amounts of FKBP52, FKBP51, CyP40 and PP5 vary somewhat among different steroid receptor heterocomplexes (Ratajczak et al. 2003), probably according to immunophilin interaction with the receptor itself (reviewed in Pratt et al. 1999).

No clear role for *cis-trans* isomerization of peptidylprolyl bonds has emerged in steroid hormone action. FKBP52 is a positive regulator of GR and AR signaling (Riggs et al. 2003; Cheung-Flynn et al. 2005), but although the PPIase domain is required, PPIase activity is not required for the regulation (Riggs et al. 2007). Thus, here we will consider the PPIase domain as it functions as a protein-protein interaction domain that links the immunophilins to the dynein motor system. Although it does not have an NLS, the majority of FKBP52 is localized to the nucleus where it colocalizes with the GR (Czar et al. 1994), suggesting that it moves to and into the nucleus in association with NLS-containing proteins. As mentioned above, the cytoplasmic portion of FKBP52 co-localizes with microtubules (Ruff et al. 1992; Czar et al. 1994; Perrot-Applanat et al. 1995). Like FKBP52, the majority of PP5 is localized in the nucleus, but the portion that is cytoplasmic co-localizes with microtubules and cytoplasmic dynein (Galigniana et al. 2002). Because FKPB52 contains a conserved negatively charged domain that could serve as a nuclear localization recognition sequence, we injected antibody against this sequence into L cells and showed that it impeded steroid-induced trafficking from the cytoplasm to the nucleus (Czar et al. 1995). Also, consistent with a role in targeting receptor trafficking is the finding that FKBP52 binds directly to the GR and that a 35-amino acid segment of the receptor that spans the protosignals comprising the NL1 nuclear localization signal is sufficient for binding (Silverstein et al. 1999).

The PPIase domain of the Hsp90-binding immunophilins is the domain that links to the dynein motor complex. This was first shown by immuoadsorbing FKBP52 from reticulocyte lysate and showing co-immunoadsorption of dynein that was competed by a fragment of FKBP52 containing its PPIase domain but not by a TPR fragment that blocks FKBP52 binding to Hsp90 (Silverstein et al. 1999). Also, GR•Hsp90 heterocomplexes assembled in reticulocyte lysate contain cytoplasmic dynein in a manner that is competed by the PPIase domain of dynein (Galigniana et al. 2001). Subsequently, it was shown that immunoadsorption of FKBP52, CyP40 or PP5 from rabbit brain cytosol yielded co-immunoadsorption of cytoplasmic dynein that was competed by the PPIase domain fragment, establishing the PPIase domain as the linker to the motor system for all three of these components of GR•Hsp90 heterocomplexes (Galigniana et al. 2002). The PPIase domain of plant immunophilins also links to the dynein motor system, suggesting that this is a conserved linkage throughout eukaryotes (Harrell et al. 2002). All PPIase or PPIase homology domains do not bind dynein (e.g. FKBP12), and binding occurs even when the isomerase activity is blocked with FK506 (Silverstein et al. 1999). In contrast to FKBP52, the PPIase domain of FKBP51 does not link to dynein, and overexpression of FKBP51 disrupts the GR association with dynein and impairs nuclear translocation (Wochnik et al. 2005). Overexpression of FKBP51 also favors cytoplasmic localization of the mineralocorticoid receptor (Galigniana et al. 2010). Whether such a negative regulation of movement is of physiological importance or not is unknown.

The component of the dynein motor system that interacts with the immunophilin PPIase domain is dynamitin. Cytoplasmic dynein links to vesicles and organelles indirectly through the dynein-associated dynactin complex (Hirokawa 1998). Dynamitin is a component of the dynactin complex, and we have shown that purified FKBP52 binds to purified dynamitin via its PPIase domain (Galigniana et al. 2004b). GR•Hsp90•immunophilin complexes immunoadsorbed from cytosols with anti-GR antibody contain the dynactin components dynamitin and p150^{Glued} as well as cytoplasmic dynein (Harrell et al. 2004).

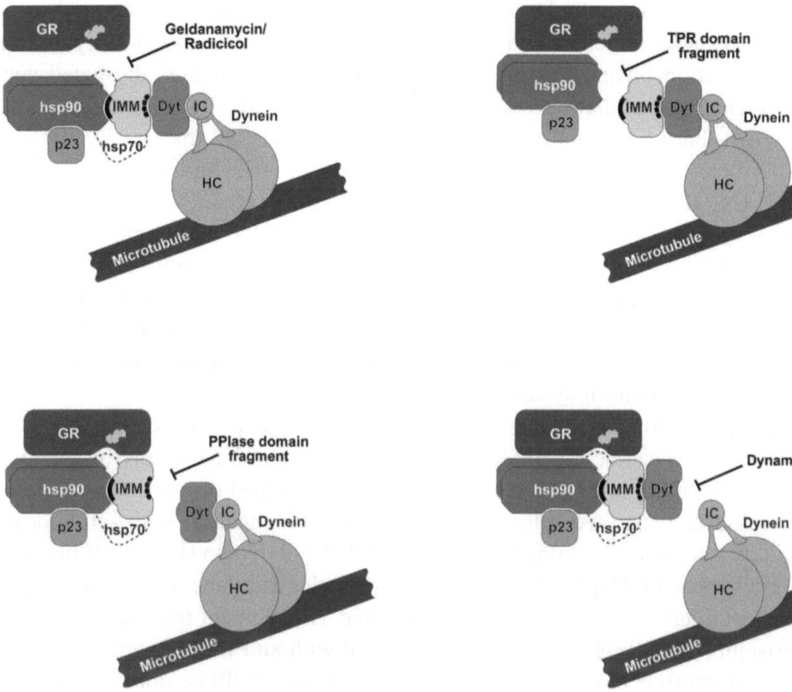

Fig. 9.3 The selective uncoupling of the GR movement system. Sites of uncoupling by the Hsp90 inhibitors geldanamycin and radicicol and by overexpression of the TPR domain fragment of PP5, the PPIase domain fragment of FKBP52, or dynamitin

9.1.4 Proof of the Hsp90-based Trafficking Model

Figure 9.3 shows the sites of uncoupling of the GR movement system by four uncoupling agents. Treatment of cells with the Hsp90 inhibitors geldanamycin or radicicol blocks Hsp90 binding to substrate. These Hsp90 inhibitors have been shown to inhibit cytoplasmic-nuclear trafficking of the GR (Czar et al. 1997; Galigniana et al. 1998), the MR (Galigniana et al. 2010), the androgen receptor (AR) (Georget et al. 2002; Thomas et al. 2004), the aryl hydrocarbon receptor (AhR) (Kazlauskas et al. 2000, 2001), and the tumor suppressor p53 (Galigniana et al. 2004b). Although the effect of overexpression of a TPR domain fragment on cytoplasmic-nuclear trafficking of the GR has not been tested, it has been shown to inhibit trafficking of the MR to the nucleus (Galigniana et al. 2010). Overexpression of the FKBP52 PPIase domain fragment has been shown to inhibit cytoplasmic-nuclear trafficking of the GR (Galigniana et al. 2001), the MR (Galigniana et al. 2010) and p53 (Galigniana et al. 2004b). Similarly, overexpression of dynamitin inhibits cytoplasmic-nuclear trafficking of the GR (Harrell et al. 2004), the MR (Galigniana et al. 2010) and p53 (Giannakakou et al. 2000; Galigniana et al. 2004b).

The tumor suppressor p53 was the first transcription factor shown to be moved to the nucleus by cytoplasmic dynein (Giannakakou et al. 2000). p53 was found to colocalize with microtubules in several human carcinoma cell lines and to be in cytosolic heterocomplexes with microtubules. Both overexpression of dynamitin and microinjection of anti-dynein antibody before DNA damage abrogated subsequent nuclear accumulation of p53 (Giannakakou et al. 2000). Subsequently, it was shown that p53 exists in heterocomplexes with Hsp90 that are identical to GR•Hsp90 heterocomplexes in that they contain Hsp70, p23, and an immunophilin (FKBP52, CyP40 or PP5), as well as dynein (Galigniana et al. 2004b). It was also shown that the p53 complex was linked to dynamitin via the immunophilin PPIase domain. Inhibition of movement through overexpression of dynamitin is powerful evidence that movement occurs through attachment to cytoplasmic dynein (Hirokawa 1998; Burkhardt et al. 1997). The free dynamitin resulting from overexpression competes with dynamitin in the dynein/dynactin complex for binding immunophilin, thus dissociating the motor system from its cargo.

All of the components of the trafficking complex exist in reticulocyte lysate, and the complexes can be assembled under cell-free conditions simply by incubating immunoadsorbed, unliganded GR that has been stripped of associated proteins with reticulocyte lysate. As illustrated in Fig. 9.4a, incubation with reticulocyte lysate yields a complex containing Hsp90, FKBP52 and dynein (lane 3). When geldanamycin is present, Hsp90 binding to the GR is blocked (lane 4) as illustrated in the upper left diagram in Fig. 9.3. When a purified TPR domain fragment is present during assembly, a GR•Hsp90 complex is formed without FKBP52 or dynein (lane 5) as in the upper right diagram of Fig. 9.3. When a purified PPIase domain fragment is present, a GR•Hsp90•FKBP52 complex is formed without any dynein (lane 6) as illustrated in the lower left diagram of Fig. 9.3.

The rapid, steroid-dependent trafficking of GFP-GR to the nucleus is shown in Fig. 9.4b (open squares). The rate of nuclear movement is inhibited by uncoupling the receptor from the movement system by treating the cells with geldanamycin (closed squares) or by overexpression of a PPIase domain fragment (open circles). The fact that movement is slowed but not blocked shows that the Hsp90-dependent trafficking is required for rapid movement, but in the presence of an intact cytoskeleton, slower movement occurs. This probably reflects receptor diffusion in the cytoplasm, with trapping for nuclear import by binding of the GR NLS to importin-α (Pratt et al. 2004). The rapid Hsp90- and immunophilin-dependent movement defined by the open squares in Fig. 9.4b requires an intact cytoskeletal system (Galigniana et al. 1998).

9.2 Perspectives on Hsp90-Dependent Trafficking

9.2.1 Essential or Non-essential or Both?

The notion that steroid receptors traffic through the cytoplasm along microtubular tracks has been resisted, because it was shown quite early that receptors undergo a ligand-dependent translocation from the cytoplasm to nucleus in cells treated with

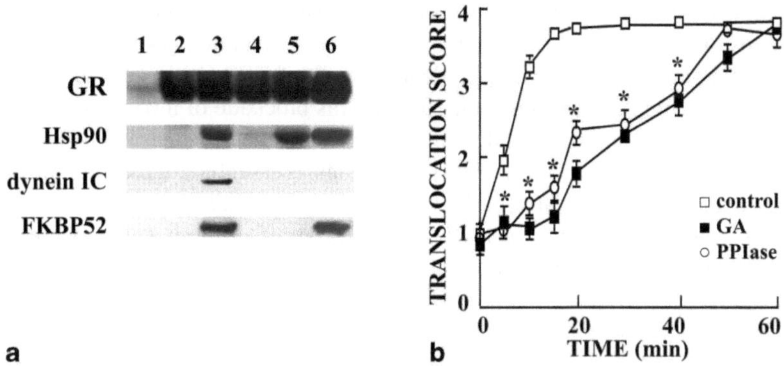

Fig. 9.4 Inhibition of GR heterocomplex assembly and cytoplasmic-nuclear translocation. **a** GR heterocomplex assembly. Immunoadsorbed GR pellets were stripped of Hsp90 by incubation in buffer containing 0.7 M NaCl and subsequent washing in buffer. The stripped GR immune pellets were then incubated for 30 min at 30 °C with 50 μl rabbit reticulocyte lysate and 5 μl of an ATP regenerating system. After three washes, the samples were boiled in SDS sample buffer and immunoblotted for GR, Hsp90, dynein intermediate chain and FKBP52. Lane 1, nonimmune pellet incubated with reticulocyte lysate (RL); lanes 2–6, stripped GR immune pellets incubated with buffer (lane 2), with RL preincubated with buffer (lane 3), with RL preincubated with 10 μM geldanamycin (lane 4), with RL preincubated with lysate of Sf9 cells expressing the TPR domain of rat PP5 (lane 5), or with RL preincubated with lysate of Sf9 cells expressing the PPIase domain of rabbit FKBP52 (lane 6). **b** Inhibition of GR cytoplasmic-nuclear translocation. 3T3 cells transfected with GFP-GR and either FKBP52 PPIase domain or vector plasmid were grown for 2 days and then placed on ice for 90 min with 1 μM dexamethasone and for 30 min with 10 μM geldanamycin (GA). The temperature was then shifted to 37 °C to permit nuclear translocation of GFP-GR, which was scored from 0 (cytoplasmic fluorescence much greater than nuclear fluorescence) to 4 (nuclear fluorescence much greater than cytoplasmic). □, vector control; ■, geldanamycin; ○, expressing PPIase domain. * difference from vector control significant at $p < 0.01$. Trafficking data are from Galigniana et al. 2001

microtubule disrupting agents (Perrot-Applanat et al. 1992; Czar et al. 1995; Galigniana et al. 1998). Also, steroid-dependent transcriptional activation by the GR is unaffected by microtubular disruption (Szapary et al. 1994) and steroid-dependent transcriptional activation occurs in yeast expressing GR but lacking dynein (Riggs et al. 2003), showing that Hsp90-dependent movement is nonessential for receptor action. Interestingly, when the cytoskeleton is disrupted, the GR moves to the nucleus in a ligand-dependent manner as rapidly as it does by Hsp90-dependent trafficking when the cytoskeleton is intact, but the rapid movement in the presence of a disrupted cytoskeleton is not affected by geldanamycin (Galigniana et al. 1998). Because GR movement by diffusion is slow when the cytoskeleton is intact (e.g., geldanamycin condition in Fig. 9.4b), and rapid when the cytoskeleton is disrupted, it would seem that the presence of an intact cytoskeleton reduces the rate of GR diffusion through the cytoplasm.

Taken together, these observations tell us that Hsp90-dependent trafficking is not essential in most cells. However, such a nonessential movement system had to exist in primitive eukaryotes for higher organisms with nervous systems to develop. Axons and dendrites are specialized cytoplasmic extensions where proteins cannot move by diffusion. In contrast to the slowing of GR movement that is seen in 3T3 cells in Fig. 9.4b, ligand-dependent retrograde movement of GFP-GR in neurites is blocked by geldanamycin (Galigniana et al. 2004a). This blockade of movement in neurites suggests that the Hsp90-dependent movement machinery, which is nonessential in non-neuronal cells and in the cell body of neurons, is essential for receptor movement in axons and dendrites. Thus, in the integrated multicellular organism possessing a nervous system, a function of Hsp90 in protein trafficking that is nonessential in lower organisms is probably essential.

9.2.2 Hsp90-binding TPR Proteins and Protein Targeting

As part of a general model of targeted protein movement we proposed that Hsp90-binding TPR proteins associated with organelles, such as mitochondria, might serve to accept preproteins in a "hand off" from the Hsp90-dependent movement machinery (Owens-Grillo et al. 1996a). Subsequent research on the mechanism of preprotein import into organelles could be consistent with such a model. For example, before import, mitochondrial preproteins are bound to Hsp70 and Hsp90, which deliver the preproteins to the mitochondrial import receptor. The TOM (*t*ranslocase of the mitochondrial *o*uter *m*embrane) complex contains import receptors, TOM70 and its homolog TOM71, that each contain seven TPR domains and bind to the TPR acceptor sites at the C-termini of Hsp70 and Hsp90. Both chaperones play an important role in delivery of the preproteins to these mitochondrial import receptors, as well as in the subsequent translocation of the proteins through the general import pore (Young et al. 2003; Fan et al. 2006; Li et al. 2009).

Mitochondrial import has been studied in cell-free systems where there is no Hsp90-dependent trafficking, and interest has focused on the role of Hsp90 and Hsp70 in maintaining the preproteins in a state that is competent for import. In early studies of import, it was found that a hybrid preprotein of dihydrofolate reductase fused to a mitochondrial matrix targeting signal lost import competence unless components of rabbit reticulocyte lysate were present (Sheffield et al. 1990). The import competence activity resided in a large (200–250 kDa) complex that was partially purified and shown to contain Hsp70, but import competence was not replaced by Hsp70 itself. This partially purified import competence complex was then shown to contain Hsp90 in a complex with Hsp70, and it formed GR•Hsp90 heterocomplexes with high affinity steroid binding activity (Scherrer et al. 1992). It was subsequently found that the GR•Hsp90 heterocomplex assembly activity was markedly increased by adding the p23 component of the receptor•Hsp90 heterocomplex assembly machinery. It is important to realize that the same Hsp90/Hsp70-based multichaperone machinery that makes client protein•Hsp90 heterocomplexes determines protein activity (high affinity steroid binding), Hsp90-dependent trafficking (when the Hsp90 binds TPR domain immunophilins), and preprotein import competence (when Hsp90 interacts with TPR domain import receptor). Because these all appear to result from the same heterocomplex assembly, in the cell the preproteins undergoing Hsp90-dependent trafficking are likely to arrive at mitochondria in the import competent state, and the Hsp90-associated targeted protein is "handed off" from the TPR domain immunophilin to the TPR domain import receptor.

Similar systems have evolved for proteins that are targeted for import into chloroplasts and into peroxisomes. The chloroplast protein translocon subunit Toc64 possesses a TPR domain that binds to Hsp90, and the Hsp90-bound preprotein is transferred to the Toc (translocon of the outer envelope of chloroplasts) core complex via a GTP-dependent association of Toc64 with Toc34 (Qbadou et al. 2006). Peroxins (Pex) are required for import of proteins with peroxisome targeting signals (PTSs) into peroxisomes. Most peroxisomal matrix proteins are targeted by PTS1, and their import is determined by Pex5p (Crookes and Olsen 1999). Pex5p contains seven TPRs, and an intact TPR domain is necessary for interaction with PTS1 targeted proteins (Klein et al. 2001). Import of a PTS1 protein into peroxisomes was inhibited by antibodies directed against Hsp70 and Hsp90 (Crookes and Olsen 1998), suggesting that similar chaperone interactions may be involved in precursor recognition by mitochondria, chloroplasts and peroxisomes.

Other TPR proteins are involved in the trafficking of both protein solutes and vesicles. Like the steroid receptors, the AhR shuttles in and out of the nucleus, and its ligand-induced nuclear accumulation is inhibited by geldanamycin (Kazlauskas et al. 2000, 2001). Thus, it utilizes an Hsp90-dependent trafficking system, and although the model that has developed is somewhat different from steroid receptor movement, there are similarities (reviewed in Pratt et al. 2004). The AhR is associated with a TPR protein called ARA9 (also called AIP or XAP2), which contains three TPRs in its C-terminus and a PPIase homology domain without PPIase activity in its N-terminus (Ma and Whitlock 1997; Carver and Bradfield 1997; Meyer et al. 1998). The TPR domain binds Hsp90 and the N-terminal region is essential for ARA9 to regulate the intracellular localization of the AhR (Kazlauskas et al. 2002). The peroxisome proliferator-activated receptor α has also been recovered from cytosols in a PPARα•Hsp90•ARA9 complex (Sumanasekera et al. 2003).

Hsp90 and a TPR protein called TPR1 are involved in the vesicular trafficking of vesicular stomatitis virus glycoprotein (VSVG) from the Golgi apparatus to the cell membrane. In this case TPR1 binds to Hsp90 and links it to the membrane-bound protein VAP-33 (Lotz et al. 2008). Depletion of TPR1, or inhibition of Hsp90 by geldanamycin or radicicol, inhibit the transport of the VSVG cargo protein. Intact adeno-associated virus 2 (AAV) particles have been shown to interact with FKBP52 and dynein, and expression of FKBP52 in cells from FKBP52 knockout mice improves AAV trafficking to the nucleus (Zhao et al. 2006). In addition to being blocked by geldanamycin and radicicol, the synaptic cycling of AMPA receptors was blocked by expression of the TPR domain fragment of PP5, suggesting the participation of Hsp90 and a TPR protein in this form of vesicular trafficking (Gerges et al. 2004). Taken together, these observations suggest that Hsp90-binding TPR proteins play a broader role in trafficking than just the retrograde trafficking of transcription factors that has been emphasized here.

9.2.3 Hsp90-dependent Trafficking and Protein Quality Control

To understand the last section of this chapter, which is concerned with chaperone regulation of the polyglutamine androgen receptor (polyQ AR), it is important to understand how the Hsp90/Hsp70-based chaperone machinery plays a key role in the triage of damaged and aberrant (e.g. polyQ AR) proteins for degradation via the ubiquitin-proteasome pathway. Until the discovery of CHIP, a chaperone-binding E3 ubiquitin ligase, it was not understood how proteins that are damaged and unfolding are selected for ubiquitination. The prevailing view was that E3 ubiquitin ligases perform the role of protein substrate recognition and bring the ubiquitin-charged E2 to the substrate for ubiquitin transfer (reviewed in Hershko and Ciechanover 1998; Pickart 2004). However, in the case of proteins that are unfolding, chaperones appear to be responsible for substrate recognition, and Hsp70-dependent ubiquitin ligases, such as CHIP, then target the E2 enzyme to the Hsp70-bound substrate (Cyr et al. 2002).

In the mid-1990s, it was shown that Hsp70 and its co-chaperone Hsp40 are required for ubiquitin-dependent degradation of short-lived and abnormal proteins (Lee et al. 1996; Bercovich et al. 1997). It was also shown that ansamycins, like geldanamycin, disrupt client protein•Hsp90 complexes (Whitesell et al. 1994), and that ansamycin-induced degradation of the Hsp90 substrate occurs via the ubiquitin-proteasome pathway (Sepp-Lorenzino et al. 1995). This general notion that Hsp90 binding to a client protein inhibits its degradation and inhibitors like geldanamycin relieve that inhibition by preventing cycling with Hsp90 has stood the test of time. The observations of the chaperone effects on client protein turnover are consistent with the two essential components of the chaperone machinery having opposing effects, with Hsp70 promoting degradation and Hsp90 stabilizing the protein against degradation.

Studies of the ubiquitination of neuronal NOS (nNOS) have led to a model of how Hsp90 and Hsp70 function in protein triage (reviewed in Pratt et al. 2010). Treatment of cells with geldanamycin leads to nNOS degradation via the ubiquitin-proteasome pathway (Bender et al. 1999; Noguchi et al. 2000). CHIP overexpression decreases nNOS protein levels, it functions as an E3 ligase for nNOS ubiquitination, and CHIP accounts for all of the nNOS ubiquitinating activity in reticulocyte lysate (Peng et al. 2004; Clapp et al. 2010). A number of nNOS substrates cause mechanism-based inactivation in the heme/substrate binding cleft to yield a more unfolded state of the enzyme, triggering CHIP-dependent ubiquitination and proteasomal degradation (Noguchi et al. 2000; Peng et al. 2009). Hsp70 and Hsp90 have opposing effects on CHIP-dependent ubiquitination of nNOS, both in a purified ubiquitinating system (Peng et al. 2009) and in the cell (Peng, work in progress), with Hsp70 promoting ubiquitination and Hsp90 inhibiting ubiquitination. Thus, although CHIP binds via an amino-terminal TPR domain to both Hsp70 and Hsp90 (Ballinger et al. 1999; Connell et al. 2001), it is the binding to Hsp70 that is critical for substrate ubiquitination.

The model of triage that has evolved is one in which Hsp90 and Hsp70 cooperate in making the triage decision for degradation via the ubiquitin-proteasomal pathway

(Pratt et al. 2010). The effects of mechanism-based inactivation on nNOS serve as examples of toxic damage that is targeted to the ligand binding cleft and triggers ubiquitination of the enzyme. As nNOS undergoes such toxic damage, the ligand binding cleft opens as the initial step in unfolding of the enzyme. As long as Hsp90 can interact even transiently with the opening cleft, ubiquitination by Hsp70-dependent ubiquitin ligases, like CHIP, is inhibited. But a point is reached where unfolding of the cleft progresses to a state that cannot cycle with Hsp90, and ubiquitination by Hsp70-dependent E3 ligases is unopposed. Inasmuch as CHIP co-immunoadsorbs with nNOS•Hsp70 complexes undergoing normal dynamic cycling with Hsp90 (Peng et al. 2004), it appears that the CHIP TPR interaction with substrate-bound Hsp70 is not affected but that Hsp90 inhibits the subsequent ubiquitination step.

While studying ligand-dependent GR movement by the Hsp90-dependent trafficking system in neurites, we found, serendipitously, a link between the trafficking and protein quality control functions of Hsp90 (Galigniana et al. 2004a). In neuronal cells treated with both dexamethasone to trigger movement and geldanamycin to block movement, the GFP-GR becomes concentrated in fluorescent puncta that are located periodically along the neurites. Both CHIP and Hsp70 also concentrate in the same loci in a steroid-dependent and geldanamycin-dependent manner, suggesting that they move to the puncta in a complex with the overexpressed GFP-GR. If geldanamycin treatment is continued, the GFP-GR undergoes proteasomal degradation, but if geldanamycin is removed shortly after puncta formation, the GFP-GR exits the puncta and continues its retrograde movement. Because CHIP and Hsp70 move in an Hsp90-dependent manner with the GR, it is clear that they are not prelocated in protein quality control centers located periodically along the neurite. But proteasomes may exist in quality control centers that the GR passes through during Hsp90-dependent trafficking.

One of the earliest examples of linkage between protein trafficking and protein quality control was established with the CFTR (cystic fibrosis transmembrane regulator). The CFTR is a chloride ion channel in the plasma membrane that in the wild-type form is inefficiently folded, with only ~ 40 % of nascent chains being able to mature beyond the ER (Ward and Kopito 1994). Nascent CFTR is in heterocomplex with Hsp90 and geldanamycin promotes its proteasomal degradation (Loo et al. 1998). The most common allele causing the autosomal recessive disease of cystic fibrosis, ΔF508, interferes with the CFTR folding such that it is effectively not functionally expressed and is eliminated by the ubiquitin-proteasome pathway (Ward et al. 1995). Inhibition of CFTR-Δ508 degradation by a proteasome inhibitor, such as lactacystin, leads to the intracellular accumulation of ubiquitinated CFTR aggregates. The undegraded, misfolded CFTR molecules accumulate at a single pericentriolar structure that is called the aggresome (Johnston et al. 1998). It is clear that aggregated CFTR particles form in the cytoplasm and move retrograde along microtubules in a dynein -dependent manner to form a single large aggresome at the microtubule organizing center. Disruption of microtubules or inhibition of dynein-mediated transport by overexpression of dynamitin inhibits aggresome formation, and the particles of misfolded CFTR remain distributed throughout the cytoplasm (Johnston et al. 1998; Garcia-Mata et al. 1999; Johnston et al. 2002). In addition to misfolded CFTR, the aggresomes contain Hsp70, Hsp40, proteasomes, ubiquitin and cytoplasmic dynein (Kopito 2000; Garcia-Mata et al. 2002).

It is generally regarded that aggresomes form when a coordinated system involving chaperones, dynein-dependent trafficking, and the ubiquitin-proteasome system of protein degradation becomes overwhelmed. Normally, this system constitutes a cytoprotective mechanism that protects cells from protein aggregate toxicity (Kopito 2000; Garcia-Mata et al. 2002). As Kopito (2000) notes, a major unresolved question is "How are aggregated proteins recognized by and attached to retrograde motors such as dynein?" There is good reason to propose that the Hsp90/immunophilin machinery of Fig. 9.1 is involved. Cells that are deficient in HDAC6, a microtubule-associated deacetylase, fail to clear CFTR-Δ508 aggregates from the cytoplasm and cannot form aggresomes properly (Kawaguchi et al. 2003). In wild-type HDAC6 cells, the CFTR-Δ508 aggresomes contain HDAC6 (Kawaguchi et al. 2003). It was subsequently shown that HDAC6 binds to Hsp90 and deacetylates it, and in HDAC6 knockdown cells where Hsp90 is acetylated, the GR is not bound to Hsp90 and does not have high affinity steroid binding activity (Kovacs et al. 2005; Murphy et al. 2005). This inactivation of both systems by acetylation of Hsp90 would be consistent with the same chaperone machinery being involved in dynein-dependent CFTR movement and GR function.

9.3 Protein Trafficking and Quality Control in SBMA (Kennedy Disease)

Among the adult onset neurodegenerative disorders are nine disorders caused by CAG/glutamine tract expansions (reviewed by Zoghbi and Orr 2000; Orr and Zoghbi 2007). These polyglutamine expansion disorders are characterized by the accumulation of aberrant proteins, and they include Huntington disease (HD), spinal and bulbar muscular atrophy (SBMA), and several autosomal-dominant spinocerebellar ataxias (e.g. SCA1, SCA3). In SBMA, an expanded glutamine tract near the amino terminus of the AR leads to hormone-dependent protein misfolding, aggregate formation in the cell cytoplasm and nucleus (Adachi et al. 2005) and the predominant loss of motor neurons in the brainstem and spinal cord of affected males (Lieberman and Fischbeck 2000). The insoluble polyQ AR aggregates form from soluble polyQ AR oligomers that may cause the pathology (Li et al. 2007), and the insoluble aggregates are essentially 'log jams' of oligomers along with components of the ubiquitin-proteasome machinery that form during trafficking along microtubules in a manner akin to the formation of aggresomes (Taylor et al. 2003).

Geldanamycin prevents formation of protein aggregates in models of Parkinson disease (PD), HD and SBMA (Auluck and Bonini 2002; Hay et al. 2004; Sittler et al. 2001; Waza et al. 2005). Because Hsp90 binding to heat shock factor 1 (HSF1) maintains it in an inactive state and treatment of cells with geldanamycin induces an HSF1-dependent stress response (Zou et al. 1998; Bagatell et al. 2000), it is often proposed that geldanamycin alleviates the phenotype and accumulation of misfolded proteins in neurodegenerative disease models by inducing a stress response (Auluck and Bonini 2002; Hay et al. 2004; Sittler et al. 2001; Muchowski and Wacker 2005). However, because geldanamycin promotes proteasomal degradation of polyQ AR aggregates in *Hsf1*$^{-/-}$ cells that cannot mount a stress response, this explanation is not valid (Thomas et al. 2006). The simple explanation is that huntingtin and the polyQ AR are client proteins of Hsp90 that undergo degradation by the ubiquitin-proteasome pathway when Hsp90 is inhibited.

Like the wild-type AR, the polyQ AR undergoes ligand-dependent trafficking from the cytoplasm to the nucleus utilizing the Hsp90-dependent movement machinery. Dynein, FKBP52 and PP5 co-localize with polyQ AR aggregates in cell culture, consistent with the entry of polyQ AR into aggregates as trafficking complexes (Thomas et al. 2006). In a knock-in mouse model a subset of intranuclear aggregates of the expanded glutamine AR in skeletal muscle contain PP5 and Hsp90. Geldanamycin and radicicol inhibit Hsp90-dependent AR movement from cytoplasm to the nucleus (Thomas et al. 2004), and they inhibit the formation of polyQ AR aggregates (Thomas et al. 2006). The Hsp90 cochaperone p23 is required for stable cycling of steroid receptors with Hsp90, and its overexpression both slows the rate of ligand-dependent AR trafficking and reduces polyQ AR aggregation (Thomas et al. 2006). Thus, it seems that Hsp90-dependent trafficking is required for insoluble aggregates to form, and the soluble oligomers are more readily eliminated by the ubiquitin-proteasome system.

As there is a clear linkage between polyQ AR trafficking and aggregate formation, inhibition of trafficking may be a productive therapeutic approach to SBMA. One approach would be to exploit the Hsp90 inhibitors that are being developed as anti-cancer drugs. For example, the geldanamycin derivative 17-AAG (17-allylamino-17-demethoxygeldanamycin) has proven effective in a SBMA transgenic mouse model (Waza et al. 2005). Another approach would be to use HDAC inhibitors. As mentioned above, CFTR-Δ508 aggresomes do not form in HDAC6 knockdown cells because of failure to load misfolded proteins onto the dynein motor for transport to aggresomes (Kawaguchi et al. 2003). As HDAC6 inhibition leads to acetylation of Hsp90 and disruption of its function, specific HDAC6 inhibitors that are being developed as anti-cancer drugs may prove useful in treatment of SBMA. Another approach is to disrupt the immunophilin connection to dynamitin. The addition of bulky substituents to FK506 that permit the drug to bind to the PPIase site on the immunophilin but block interaction with dynamitin would uncouple the trafficking machinery and aggregate formation. Both FK506 derivatives of this sort (Gestwicki et al. 2004) and bivalent FK506 compounds (Amara et al. 1997) are being developed, albeit for different purposes and with different rationales for their use. Inasmuch as FKBP52 is a positive regulator of AR signaling in cells and animal models (Yong et al. 2007), a library of compounds was screened for inhibitors of FKBP52-enhanced AR function in yeast (De Leon et al. 2011). The lead compound, MJC13, was found to prevent hormone-induced AR•Hsp90 complex dissociation when FKBP52 was present. MJC13 apparently increases the affinity of interaction between FKBP52 and the AR to inhibit receptor activation and nuclear transfer, thus inhibiting AR-dependent gene expression (De Leon et al. 2011). Although the goal was to screen for compounds that could be used to treat castrate-resistant prostate cancer, such compounds would have direct application in the treatment of SBMA.

It is clear that Hsp70/CHIP-mediated ubiquitination plays a major role in the degradation of aberrant proteins in a variety of adult onset neurodegenerative diseases. Overexpression of Hsp70 or its cochaperone Hsp40 decreases the level of aberrant proteins in cellular models of SBMA, HD and PD (Jana et al. 2000; Bailey et al. 2002; Klucken et al. 2004). Overexpression of these chaperones also ameliorates the disease phenotype in *Drosophila* and mouse models of several of these diseases, including a mouse model of SBMA (Warrick et al. 1999; Chan et al. 2000; Auluck et al. 2002; Adachi et al. 2003; reviewed in Muchowski and Wacker 2005).

Overexpression of CHIP has been shown to increase ubiquitination and degradation of many established Hsp90 client proteins, such as the GR (Connell et al. 2001), p53 (Esser et al. 2005), and ErbB-2 (Zhou et al. 2003). CHIP is found in aggregates of aberrant proteins involved in neurodegenerative diseases, such as alpha-synuclein (PD) and polyglutamine proteins, including the polyQ AR, huntingtin, ataxin-1 and ataxin-3 (Jana et al. 2005; Miller et al. 2005; Shin et al. 2005; Al-Ramahi et al. 2006; Morishima et al. 2008). Overexpression of CHIP suppresses aggregation and protein levels in cellular disease models (Jana et al. 2005; Miller et al. 2005; Shin et al. 2005; Al-Ramahi et al. 2006). Importantly, overexpression of CHIP in *Drosophila* and mouse polyglutamine disease models, including SBMA transgenic mice, suppresses toxicity (Jana et al. 2005; Adachi et al. 2007), whereas haploinsufficiency for CHIP in Huntington disease transgenic mice accelerates the disease phenotype (Al-Ramahi et al. 2006).

If one could modulate Hsp70/CHIP action to promote polyQ AR degradation, it would provide an alternative approach to SBMA treatment. That this might be possible is suggested by experiments focusing on the effect of Hip (Hsp70 interacting protein) on polyQ AR aggregation. Hip is an Hsp70-binding cochaperone that stabilizes the ADP-bound state of Hsp70 (Hohfeld et al. 1995), thus stabilizing the binding of Hsp70 to unfolded substrate. Adenoviral-mediated overexpression of Hip was shown to decrease the formation of polyQ aggregates in a cellular model of SBMA (Howarth et al. 2009). Because they did not see any effect of Hip overexpression on polyQ AR ubiquitination, the authors argued against any effect on proteasomal degradation and suggested that Hip may prevent inclusion formation by facilitating the constitutive Hsc70 refolding cycle and possibly by preventing aggregation. However, we have found that overexpression of Hip promotes ubiquitination while confirming that it reduces polyQ AR aggregates in a cellular model of SBMA (A.M. Wang, thesis work in progress).

The availability of a small molecule that functions like Hip to promote Hsp70/CHIP-dependent ubiquitination could provide a novel approach to the treatment of several neurodegenerative diseases. As part of a chemical screening program for modulators of Hsp70 action (reviewed in Patury et al. 2009), the Gestwicki laboratory has identified both inhibitors and promoters of Hsp70 action. Hsp70 inhibitors impair degradation of the polyQ AR and increase aggregate formation in a cellular model of SBMA (Wang et al. 2010). In contrast, two rhodacyanine dye analogs, MKT-077 and YM-1, act like Hip in stabilizing Hsp70 association with unfolded protein and function like Hip in a biological assay (Morishima et al. 2011). Like Hip overexpression, YM-1 promotes ubiquitination and reduces polyQ AR aggregates when it is added to cells in culture, and it also ameliorates the disease phenotype when administered to a fly model of SBMA (A.M. Wang, thesis work in progress). Thus, YM-1 may function as a lead compound for a new class of drugs that function by stimulating Hsp70/CHIP-dependent ubiquitination and degradation via the proteasome.

Acknowledgments The work of the authors described herein was supported by National Institutes of Health grants GM077430 (to Y.O.) and NS055746 (to A.P.L.) and by a McKnight Foundation Neuroscience of Brain Disorders Award (to A.P.L.).

References

Adachi H, Katsuno M, Minamiyama M et al (2003) Heat shock protein 70 chaperone overexpression ameliorates phenotypes of the spinal and bulbar muscular atrophy transgenic mouse model by reducing nuclear-localized mutant androgen receptor protein. J Neurosci 23:2203–2211

Adachi H, Katsuno M, Minamiyama M et al (2005) Widespread nuclear and cytoplasmic accumulation of mutant androgen receptor in SBMA patients. Brain 128:659–670

Adachi H, Waza M, Tokui K et al (2007) CHIP overexpression reduces mutant androgen receptor protein and ameliorates phenotypes of the spinal and bulbar muscular atrophy transgenic mouse model. J Neurosci 27:5115–5126

Al-Ramahi I, Lam YC, Chen HK et al (2006) CHIP protects from the neurotoxicity of expanded and wild-type ataxin-1 and promotes their ubiquitination and degradation. J Biol Chem 281:26714–26724

Amara JF, Clackson T, Rivera VM et al (1997) A versatile synthetic dimerizer for the regulation of protein-protein interactions. Proc Natl Acad Sci USA 94:10618–10623

Auluck PK, Bonini NM (2002) Pharmacological prevention of Parkinson disease in *Drosophila*. Nat Med 8:1185–1186

Auluck PK, Chan HY, Trojanowski JQ, Lee VM, Bonini NM (2002) Chaperone suppression of alpha-synuclein toxicity in a *Drosophila* model for Parkinson's disease. Science 295:865–868

Bagatell R, Paine-Murrieta GD, Taylor CW et al (2000) Induction of a heat shock factor 1-dependent stress response alters the cytotoxicity activity of hsp90-binding agents. Clin Cancer Res 6:3312–3318

Bailey CK, Andriola IF, Kampinga HH, Merry DE (2002) Molecular chaperones enhance the degradation of expanded polyglutamine repeat androgen receptor in a cellular model of spinal and bulbar muscular atrophy. Hum Mol Genet 11:515–523

Ballinger CA, Connell P, Wu Y et al (1999) Identification of CHIP, a novel tetratricopeptide repeat-containing protein that interacts with heat shock proteins and negatively regulates chaperone functions. Mol Cell Biol 19:4535–4545

Bender AT, Silverstein AM, Demady DR et al (1999) Neuronal nitric oxide synthase is regulated by the hsp90-based chaperone system *in vivo*. J Biol Chem 274:1472–1478

Bercovich B, Stancoviski I, Mayer A et al (1997) Ubiquitin-dependent degradation of certain protein substrates *in vitro* requires the molecular chaperone Hsp70. J Biol Chem 272:9002–9010

Brinker A, Scheufler C, von der Mulbe F et al (2002) Ligand discrimination by TPR domains. Relevance and selectivity of EEVD-recognition in Hsp70•Hop•Hsp90 complexes. J Biol Chem 277:19265–19275

Burkhardt JK, Escheverri CJ, Nilsson T, Vallee RB (1997) Overexpression of the dynamitin (p50) subunit of the dynactin complex disrupts dynein-dependent maintenance of membrane organelle distribution. J Cell Biol 139:469–484

Carrello A, Ingley E, Minchin RF, Tsai S, Ratajczak T (1999) The common tetratricopeptide repeat acceptor site for steroid receptor-associated immunophilins and hop is located in the dimerization domain of Hsp90. J Biol Chem 274:2682–2689
Carver LA, Bradfield CA (1997) Ligand-dependent interaction of the aryl hydrocarbon receptor with a novel immunophilin homolog in vivo. J Biol Chem 272:11452–11456
Chan HY, Warrick JM, Gray-Board GL, Paulson HL, Bonini NM (2000) Mechanisms of chaperone suppression of polyglutamine disease: selectivity, synergy and modulation of protein solubility in *Drosophila.* Hum Mol Genet 9:2811–2820
Chen CY, Balch WE (2006) The Hsp90 chaperone complex regulates GDI-dependent Rab recycling. Mol Biol Cell 17:3494–3507
Chen S, Sullivan WP, Toft DO, Smith DF (1998) Differential interactions of p23 and the TPR-containing proteins Hop, Cyp40, FKBP52 and FKB51 with Hsp90 mutants. Cell Stress Chaperones 3:118–129
Chen L, Hamada S, Fujiwara M et al (2010) The Hop/Sti1-Hsp90 chaperone complex facilitates the maturation and transport of a PAMP receptor in rice innate immunity. Cell Host Microbe 7:185–196
Cheung-Flynn J, Prapapanich V, Cox MB, Riggs DL, Suarez-Quian C, Smith DF (2005) Physiological role for the cochaperone FKBP52 in androgen receptor signaling. Mol Endocrinol 19:1654–1666
Citri A, Alroy I, Lavi S et al (2002) Drug-induced ubiquitylation and degradation of ErbB receptor tyrosine kinases: implications for cancer therapy. EMBO J 21:2407–2417
Clapp KM, Peng HM, Morishima Y et al (2010) C331A mutant of neuronal nitric-oxide synthase is labilized for Hsp70/CHIP (C-terminus of Hsc70-interacting protein)-dependent ubiquitination. J Biol Chem 285:33642–33651
Connell P, Ballinger CA, Jiang J et al (2001) The co-chaperone CHIP regulates protein triage decisions mediated by heat shock proteins. Nat Cell Biol 3:93–96
Crookes WJ, Olsen LJ (1998) The effects of chaperones and the influence of protein assembly on peroxisomal protein import. J Biol Chem 273:17236–17242
Crookes WJ, Olsen LJ (1999) Peroxin puzzles and folded freight. Naturwissenschaften 86:51–61
Cyr DM, Hohfeld J, Patterson C (2002) Protein quality control: U-box-containing E3 ubiquitin ligases. Trends Biochem Sci 27:368–375
Czar MJ, Owens-Grillo JK, Yem AW et al (1994) The hsp56 immunophilin component of untransformed steroid receptor complexes is localized both to microtubules in the cytoplasm and to the same nonrandom regions within the nucleus as the steroid receptor. Mol Endocrinol 8:1731–1741
Czar MJ, Lyons RH, Welsh MJ, Renoir JM, Pratt WB (1995) Evidence that the FK506-binding immunophilin heat shock protein 56 is required for trafficking of the glucocorticoid receptor from the cytoplasm to the nucleus. Mol Endocrinol 9:1549–1560
Czar MJ, Galigniana MD, Silverstein AM, Pratt WB (1997) Geldanamycin, an hsp90-binding benzoquinone ansamycin, inhibits steroid-dependent translocation of the glucocorticoid receptor from the cytoplasm to the nucleus. Biochemistry 36:7776–7785
DeFranco DB, Madan AP, Tang Y, Chandran UR, Xiao N, Yang J (1995) Nucleocytoplasmic shuttling of steroid receptors. Vitam Horm 51:315–338
DeFranco DB, Ramakrishnan C, Tang Y. (1998) Molecular chaperones and subcellular trafficking of steroid receptors. J Steroid Biochem Mol Biol 65:51–58
De Leon JT, Iwai A, Feau C et al (2011) Targeting the regulation of androgen receptor signaling by the heat shock protein 90 cochaperone FKBP52 in prostate cancer cells. Proc Natl Acad Sci USA 108 (in press)
Elbi C, Walker DA, Romero G et al (2004) Molecular chaperones function as steroid receptor nuclear mobility factors. Proc Natl Acad Sci USA 101:2876–2881
Esser C, Scheffner M, Hohfeld J (2005) The chaperone-associated ubiquitin ligase CHIP is able to target p53 for proteasomal degradation. J Biol Chem 280:27443–27448
Fan ACY, Bhangoo MK, Young JC (2006) Hsp90 functions in the targeting and outer membrane translocation steps of TOM70-mediated mitochondrial import. J Biol Chem 281:33313–33324

Fontana J, Fulton D, Chen Y et al (2002) Domain mapping studies reveal that the M domain of hsp90 serves as a molecular scaffold to regulate Akt-dependent phosphorylation of endothelial nitric oxide synthase and NO release. Circ Res 90:866–873

Galigniana MD, Scruggs JL, Herrington J et al (1998) Heat shock protein 90-dependent (geldanamycin-inhibited) movement of the glucocorticoid receptor through the cytoplasm to the nucleus requires intact cytoskeleton. Mol Endocrinol 12:1903–1913

Galigniana MD, Housley PR, DeFranco DB, Pratt WB (1999) Inhibition of glucocorticoid receptor nucleocytoplasmic shuttling by okadaic acid requires intact cytoskeleton. J Biol Chem 274:16222–16227

Galigniana MD, Radanyi C, Renoir JM, Housley PR, Pratt WB (2001) Evidence that the peptidyl-prolyl isomerase domain of the hsp90-binding immunophilin FKBP52 is involved in both dynein interaction and glucocorticoid receptor movement to the nucleus. J Biol Chem 276:14884–14889

Galigniana MD, Harrell JM, Murphy PJM et al (2002) Binding of Hsp90-associated immunophilins to cytoplasmic dynein: direct binding and in vivo evidence that the peptidylprolyl isomerase domain is a dynein interaction domain. Biochemistry 41:13602–13610

Galigniana MD, Harrell JM, Housley PR, Patterson C, Fisher SK, Pratt WB (2004a) Retrograde transport of the glucocorticoid receptor in neurites requires dynamic assembly of complexes with the protein chaperone hsp90 and is linked to the CHIP component of the machinery for proteasomal degradation. Mol Brain Res 123:27–36

Galigniana MD, Harrell JM, O'Hagen HM, Ljungman M, Pratt WB (2004b) Hsp90-binding immunophilins link p53 to dynein during p53 transport to the nucleus. J Biol Chem 279:22843–22849

Galigniana MD, Erlejman AG, Monte M, Gomez-Sanchez C, Piwien-Pilipuk G (2010) The hsp90-FKBP52 complex links the mineralocorticoid receptor to motor proteins and persists bound to the receptor in early nuclear events. Mol Cell Biol 30:1285–1298

Garcia-Cardena G, Fan R, Shah V et al (1998) Dynamic activation of endothelial nitric oxide synthase by Hsp90. Nature 392:821–824

Garcia-Mata R, Bebok Z, Sorscher EL, Sztul ES (1999) Characterization and dynamics of aggresome formation by a cytosolic GFP-chimera. J Cell Biol 146:1239–1254

Garcia-Mata R, Gao YS, Sztul E (2002) Hassles with taking out the garbage: aggravating aggresomes. Traffic 3:388–396

Georget V, Terouanne B, Nicolas JC, Sultan C (2002) Mechanism of antiandrogen action: key role of hsp90 in conformational change and transcriptional activity of the androgen receptor. Biochemistry 41:11824–11831

Gerges NZ, Tran IC, Backos DS et al (2004) Independent functions of hsp90 in neurotransmitter release and in continuous synaptic cycling of AMPA receptors. J Neurosci 24:4758–4766

Gestwicki JE, Crabtree GR, Graef IA (2004) Harnessing chaperones to generate small-molecule inhibitors of amyloid beta aggregation. Science 306:865–869

Giannakakou P, Sackett DL, Ward Y, Webster KR, Blagosklonny MV, Fojo T (2000) p53 is associated with microtubules and is transported to the nucleus by dynein. Nat Cell Biol 2:709–717

Giannoukos G, Silverstein AM, Pratt WB, Simons SS (1999) The seven amino acids (547–553) of rat glucocorticoid receptor required for steroid and Hsp90 binding contain a functionally independent LXXLL motif that is critical for steroid binding. J Biol Chem 274:36527–36536

Harrell JM, Kurek I, Breiman A et al (2002) All of the protein interactions that link steroid receptor•hsp90•immunocomplexes to cytoplasmic dynein are common to plant and animal cells. Biochemistry 41:5581–5587

Harrell JM, Murphy PJM, Morishima Y et al (2004) Evidence for glucocorticoid receptor transport on microtubules by dynein. J Biol Chem 279:54647–54654

Hay DG, Sathasivam K, Tobaben S et al (2004) Progressive decrease in chaperone levels in a mouse model of Huntington's disease and induction of stress protein as a therapeutic approach. Hum Mol Genet 13:1389–1405

Hershko A, Ciechanover A (1998) The ubiquitin system. Annu Rev Biochem 67:425–479

Hirokawa N (1998) Kinesin and dynein superfamily proteins and the mechanism of organelle transport. Science 279:519–526

Hoffman K, Handschumacher RE (1995) Cyclophilin-40: evidence for a dimeric complex with hsp90. Biochem J 307:5–8

Hohfeld J, Minami Y, Hartl FU (1995) Hip, a novel cochaperone involved in the eukaryotic Hsc70/Hsp40 reaction cycle. Cell 83:589–596

Howarth JL, Glover CPJ, Uney JB (2009) Hsp70 interacting protein prevents the accumulation of inclusions in polyglutamine disease. J. Neurochem. 108:945–951

Isaacs JS, Xu W, Neckers L (2003) Heat shock protein 90 as a molecular target for cancer therapeutics. Cancer Cell 3:213–217

Jana NR, Tanaka M, Wang G, Nukina N (2000) Polyglutamine length-dependent interaction of Hsp40 and Hsp70 family chaperones with truncated N-terminal huntingtin: their role in suppression of aggregation and cellular toxicity. Hum Mol Genet 9:2009–2018

Jana NR, Dikshit P, Goswami A et al (2005) Co-chaperone CHIP associates with expanded polyglutamine protein and promotes their degradation by proteasomes. J Biol Chem 280:11635–11640

Johnston JA, Ward CL, Kopito RR (1998) Aggresomes: a cellular response to misfolded proteins. J Cell Biol 143:1883–1898

Johnston JA, Illing ME, Kopito RR (2002) Cytoplasmic dynein/dynactin mediates the assembly of aggresomes. Cell Motil Cytoskel 53:26–38

Kawaguchi Y, Kovacs JJ, McLaurin A, Vance JM, Ito A, Yao TP (2003) The deacetylase HDAC6 regulates aggresome formation and cell viability in response to misfolded protein stress. Cell 115:727–738

Kazlauskas A, Poellinger L, Pongratz I. (2000) The immunophilin-like protein XAP2 regulates ubiquitination and subcellular localization of the dioxin receptor. J Biol Chem 275:41317–41324

Kazlauskas A, Sundstrom S, Poellinger L, Pngratz I (2001) The hsp90 chaperone complex regulates intracellular localization of the dioxin receptor. Mol Cell Biol 21:2594–2607

Kazlauskas A, Poellinger L, Pongratz I (2002) Two distinct regions of the immunophilin-like protein XAP2 regulate dioxin receptor function and interaction with Hsp90. J Biol Chem 277:11795–11801

Klein AT, Barnett P, Bottger G, Konings D, Tabak HF, Distel B (2001) Recognition of peroxisomal targeting signal type 1 by the import receptor Pex5p. J Biol Chem 276:15034–15041

Klucken J, Shin Y, Masliah E, Hyman BT, McLean PJ (2004) Hsp70 reduces alpha-synuclein aggregation and toxicity. J Biol Chem 279:25497–25502

Kopito RR (2000) Aggresomes, inclusion bodies and protein aggregation. Trends Cell Biol 10:524–530

Kovacs JJ, Murphy PJM, Gaillard S et al (2005) The deacetylase HDAC6 regulates Hsp90 acetylation and chaperone-dependent activation of the glucocorticoid receptor. Mol Cell 18:601–607

Lee DH, Sherman MY, Goldberg AL (1996) Involvement of the molecular chaperone Ydj1 in the ubiquitin-dependent degradation of short-lived and abnormal proteins in *Saccharomyces cerevisiae*. Mol Cell Biol 16:4773–4781

Li M, Chevalier-Larson ES, Merry DE, Diamond MI (2007) Soluble androgen receptor oligomers underlie pathology in a mouse model of spinobulbar muscular atrophy. J Biol Chem 282:3157–3164

Li J, Qian X, Hu J, Sha B (2009) Molecular chaperone Hsp70/Hsp90 prepares the mitochondrial outer membrane translocon receptor TOM71 for preprotein loading. J Biol Chem 284:23852–23859

Lieberman AP, Fischbeck KH (2000) Triplet repeat expansion in neuromuscular disease. Muscle Nerve 23:843–850

Lieberman AP, Pratt WB (2011) Regulation of the polyglutamine androgen receptor by the Hsp90/Hsp70-based chaperone machinery. In: S Witt (ed) Protein Chaperones and Protection from Neurodegeneration. Wiley, Hoboken, p 211–233

Loo ML, Jensen TJ, Cui L, Hou Y, Chang XB, Riordan JR (1998) Perturbation of Hsp90 interaction with nascent CFTR prevents its maturation and accelerates its degradation by the proteasome. EMBO J 17:6879–6887

Lotz GP, Brychzy A, Heinz S, Oberman WMJ (2008) A novel HSP90 chaperone complex regulates intracellular vesicle transport. J Cell Sci 121:717–723

Ma Q, Whitlock JP (1997) A novel cytoplasmic protein that interacts with the Ah receptor, contains tetratricopeptide repeat motifs, and augments transcriptional response to 2,3,7,8-tetrachlorodibenzo-*p*-dioxin. J Biol Chem 272:8878–8884

Meyer BK, Pray-Grant MG, Vanden Heuvel JP, Perdew GH (1998) Hepatitis B virus X-associated protein 2 is a subunit of the unliganded aryl hydrocarbon receptor core complex and exhibits transcriptional enhancer activity. Mol Cell Biol 18:978–988

Miller VM, Nelson RF, Gouvion CM et al (2005) CHIP suppresses polyglutamine aggregation and toxicity in vitro and in vivo. J Neurosci 25:9152–9161

Morishima Y, Wang AM, Yu Z, Pratt WB, Osawa Y, Lieberman AP (2008) CHIP deletion reveals functional redundancy of E3 ligases in promoting degradation of both signaling proteins and expanded glutamine proteins. Hum Mol Genet 17:3942–3952

Morishima Y, Lau M, Peng HM et al (2011) Heme-dependent activation of neuronal nitric-oxide synthase by cytosol is due to an Hsp70-dependent, thioredoxin-mediated thiol-disulfide interchange in the heme/substrate binding cleft. Biochemistry 50:7146–7156

Muchowski PJ, Wacker JL (2005) Modulation of neurodegeneration by molecular chaperones. Nat Rev Neurosci 6:11–22

Murphy PJM, Morishima Y, Kovacs JJ, Yao TP, Pratt WB (2005) regulation of the dynamics of hsp90 action on the glucocorticoid receptor by acetylation/deacetylation of the chaperone. J Biol Chem 280:33792–33799

Noguchi S, Jianmongkol S, Bender AT, Kamada Y, Demady DR, Osawa Y (2000) Guanabenz-mediated inactivation and enhanced proteolytic degradation of neuronal nitric oxide synthase. J Biol Chem 275:2376–2380

Orr HT, Zoghbi HY (2007) Trinuclotide repeat disorders. Annu Rev Neurosci 30:575–621

Owens-Grillo JK, Hoffman K, Hutchison KA et al (1995) The cyclosporine A-binding immunophilin Cyp-40 and the FK506-binding immunophilin hsp56 (FKBP52) bind to a common site on Hsp90 and exist in independent cytosolic heterocomplexes with the untransformed glucocorticoid receptor. J Biol Chem 270:20479–20484

Owens-Grillo JK, Czar MJ, Hutchison KA, Hoffmann K, Perdew GH, Pratt WB (1996a) A model of protein targeting mediated by immunophilins and other proteins that bind to hsp90 via tetratricopeptide repeat domains. J Biol Chem 271:13468–13475

Owens-Grillo JK, Stancato LF, Hoffman K, Pratt WB, Krishna P (1996b) Binding of immunophilins to the 90 kDa heat shock protein (hsp90) via a tetratricopeptide repeat domain is a conserved interaction in plants. Biochemistry 35:15249–15255

Patury S, Miyata Y, Gestwicki JE (2009) Pharmacological targeting of the Hsp70 chaperone. Curr Top Med Chem 9:1337–1351

Peng HM, Morishima Y, Jenkins GJ et al (2004) Ubiquitination of neuronal nitric-oxide synthase by CHIP, a chaperone-dependent E3 ligase. J Biol Chem 279:52970–52977

Peng HM, Morishima Y, Clapp KM, Lau M, Pratt WB, Osawa Y (2009) Dynamic cycling with Hsp90 stabilizes neuronal nitric oxide synthase through calmodulin-dependent inhibition of ubiquitination. Biochemistry 48:8483–8490

Perrot-Applanat M, Lescop P, Milgram E (1992) The cytoskeleton and the cellular traffic of the progesterone receptor. J Cell Biol 119:337–348

Perrot-Applanat M, Cibert C, Geraud G, Renoir JM, Baulieu EE (1995) The 59 kDa FK506-binding protein, a 90 kDa heat shock protein binding immunophilin (FKBP59-HBI), is associated with the nucleus, the cytoskeleton and mitotic apparatus. J Cell Sci 108:2037–2051

Pickart CM (2004) Back to the future with ubiquitin. Cell 116:181–190

Pratt WB (1993) The role of heat shock proteins in regulating the function, folding, and trafficking of the glucocorticoid receptor. J Biol Chem 268:21455–21458

Pratt WB, Toft DO (1997) Steroid receptor interactions with heat shock protein and immunophilin chaperones. Endocrine Rev 18:306–360

Pratt WB, Toft DO (2003) Regulation of signaling protein function and trafficking by the hsp90/hsp70-based chaperone machinery. Exp Biol Med 228:111–133

Pratt WB, Silverstein AM, Galigniana MD (1999) A model for the cytoplasmic trafficking of signalling proteins involving the hsp90-binding immunophilins and p50^{cdc37}. Cell Signal 11:839–851

Pratt WB, Galigniana MD, Harrell JM, DeFranco DB (2004) Role of Hsp90 and the Hsp90-binding immunophilins in signalling protein movement. Cell Signal 16:857–872

Pratt WB, Morishima Y, Osawa Y (2008) The Hsp90 chaperone machinery regulates signaling by modulating ligand binding clefts. J Biol Chem 283:22885–22889

Pratt WB, Morishima Y, Peng HM, Osawa Y (2010) Proposal for a role of the Hsp90/Hsp70-based chaperone machinery in making triage decisions when proteins undergo oxidative and toxic damage. Exp Biol Med 235:278–289

Qbadou S, Becker T, Mirus O, Teus I, Soll J, Schleiff E (2006) The molecular chaperone Hsp90 delivers precursor proteins to the chloroplast import receptor Toc64. EMBO J 25:1836–1847

Radanyi C, Chambraud B, Baulieu EE (1994) The ability of the immunophilin FKBP59-HBI to interact with the 90-kDa heat shock protein is encoded by its tetratricopeptide repeat domain. Proc Natl Acad Sci USA 91:11197–11201

Ratajczak T, Carrello A (1996) Cyclophilin 40 (CyP-40), mapping of it hsp90 binding domain and evidence that FKBP52 competes with CyP-40 for hsp90 binding. J Biol Chem 271:2961–2965

Ratajczak T, Ward BK, Minchin RF (2003) Immunophilin chaperones in steroid receptor signalling. Curr Top Med Chem 3:1348–1357

Reddy RK, Kurek I, Silverstein AM, Chinkers M, Breiman A, Krishna P (1998) High molecular-weight FK506-binding proteins are components of heat-shock protein 90 heterocomplexes in wheat germ lysates. Plant Physiol 118:1395–1401

Renoir JM, Mercier-Bodard C, Hoffman K et al (1995) Cyclosporin A potentiates the dexamethasone-induced mouse mammary tumor virus-chloramphenicol acetyltransferase activity in LMCAT cells: a possible role for different heat shock protein-binding immunophilins in glucocorticoid receptor-mediated gene expression. Proc Natl Acad Sci USA 92:4977–4981

Riggs DL, Roberts PJ, Chirillo SC et al (2003) The Hsp90-binding peptidylprolyl isomerase FKBP52 potentiates glucocorticoid signaling *in vivo*. EMBO J 22:1158–1167

Riggs DL, Cox MB, Tardif HL, Hessling M, Buchner J, Smith DF (2007) Noncatalytic role of the FKBP52 peptidyl-prolyl isomerase domain in the regulation of steroid hormone signaling. Mol Cell Biol 27:8658–8669

Ruff VA, Yem AW, Munns PL et al (1992) Tissue distribution and cellular localization of hsp56, an FK506-binding protein. Characterization using a highly specific polyclonal antibody. J Biol Chem 267:21285–21288

Russell LC, Whitt SR, Chen MS, Chinkers M (1999) Identification of conserved residues required for the binding of a tetratricopeptide repeat domain to heat shock protein 90. J Biol Chem 274:20060–20063

Sakisaka T, Meerlo T, Matteson J, Plutner H, Balch WE (2002) Rab-α GDI activity is regulated by a Hsp90 chaperone complex. EMBO J 21:6125–6135

Sanchez ER, Redmond T, Scherrer LC, Bresnick EH, Welsh MJ, Pratt WB (1988) Evidence that the 90-kilodalton heat shock protein is associated with tubulin-containing complexes in L cell cytosol and in intact PtK cells. Mol Endocrinol 2:756–760

Sanchez ER, Hirst M, Scherrer LC et al (1990) Hormone-free glucocorticoid receptors overexpressed in Chinese hamster ovary cells are localized to the nucleus and are associated with both hsp70 and hsp90. J Biol Chem 265:20123–20130

Savory JGA, Hsu B, Laquian IR et al (1999) Discrimination between NL1- and NL2-mediated nuclear localization of the glucocorticoid receptor. Mol Cell Biol 19:1025–1037

Scherrer LC, Hutchison KA, Sanchez ER, Randall SK, Pratt WB (1992) A heat shock protein complex isolated from rabbit reticulocyte lysate can reconstitute a functional glucocorticoid receptor-Hsp90 complex. Biochemistry 31:7325–7329

Sepp-Lorenzino L, Ma Z, Lebwhol DE, Vinitsky A, Rosen N. (1995) Herbimycin A induces 20 S proteasome- and ubiquitin-dependent degradation of receptor tyrosine kinases. J Biol Chem 270:16580–16587

Sheffield WP, Shore GC, Randall SK (1990) Mitochondrial precursor protein. Effects of 70-kilodalton heat shock protein on polypeptide folding, aggregation and import competence. J Biol Chem 265:11069–11076

Scheufler C, Brinker A, Bourenkov G et al (2000) Structure of TPR domain-peptide complexes: critical elements in the assembly of the Hsp70-Hsp90 multichaperone machine. Cell 101:199–210

Shin Y, Klucken J, Patterson C, Hyman BT, McLean PJ (2005) The co-chaperone carboxyl terminus of Hsp70-interacting protein (CHIP) mediates alpha-synuclein degradation decisions between proteasomal and lysosomal pathways. J Biol Chem 280:23727–23734

Sikorski RS, Boguski MS, Goebl M, Hieter P (1990) A repeating amino acid motif in CDC23 defines a family of proteins and a new relationship among genes required for mitosis and RNA synthesis. Cell 60:307–317

Silverstein AM, Galigniana MD, Chen MS, Owens-Grillo JK, Chinkers M, Pratt WB (1997) Protein phosphatase 5 is a major component of glucocorticoid receptor•hsp90 complexes with properties of an FK506 binding immunophilin. J Biol Chem 272:16224–16230

Silverstein AM, Galigniana MD, Kanelakis KC, Radanyi C, Renoir JM, Pratt WB (1999) Different regions of the immunophilin FKBP52 determine its association with the glucocorticoid receptor, hsp90 and cytoplasmic dynein. J Biol Chem 274:36980–36986

Sittler A, Lurz R, Lueder G et al (2001) Geldanamycin activates a heat shock response and inhibits huntington aggregation in a cell culture model of Huntington's disease. Hum Mol Genet 10:1307–1315

Song Y, Zweir JL, Xia Y (2001) Heat shock protein 90 augments neuronal nitric oxide synthase by enhancing Ca^{2+}/calmodulin binding. Biochem. J. 355:357–360

Stancato LF, Chow YH, Hutchison KA, Perdew GH, Jove R, Pratt WB (1993) Raf exists in a native heterocomplex with hsp90 and p50 that can be reconstituted in a cell-free system. J Biol Chem 268:21711–21716

Sumanasekera WK, Tien ES, Tupey R, Vanden Heuvel JP, Perdew GH (2003) Evidence that peroxisome proliferator-activated receptor alpha is complexed with the 90-kDa heat shock protein and the hepatitis virus B X-associated protein 2. J Biol Chem 278:4467–4473

Szapary D, Barber T, Dwyer NK, Blanchette-Mackie EJ, Simons SS (1994) Microtubules are not required for glucocorticoid receptor mediated gene induction. J Steroid Biochem Mol Biol 51:143–148

Taylor JP, Tanaka F, Robitschek J et al (2003) Aggresomes protect cells by enhancing the degradation of toxic poly-glutamine-containing protein. Hum Mol Genet 12:749–757

Thomas M, Dadgar N, Aphale A et al (2004) Androgen receptor acetylation site mutations cause trafficking defects, misfolding, and aggregation similar to expanded glutamine tracts. J Biol Chem 279:8389–8395

Thomas M, Harrell JM, Morishima Y, Peng HM, Pratt WB, Lieberman AP (2006) Pharmacologic and genetic inhibition of Hsp90-dependent trafficking reduces aggregation and promotes degradation of the expanded glutamine androgen receptor without stress protein induction. Hum Mol Genet 15:1876–1883

Wang AM, Morishima Y, Clapp KM et al (2010) Inhibition of Hsp70 by methylene blue affects signaling protein function and ubiquitination and modulates polyglutamine protein degradation. J Biol Chem 285:15714–15723

Ward CL, Kopito RR (1994) Intracellular turnover of cystic fibrosis transmembrane conductance regulator. Inefficient processing and rapid degradation of wild-type and mutant proteins. J Biol Chem 269:25710–25718

Ward CL, Omura S, Kopito RR (1995) Degradation of CFTR by the ubiquitin-proteasome pathway. Cell 83:121–127

Warrick JM, Chan HY, Gray-Board GL, Chai Y, Paulson HL, Bonini NM (1999) Suppression of a polyglutamine-mediated neurodegeneration in *Drosophila* by the molecular chaperone HSP70. Nat Genet 23:425–428

Waza M, Adachi H, Katsuno M et al (2005) 17-AAG, an Hsp90 inhibitor, ameliorates polyglutamine-mediated motor neuron degeneration. Nat Med 11:1088–1095

Whitesell L, Mimnaugh EG, De Costa B, Myers CE, Neckers LM (1994) Inhibition of heat shock protein HSP90-pp60v-src heteroprotein complex formation by benzoquinone ansamycins: essential role for stress proteins in oncogenic transformation. Proc Natl Acad Sci USA 91:8324–8328

Wochnik GM, Ruegg J, Abel GA, Schmidt U, Holsboer F, Rein T (2005) FK506-binding proteins 51 and 52 differentially regulate dynein interaction and nuclear translocation of the glucocorticoid receptor in mammalian cells. J Biol Chem 280:4609–4616

Xu M, Dittmar KD, Giannoukos G, Pratt WB, Simons SS (1998) Binding of hsp90 to the glucocorticoid receptor requires a specific 7-amino acid sequence at the amino terminus of the hormone binding domain. J Biol Chem 273:13918–139294

Xu W, Mimnaugh EG, Rosser MFN et al (2001) Sensitivity of mature ErbB2 to geldanamycin is conferred by its kinase domain and is mediated by the chaperone protein Hsp90. J Biol Chem 276:3702–3708

Xu W, Mimnaugh EG, Kim JS, Trepel JB, Neckers LM (2002) Hsp90, not Grp94, regulates the intracellular trafficking and stability of nascent ErbB2. Cell Stress Chaperones 7:91–96

Xu H, Shi Y, Wang J et al (2007) A heat shock protein 90 binding domain in endothelial nitric oxide synthase influences enzyme function. J Biol Chem 282:37567–37574

Yong W, Yang Z, Periyasamy S et al (2007) Essential role for co-chaperone FKBP52 but not FKBP51 in androgen receptor-mediated signaling and physiology. J Biol Chem 282:5026–5036

Young JC, Obermann WM, Hartl FU (1998) Specific binding of tetratricopeptide repeat domains to the C-terminal 12-kDa domain of hsp90. J Biol Chem 273:18007–18010

Young JC, Hoogenraad NJ, Hartl FU (2003) Molecular chaperones Hsp90 and Hsp70 deliver preproteins to the mitochondrial import receptor Tom70. Cell 112:41–50

Zhao W, Zhong L, Wu J et al (2006) Role of cellular FKBP52 protein in intracellular trafficking of recombinant adeno-associated virus 2 vectors. Virology 353:283–293

Zhou P, Fernandes N, Dodge IL et al (2003) ErbB2 degradation mediated by the co-chaperone protein CHIP. J Biol Chem 278:13829–13837

Zoghbi HY, Orr HT (2000) Glutamine repeats and neurodegeneration. Annu Rev Neurosci 23:217–247

Zou J, Guo Y, Guettouche T, Smith DF, Voellmy R (1998) Repression of heat shock transcription factor HSF1 activation by HSP90 (HSP90 complex) that forms a stress-sensitive complex with HSF1. Cell 94:471–480

Chapter 10
Cell Surface Molecular Chaperones and the LPS Receptor

Martha Triantafilou and Kathy Triantafilou

Abstract Heat shock proteins are ancient, highly conserved, molecular chaperones essential for maintaining cellular functions. Although originally thought to be only intracellular, it has become apparent that they exist on the surfaces of cells and in the extracellular space and are able to modulate both innate and adaptive immune responses. They have been shown to have the ability to trigger inflammatory responses similar to pathogen-associated molecular patterns (PAMPs), to act as "danger" signals" for the innate immune response (DAMPs) and to inhibit or resolve inflammation (so-called—resolution-associated molecular patterns—RAMPs). Therefore their role in innate immunity seems to be multifunctional. In this chapter, we will focus on how these proteins are involved in innate immune responses and how they interact with pattern recognition receptors in the innate immune recognition of microbial pathogens.

10.1 Introduction

The use of heat shock as a treatment has been observed in many cultures with references in Greek and Indian medical texts dating back to 500 BC. Saunas and hot springs are used by other cultures as places of healing.

Heat Shock Proteins or Hsps were discovered in 1962 by Ferruccio M Ritossa as a result of the accidental heat shocking of *Drosophila melongaster* salivary cells causing chromosomal puffs to appear at discrete loci, indicating new gene transcription was occurring (Ritossa 1962). This knowledge has opened up many new avenues to further understand innate immunity and the initiation of acquired immunity, and how cells respond to a variety of stressors.

The role of heat shock proteins (hsps) is more than just the cellular response to heat. For example, up regulation of hsps has been reported in response to glucose

K. Triantafilou (✉) · M. Triantafilou
Cardiff University, School of Medicine, University Hospital of Wales,
Heath Park, Cardiff CF14 4XN, UK
e-mail: TriantafilouK@cardiff.ac.uk

B. Henderson, A. G. Pockley (eds.), *Cellular Trafficking of Cell Stress Proteins in Health and Disease*, Heat Shock Proteins 6, DOI 10.1007/978-94-007-4740-1_10,

deprivation and the presence of heavy metals and other toxic chemicals (Srivastava et al. 1998a), bacterial and viral infections and trauma both physiological and psychological (Hartl and Hayer-Hartl 2002). Recent research, however has suggested that hsps are involved in modulating both the innate and acquired immune response in response to bacterial, fungal, viral and even cancer challenges.

The primary role of hsps within the cell during the normal cell cycle is thought to be to ensure the correct folding of newly formed polypeptides into their native state, to prevent the formation of protein aggregates which can lead, for example to amyloid diseases in neurons (Hartl and Hayer-Hartl 2002). Their role was thought to be to triage proteins that are in danger of unfolding by ensuring either that they refold correctly or that they are broken down in the proteosome. As a result of this crucial function they are amongst the proteins most widely expressed in the cell and most highly conserved between species (Hartl et al. 2011).

In the last couple of decades it has become increasingly evident that the role of hsps is not only intracellular. Hsps are found on the cell surface and in the extracellular space, modulating innate and acquired immune responses. These hsps are more than "chaperones" of misfolded proteins, but rather they are "sensors" of damage and danger for the host, an ancient "immune surveillance" system. In this chapter we will look at the involvement of hsps in the innate immune response and how they are involved in bacterial recognition and discuss their increasingly understood, but unconventional, roles on the cell surface as well as their role as modulators of the innate immune response.

10.2 The Intracellular Roles of Hsps

The intracellular fluid is a densely packed environment with a high concentration of ions, molecules, membranes and proteins, both soluble and structural. It is this crowded environment that seems to promote protein misfolding and aggregation. It was the groups of Laskey, Ellis and Georgopoulos that demonstrated this critical relationship between correctly assembled macromolecules and hsps (Ellis 1987, 2007; Ellis et al. 1989; Hemmingsen et al. 1978; Laskey et al. 1978). Hsps counter this problem by functioning as intracellular chaperones that bind to the hydrophobic residues, which are essential for the structure of proteins but can become exposed as a result of misfolding and stress, and assist the proteins in reaching their native state (Hemmingsen et al. 1978). The use of proteins as molecular chaperones was identified by Laskey and co-workers (Laskey et al. 1978) with the first protein-folding chaperone to be identified being Cpn60 (Hemminsen et al. 1978). Thus the conventional idea arose that molecular chaperones (hsps) act as molecular scaffolds, shoring up at risk proteins, holding new ones together while they are correctly organised and transporting these within the intracellular environment. Initially it was believed that this only occurred within the cell, but then it was discovered that hsps are also found in the extracellular environment (see Chap. 1). This raises the question as to how these proteins leave the cell and what their functions are in the extracellular environment.

10.3 The Extracellular Roles of Hsps

The role of hsps in the intracellular environment now seems to be an essential part of the housekeeping actions of the cell, but research has shown that this role is not limited to the intracellular compartment, with an increasing number of chaperones being shown to exist in the extracellular environment, and that their role outside of the cell is somewhat different to that inside. One of the earliest extracellular hsps detected was thioredoxin, which was found to be secreted from T-cells in patients with a specific type of leukaemia (Tagaya et al. 1994). This suggested that these secreted proteins acted as intercellular signalling proteins for leukocyte cells and regulated their cellular functions.

Since then, several heat shock proteins including Hsp10, Hsp60, Hsp70, Bip and Hsp90 have been identified in the a range of body fluids (see Chaps. 3 and 18 for more details). The role of these extracellular chaperones seems to be varied and occasionally antagonistic. The earliest hsp identified outside of the cell was Hsp10 which was originally given the name of early pregnancy factor (EPF) as it was primarily seen during the first trimester of pregnant women (Morton et al. 1977; Noonan et al. 1979). Hsp10 was shown to have an immunosuppressive function outside of the cell with soluble Hsp10 being able to down regulate the release of cytokines including TNF-α, IL-1β, IL-6 and IL-10 in order to prevent rejection of the foetus (Johnson et al. 2005). The co-chaperone of Hsp10, Hsp60 has also been shown to circulate in the blood stream. It is seen to be elevated in patients with cardiovascular disease, atherosclerosis and the early stages of vascular injury and this has been reviewed in Chap. 3 (see Henderson and Pockley 2012). In contrast, elevated levels of Hsp70 and Bip in the blood stream seem to show a link to a lower risk of cardiovascular disease and atherosclerosis. The potential mechanism indicated for this is that it is acting in an anti-inflammatory and cytoprotective manner (Henderson and Pockley 2012). The circulating serum levels of Hsp70 are lowered in pregnant women compared to non-pregnant individuals, however in pre-eclamptic and eclamptic patients their serum levels are increased in proportion to the severity of haemolytic and hepatocellular tissue damage (Molvarec et al. 2009). Hsp70 has also been observed to be released into the cerebro-spinal fluid in response to ischemic reperfusion as a result of trauma to the head and neck and the subsequent corrective surgery (Hecker and McGarvey 2011). Hsp70 is also released in a similar manner into the blood stream as a result of coronary artery bypass grafting, it is indicated that this triggers a subsequent high level of the pro-inflammatory cytokine IL-6 to be released (Szerafin et al. 2008). Hsp10 is also known to have contradicting levels in response to disease; research has recently shown that circulating levels within the blood stream are decreased in response to the increasing severity of periodontal disease (Shamaei-Tousi et al. 2007). This contradictory evidence seems to indicate that the role of extracellular chaperones is a complex one with the same protein showing both pro- and anti- inflammatory functions. This suggests that they are able to control the immune system in a context sensitive manner.

10.4 Cell Surface Expression of Hsps

Research has shown that some hsps are localised to the cell surface in response to cellular stress (Gupta et al. 2008). Heat shock proteins have also been detected on the surface of cells that have had their intracellular environment altered in someway, in particular they have been seen on the surface of tumour cells (Multhoff et al. 1997) on cells infected with respiratory syncytial virus (Anderson et al. 1992) and Dengue virus (Reys-Del Valle et al. 2005) and apoptotic cells (Sapozhnikov et al. 1999). Hsps that are associated with the membrane have been shown to be an important part of the immune response with their presence triggering cytokine release and immune receptor clustering (Asea et al. 2000; Singh-Jasuja et al. 2000) with a particular relevance to the recognition of LPS (Triantafilou et al. 2001, 2002; Triantafilou and Triantafilou 2004). The mechanism by which hsps are released outside of the cell remains elusive, but evidence suggests that hsps are secreted by the cell before they associate with the plasma membrane (Calderwood et al. 2007).

The modulation of the immune system indicates that the hsp family in response to trauma localise on the cell surface in lipid rafts and that one of the release mechanisms proposed for hsp release is lipid raft dependent (Gangalum et al. 2011).

10.5 Hsps and Antigen Presentation

Hsps are evolutionarily ancient and highly conserved. They are defined as molecular chaperones that non-covalently bind exposed hydrophobic surfaces of non-native proteins (Hartl and Hayer-Hartl 2002; Hartl et al. 2011). The ability of gp96 and hsp70 to bind peptides was first postulated, and then demonstrated, by Srivastava and coworkers (Srivastava 2002; Srivastava et al. 1998b). Srivastava et al reported that immunisation of mice with tumour-derived gp96 induced an anti-tumour immune response sufficient for tumour rejection (Srivastava et al. 1986). The specificity of this response was based on the binding of endogenous antigenic peptides by hsps and transporting them to the MHC class I molecules present on the cell surface triggering $CD8^+$ T-cell activation (Li et al. 2002; Ishii et al. 1999). Since then, it has been shown that Hsp70, Hsc70, Hsp90, Gp96, Grp110 and Grp170 are all able to bind and present peptides. It has been shown that hsp-associated peptides are presented by MHC molecules after receptor-mediated endocytosis by antigen-presenting cells (APCs) (Arnold et al. 1995). The receptor responsible for this endocytosis was shown to be CD91 (Basu et al. 2001).

In addition to binding and presenting endogenous antigenic peptides, hsps have also been shown to bind and present exogenous peptides. Hsps released extracellularly or hsp-peptide complexes that might be administered as a vaccine are taken up by antigen-presenting cells via receptor-mediated endocytosis (Gullo and Teoh 2004). These hsp-peptide complexes are then presented to T-cells via MHC class I. The capacity of hsps to deliver exogenous antigenic peptides into the MHC class I presentation pathway is now recognised as a process called cross-presentation

(Murshid et al. 2008). This presentation seems be orders of magnitude more sensitive in sensitizing CD8 T-cells than peptide alone. This has a tremendous application in generating hsp-peptide complexes *in vitro* and utilising them therapeutically for the treatment of different cancers (Chandawarkar et al. 1999; Tamura et al. 1997; Udono et al. 1994). The question that remains is do hsps only modulate acquired immune responses, or can they also modulate the innate immune response?

10.6 Hsps Involvement in the Innate Immune Response

In the 20th century immunologists focused much of their attention in deciphering the mechanisms of adaptive immunity. As a consequence, tremendous progress was made in the field of adaptive immunity, including tolerance, mechanisms of MHC restriction, the structure and function of MHC receptors and development and activation of B and T-cells.

In contrast, innate immunity was sidelined. It was viewed as the most archaic of the two branches of immunity, being simple and unsophisticated. It was seen as an undifferentiated system, just involved in engulfing and digesting invaders.

Charles Janeway changed our view of the innate immune system, by predicting almost twenty years ago that there would be molecules that were encoded in the germ line which would recognize the presence of molecules produced by broad classes of pathogens. He called these molecules pattern recognition receptors (PRRs) and the ligands that they recognise, pathogen-associated molecular pattern (PAMPs).

Janeway's view was justified in the late 1990s with the discovery of Toll-like receptors (TLRs). The discovery of TLRs demonstrated that the innate immune system is actually highly specific, relying on germline-encoded pattern-recognition receptors (PRRs) that have evolved to detect components of foreign pathogens. Since the discovery of TLRs, an onslaught of research has shown that the innate immune system is highly specific in recognising microbial signatures. It seems to be equally specific and elaborate as adaptive immunity, as e.g. TLR4 was found to recognise bacterial lipopolysaccharide (LPS) (Poltorak et al. 1998; Qureshi et al. 1999); TLR2 was found to recognise lipoteichoic acid (LTA) and peptidoglycan (Takeuchi et al. 1999), TLR3 was able to sense double stranded viral RNA (Alexopoulou et al. 2001), TLR5 was found to recognise bacterial flagellin (Hayashi et al. 2001), TLR7 (Lund et al. 2004) and TLR8 (Heil et al. 2004) to sense single stranded viral RNA, whereas TLR9 was found to delicate distinguish between methylated DNA from host DNA and unmethylated DNA from microorganisms. Subsequently, TLRs have been identified as operational centers for both innate and adaptive immunity.

Similarly to TLRs and other PRRs of the innate immune system, hsps are also evolutionary ancient and highly conserved molecules, thus it is highly likely that they are also part of this ancient "microbial sensing" apparatus. The question that remains is whether they are themselves PRRs, or do they act in conjunction with PRRs or are they modulators/regulators of this whole archaic sensing system?

10.7 Hsps and the LPS Receptor

Over the years, several studies have linked the ancient surveillance system of PRRs with hsps. Hsps have been shown to interact with TLRs both on the cell surface as well as intracellularly (Asea et al. 2000, 2002; Triantafilou et al. 2001; Vabulas et al. 2001, 2002a, b). It is rather perplexing that they have been shown to both trigger the innate immune response as well as dampen it.

Interestingly, in the case of bacterial lipopolysaccharide (LPS) recognition by the innate immune system, hsps seem to be part of the "sensing apparatus". LPS constitutes one of the most potent PAMPs of the innate immune response. Recognition of LPS triggers the release of pro-inflammatory cytokines and can lead to sepsis and septic shock. Understanding how LPS is recognised has been pivotal in the discovery of TLRs. The work of Bruce Beutler, which demonstrated that TLR4 is the main LPS receptor, triggered the onslaught of research that revealed the existence of other PRRs and also won him the Nobel Prize.

Although TLR4 has been shown to be essential for LPS innate immune recognition, it has been shown that accessory molecules are also required. In addition to TLR4, CD14 (Wright et al. 1990) and MD2 (Shimazu et al. 1999) are also involved. CD14 has been shown to bind LPS and transfer it to TLR4, which transduces the signal. MD-2, a secreted glycoprotein that interacts with TLR4, enhancing its sensitivity to the ligand. Without this molecule TLR4 is still capable of binding LPS, but the sensitivity to this ligand is extremely low (Re and Strominger 2003; Visintin et al. 2003). Furthermore, we have previously shown that there are other molecules involved in the LPS "sensing apparatus". Interestingly, these molecules include heat shock proteins Hsp70 and Hsp90 (Triantafilou et al. 2001, 2002), CXCR4 (Triantafilou et al. 2008) and CD55 (Heine et al. 2001). They have all been identified as being linked to this activation cluster possibly either in an LPS presentation or modulatory capacity. It has been suggested that Hsp70 and Hsp90 are essential to this cluster because neutralising antibodies against these two molecules showed a dramatic attenuation of the pro-inflammatory cytokine release in response to LPS (Triantafilou et al. 2001). Therefore the question still remains, what is the role of hsps within this LPS "sensing-apparatus"? Are they part of the LPS receptor or do they simply act as transfer molecules?

10.7.1 Hsp-LPS Interactions

Hsps contain a peptide-binding domain that binds exposed hydrophobic residues of substrate proteins. The structural features of the mammalian hsp70 binding domain and its bacterial homologue DnaK have been resolved and demonstrate that the binding pocket consists of four-stranded anti-parallel Beta-sheets and a single α helix (Zhu et al. 2003). This conformation is different from the MHC binding domain making unlikely that they bind similar peptides.

On the contrary, since it seems that the peptide-binding domain of the hsp binds hydrophobic residues, LPS, which has a lipid part and thus highly hydrophobic will have a very strong affinity for the hsp. This has been demonstrated for hsp60 and gp96, where they have been found to bind LPS tightly and enhance the immunostimulatory effects of LPS. Therefore it has been suggested that hsps might contribute to the *in vivo* recognition of Gram-negative bacteria by binding LPS and modulating the host response via TLR4 (Habich et al. 2005; Osterloh et al. 2007; Osterloh and Breloer 2008; Warger et al. 2006).

The role of hsps in LPS recognition might be similar to the role of lipopolysacchride-binding protein (LBP), which is an acute phase reactant that complexes with LPS and transports LPS to its cellular targets (Gallay et al. 1994). A possible explanation might be that hsps in the extracellular space might interact with LPS in the circulation, in the early phases of the infection. The hsp-LPS complexes are then delivered to the host's LPS "sensing-apparatus" on the cells, where hsps then deliver LPS to its cellular targets triggering the activation of the innate immune response. The inflammatory response causes damage to the surrounding area releasing more hsps in the extracellular space and thus "fueling" the LPS response, leading to sepsis and septic shock. Inhibiting the hsps with function-blocking antibodies might also inhibit LPS and thus the inflammatory response. Thus in this scenario, hsps might be acting as PRRs and enhance the innate immune responses. The question though that still remains is whether they act as modulators of the innate immune response or are they simply transfer molecules as described above?

10.8 PAMPs, DAMPs or RAMPs?

The innate immune response is our first line of defence against invading pathogens. Once believed to be a simple, unsophisticated system that ingested pathogens, it is now being shown to be highly specific and able to recognise microbial signatures, called PAMPs. PAMPs are evolutionary conserved molecules which are present on the large classes of microbes but not the host, making it easy for the host to distinct them from self-proteins. In addition to PAMPs, the innate immune system has been shown to be also triggered by the presence of damage-associated molecular patterns (DAMPs). DAMPs have been defined as endogenous molecules released during cellular stress triggering sterile inflammation and they are the "corner-stone" of the "danger theory". The "danger theory" proposes that the immune system responds to immunological danger (Matzinger 1994) and is unconcerned with the discrimination of self-non-self. Several studies suggest that hsps are the major protein species released in the extracellular space when cells die by necrosis but not by apoptosis (basu et al. 2000; Berwin et al. 2001). According to this view, hsps constitute "danger" signals and thus we can class them as DAMPs. Therefore it seems that PAMPs as well as DAMPs are able to trigger innate immune responses.

Hsp preparations from hysp60, hsp70, hsp90 and gp96 from various sources have been reported to be potent activators of the innate immune system. They seem to be recognised as PAMPs and trigger pro-inflammatory cytokine production and stimulate the production of proinflammatory cytokines such as TNF-alfa, IL-1, IL-6 and IL-12 (Asea et al. 2000; Basu et al. 2000; Singh-Jasuja et al. 2000). As already mentioned, the above hsp cytokine effects are reported to be mediated by TLR2 and TLR4 (Asea et al. 2000; Bulut et al. 2002; Ohashi et al. 2000; Vabulas et al. 2002a, b).

The induction of pro-inflammatory cytokine production by hsps may contribute to the pathogenesis of autoimmune and chronic inflammatory diseases (Pockley et al. 2000). For example, chlamydial hsp60 (Kol et al. 1998, 1999), as well as human hsp60 (Chen et al. 1999). In particular, chlamydial hsp60 has been linked with chronic inflammation in atherosclerosis (Kol et al. 1999). The mechanism of hsp60-induced activation has been shown to involve CD14 (Kol et al. 2000) as well as TLR4 and TLR2 (Ohashi et al. 2000). Similar features were observed for Gp96. Gp96 was able to stimulate DCs to release pro-inflammatory mediators (Singh-Jasuja et al. 2000). This activation seems to be TLR4, MD2 and also TLR2-dependent (Vabulas et al. 2002a, b).

These cytokine-like functions of hsps have convinced many that they might be endogenous ligands for TLRs, activating as DAMPs. A lot of questions though have been raised concerning these cytokine-like functions of hsps and some have attributed them to contamination with PAMPs. Using highly purified hsp preparations, Gao and Tsan claimed that the hsp cytokine functions previously reported were in fact a result of contaminating bacterial products (Gao and Tsan 2003). However, they also reported that their LPS preparations were heat sensitive, which is not reported by others. Thus the question remains whether indeed hsps are PAMPs themselves triggering an inflammatory response or whether the inflammatory response triggered by hsps is attributed to PAMPs that they have bound. We feel that although TLR2 and TLR4 are PRRs, they are capable of discriminating between LPS and hsps and thus some hsps might have the ability to trigger an innate immune response independently of LPS and bacterial contaminants. This is the conclusion of a recent review written by some of the leading proponents of hsps as cell signalling agonists (Henderson et al. 2010).

In the case of hsp70 and hsp90, they do not seem to be PAMPs themselves, but they do seem to contribute to the pro-inflammatory response since inhibition of these chaperones leads to downregulation of the LPS-induced response (Triantafilou et al. 2001, 2002). In this context, they seem to act as TLR accessory molecules, binding the bacterial ligand and possibly acting as LPS-transfer molecules. This action is likely to result in a more efficient anti-bacterial response.

Some hsps, such as hsp10 (Johnson et al. 2005, Vanags et al. 2006), alpha Beta-crystallin (Ousman et al. 2007), GRP78 (Corrigall et al. 2001), and hsp27 (Liu et al. 2010; Park et al. 2003) seem to be able to inhibit the inflammatory responses and modulate immunological networks when released from cells. Thus it has been proposed that these hsps are defined as resolution-associated molecular patterns (RAMPs) (Shields et al. 2011) because they are able to counterbalance the activity of PAMPs and DAMPs and help resolve the inflammation.

Thus it seems that hsps are capable of resolving inflammation. Their delivery at sites of inflammation can overcome local inflammation and balance the dyregulated inflammatory environments, providing prolonged periods of homeostasis.

10.9 An Ancient "off- and on-switch" of Inflammation?

The study of RAMPs is in its infancy, but RAMP research so far has demonstrated that when added to tissues or animal models of disease, RAMPs are able to resolve inflammation. We need to explore in detail the biology of RAMPs and explore the possibility of utilising them as therapeutic interventions for chronic inflammatory conditions.

Their capacity to control invasion by pathogens and neoplastic cells demonstrates that they are part of an immune surveillance mechanism that has been maintained through evolution. Their role might be to act as PAMPs, DAMPs and RAMPs for the innate immune system. A primitive off- and on-switch of inflammation during infection/stress.

It is possible that hsps are evolutionary ancient and highly conserved part of the innate immune system and their role is to act as regulators of the innate immune response. They can be expressed on the cell surface and have binding domains that can bind non-covalently to exposed hydrophobic surfaces, such as the surfaces of LPS and lipoproteins. In this respect they could be acting as an ancient antigen presenting system that is independent of the MHC molecules. The role of this primitive antigen presentation system can be to present PAMPs (microbial signatures) to T cells. Therefore, they might be the missing link between innate and adaptive immunity and help bridge the two systems. Their role as PAMPs might be early on in an infection, where they would bind abundant microbial products and aid in the raising of the "alarm".

In the next phase of the infection, they might contribute as DAMPs, since they would be released in the extracellular space via damage caused by the infectious agent. There would be increased DAMP release from stressed cells in the acute phase if inflammation.

Eventually in the final stages of inflammation, RAMP release could be acting as an "off-switch", switching the whole inflammation off and dampening the response via mechanisms that might direct antagonise inflammatory stimuli. Thus throughout the years, certain hsps might have been specialised to act as PAMPs, some as DAMPs and some as RAMPs.

10.10 Concluding Remarks

The involvement of hsps in pathogen recognition (i.e. LPS), their stimulatory capacity of innate immune responses by hsps (PAMPs) in conjuction with their ability to generate antigen-specific T-cell responses, has important implications at the level of immune surveillance. It demonstrates that these proteins are unique and mutlifunctional and most likely can act as modulators of the innate immune system.

Harnessing their power as PAMPs/RAMPs would be an exciting challenge for the future since they could potentially have vast therapeutic applications. They could be used to manipulate the innate and adaptive immune responses in the case of vaccines complexed with peptides. In addition, hsp-peptide complexes can be used a agents for cross-priming and as adjuvants for vaccines. They can be used in several chronic inflammatory disease conditions as therapeutic interventions, such as atheroscleoris, or hyperinflammatory responses, such as fatal sepsis syndrome. Thus by manipulating the innate immune response, we can find therapeutic interventions for all these conditions. It has already been shown that previous heat shock treatment attenuates LPS-induced hyporesponsiveness in rats (Dong et al. 2005). In addition, it has also been shown that administration of exogenous hsp70 decreases the mortality rate in rats with sepsis manifestations (Kustanova et al. 2006a, b, 2007; Nakada et al. 2005).

More exciting is the prospect of their administration as RAMPs, since they could potentially have applications in multiple chronic inflammatory conditions. The RAMP role of hsp10 has already been demonstrated in adjuvant arthritis (Agnello et al. 2002), experimental autoimmune encephalitis (Zhang et al. 2003) and allogeneic skin grafting (Morton et al. 2000), which heralds hsps as a new class of immunotherapeutics.

References

Agnello D, Scanziani E, Di Giancamillo M, Leoni F, Modena D, Mascagni P, Introna M, Ghezzi P, Villa P (2002) Preventive administration of Mycobacterium tuberculosis 10-kDa heat shock protein (hsp 10) suppresses adjuvant arthritis in Lewis rats. Int Immunopharmacol 2:463–474

Alexopoulou L, Holt AC, Medzhitov R, Flavell RA (2001) Recognition of double-stranded RNA and activation of NF-kappaB by Toll-like receptor 3. Nature 413:732–738

Anderson K, Stott EJ, Wertz GW (1992) Intracellular processing of the human respiratory syncytial virus fusion glycoprotein: amino acid substitutions affecting folding, transport and cleavage. J Gen Virol 73:1177–1188

Arnold D, Faath S, Rammensee H, Schild H (1995) Cross-priming of minor histocompatibility antigen-specific cytotoxic T cells upon immunisation with the heat shock protein gp96. J Exp Med 162:3757–3764

Asea A, Kraeft SK, Kurt-Jones EA, Stevenson MA, Chen LB, Finberg RW, Koo GC, Calderwood SK (2000) Hsp70 stimulates cytokine production through a CD14-dependent pathway, demonstrating its dual role as a chaperone and cytokine. Nature Med 6:435–442

Asea A, Rehli M, Kabingu E, Boch JA, Bare O, Auron PE, Stevenson MA, Calderwood SK (2002) Novel signal transduction pathway utilised by extracellular Hsp70: role of toll-like receptor (TLR) 2 and TLR4. J Biol Chem 277:15028–15034

Basu S, Binder RJ, Suto R, Anderson KM, Srivastava PK (2000) Necrotic but not apoptotic cell death releases heat shock proteins, which deliver a partial maturation signal to dendritic cells and activate the NF-kappa B pathway. Int Immunol 12:1539–1546

Basu S, Binder RJ, Ramalingam T, Srivastava PK (2001) CD91 is a common receptor for heat shock proteins gp96, hsp90, hsp70, and calreticulin. Immunity 14:303–313

Berwin B, Reed RC, Nicchitta CV (2001) Virally induced lytic cell death elicits the release of immunogenic GRP94/gp96. J Biol Chem 276:21083–21088

Bulut Y, Faure E, Thomas L, Karahashi H, Michelsen KS, Equils O, Morrison SG, Morrison RP, Arditi M (2002) Chlamydial heat shock protein 60 activates macrophages and endothelial cells through toll-like receptor 4 and MD2 in a MyD88-dependent pathway. J Immunol 168:1435–1440

Calderwood SK, Mambula SS, Gray PJ (2007) Extracellular heat shock proteins in cell signaling and immunity. Ann N Y Acad Sci 1113:28–39

Chandawarkar RY, Wagh MS, Srivastava PK (1999) The dual nature of specific immunological activity of tumour derived gp96 preparations. J Exp Med 189:1437–1442

Chen W, Syldath U, Bellmann K, Burkart V, Kolb H (1999) Human 60-kDa heat shock protein: A danger signal to the innate immune system. J Immunol 162:3212–3219

Corrigall VM, Bodman-Smith MD, Fife MS, Canas B, Myers LK, Wooley PH, Soh C, Staines NA, Pappin DJC, Berlo SE, van Eden W, van der Zee R, Lanchbury JS, Panayi GS (2001) The human endoplasmic reticulum molecular chaperone BiP is an autoantigen for rheumatoid arthritis and prevents the induction of experimental arthritis. J Immunol 166:1492–1498

Dong HP, Chen HW, Hsu C, Chiu HY, Lin LC, Yang RC (2005) Previous heat shock treatment attenuates lipopolysaccharide-induced hyporesponsiveness of platelets in rats. Shock 24:239–244

Ellis J (1987) Proteins as molecular chaperones. Nature 328:378–379

Ellis RJ (2007) Protein misassembly: macromolecular crowding and molecular chaperones. Adv Exp Med Biol 594:1–13

Ellis RJ, Van Der Vies SM, Hemmingsen SM (1989) The molecular chaperone concept. Biochem Soc Symp 55:145–153

Gallay P, Barras C, Tobias PS, Calandra T, Glauser MP, Heumann D (1994) Lipopolysaccharide (LPS)-binding protein in human serum determines the tumor necrosis factor response of monocytes to LPS. J Infect Dis 170:1319–1322

Gangalum RK, Atanasov IC, Zhou ZH, Bhat SP (2011) AlphaB-crystallin is found in detergent-resistant membrane microdomains and is secreted via exosomes from human retinal pigment epithelial cells. J Biol Chem 286:3261–3269

Gao B, Tsan MF (2003) Endotoxin contamination in recombinant human heat shock protein 70 (Hsp70) preparation is responsible for the induction of tumor necrosis factor a release by murine macrophages. J Biol Chem 278, 174–179

Gullo CA, Teoh G (2004) Heat shock proteins: to present or not, that is the question. Immunol Lett 94:1–10

Gupta RS, Ramachandra NB, Bowes T, Singh B (2008) Unusual cellular disposition of the mitochondrial molecular chaperones Hsp60, Hsp70 and Hsp10. Novartis Foundation Symp 291:74–95

Habich C, Kempe K, van der Zee R, Rumenapf R, Akiyama H, Kolb H, Burkart V (2005) Heat shock protein 60: Specific binding of lipopolysaccharide. J Immunol 174:1298–1305

Hartl FU, Hayer-Hartl M (2002) Molecular chaperones in the cytosol: from nascent chain to folded protein. Science 295:1852–1858

Hartl FU, Bracher A, Hayer-Hartl M (2011) Molecular chaperones in protein folding and proteostasis. Nature 475:324–332

Hayashi F, Smith KD, Ozinsky A, Hawn TR, Yi EC, Goodlett DR, Eng JK, Akira S, Underhill DM, Aderem A (2001) The innate immune response to bacterial flagellin is mediated by Toll-like receptor 5. Nature 410:1099–1103

Hecker JG, McGarvey M (2011) Heat shock proteins as biomarkers for the rapid detection of brain and spinal cord ischemia: a review and comparison to other methods of detection in thoracic aneurysm repair. Cell Stress Chaperones 16:119–131

Heil F, Hemmi H, Hochrein H, Ampenberger F, Kirschning C, Akira S, Lipford G, Wagner H, Bauer S (2004) Species-specific recognition of single-stranded RNA via Toll-like receptor 7 and 8. Science 303:1526–1529

Heine H, Ulmer AJ, El-Samalouti VT, Lentschat A, Hamann L (2001) Decay-accelerating factor (DAF/CD55) is a functional active element of the LPS receptor complex. J Endotoxin Res 7:227–231

Hemmingsen SM, Woolford C, van der Vies SM, Tilly K, Dennis DT, Georgopoulos CP, Hendrix RW, Ellis RJ (1978). Homologous plant and bacterial proteins chaperone oligomeric protein assembly. Nature 333:330–334

Henderson B, Pockley AG (2012) Proteotoxic Stress and Circulating Cell Stress Proteins in the Cardiovascular Diseases. Cell Stress Chaperones 17:303–311

Henderson B, Calderwood S, Coates ARM, Cohen IR, van Eden W, Lehner T, Pockley AG (2010) Caught with their PAMPs down? The extracellular signalling actions of molecular chaperones are not due to microbial contaminants. Cell Stress Chaperones 15:123–141

Ishii T, Udono H, Yamano T, Ohta H, Uenaka A, Ono T, Hizuta A, Tanaka N, Srivastava PK, Nakayama E (1999) Isolation of MHC class I-restricted tumour antigen peptide and its precursors associated with heat shock proteins hsp 70, hsp90, and gp96. J Immunol 162:1303–1309

Johnson BJ, Le TT, Dobbin CA et al (2005) Heat shock protein 10 inhibits lipopolysaccharide-induced inflammatory mediator production. J Biol Chem 280:4037–4047

Kol A, Sukhova GK, Lichtman AH, Libby P (1998) Chlamydial heat shock protein 60 localizes in human atheroma and regulates macrophage tumor necrosis factor-a and matrix metalloproteinase expression. Circulation 98:300–307

Kol A, Bourcier T, Lichtman AH, Libby P (1999) Chalmydial and human heat shock protein 60s activate human vascular endothelium, smooth muscle cells, and macrophages. J Clin Invest 103:571–577

Kol A, Lichtman AH, Finberg RW, Libby P, Kurt-Jones EA (2000) Cutting edge: heat shock protein (HSP) 60 activates the innate immune response: CD14 is an essential receptor for HSP60 activation of mononuclear cells. J Immunol 164:13–17

Kustanova GA, Murashev AN, Guzhova IV, Margulis BA, Prokhorenko IR, Grachev SV, Evgen'ev MB (2006a) Protective effect of exogenous 70-kDa heat shock protein during endotoxic shock (sepsis). Dokl Biol Sci 411:504–507

Kustanova GA, Murashev AN, Karpov VL, Margulis BA, Guzhova IV, Prokhorenko IR, Grachev SV, Evgen'ev MB (2006b) Exogenous heat shock protein 70 mediates sepsis manifestations and decreases the mortality rate in rats. Cell Stress Chaperones 11:276–286

Kustanova GA, Evgen'ev MB, Karpov VL, Margulis BA, Prokhorenko IR, Grachev SV, Murashev AN (2007) Effect of 70 kDa exogenous heat shock protein on biochemical parameters of Wistar rats in endotoxic shock. Dokl Biol Sci 415:253–256

Laskey RA, Honda BM, Mills AD, Finch JT (1978) Nucleosomes are assembled by an acidic protein which binds histones and transfers them to DNA. Nature 275:416–420

Li ZH, Menoret A, Srivastava P (2002) Roles of heat-shock proteins in antigen presentation and cross-presentation. Curr Opin Immunol 14:45–51

Liu JH, Hong SL, Feng ZY, Xin YQ, Wang Q, Fu J, Zhang C, Li GL, Luo L, Yin ZM (2010) Regulation of lipopolysaccharide-induced inflammatory response by heat shock protein 27 in THP-1 cells. Cell Immunol 264:127–134

Lund JM, Alexopoulou L, Sato A, Karow M, Adams NC, Gale NW, Iwasaki A, Flavell RA (2004) Recognition of single-stranded RNA viruses by Toll-like receptor 7. Proc Natl Acad Sci USA 101:5598–5603

Matzinger P (1994) Tolerance, danger, and the extended family. Ann Rev Immunol 12:991–1005

Molvarec A, Rigo J, Lazar L, Balogh K, Mako V, Cervenak L, Mezes M, Prohaszka Z (2009) Increased serum heat-shock protein 70 levels reflect systemic inflammation, oxidative stress and hepatocellular injury in preeclampsia. Cell Stress Chaperones 14:151–159

Morton H, Rolfe B, Clunie GJ (1977) An early pregnancy factor detected in human serum by the rosette inhibition test. Lancet 1:394–397

Morton H, Mckay D, Murphy RM, Somodevilla-Torres MJ, Swanson CE, Cassady AI, Summers KM, Cavanagh AC (2000) Production of a recombinant form of early pregnancy factor that can prolong allogeneic skin graft survival time in rats. Immunol Cell Biol 78:603–607

Multhoff G, Botzler C, Jennen L, Schmidt J, Ellwart J, Issels R (1997) Heat shock protein 72 on tumor cells: a recognition structure for natural killer cells. J Immunol 158:4341–4354

Murshid A, Gong J, Calderwood SK (2008) Heat-shock proteins in cancer vaccines: agents of antigen cross-presentation. Expert Rev Vaccines 7:1019–1030

Nakada J, Matsura T, Okazaki N, Nishida T, Togawa A, Minami Y, Inagaki Y, Ito H, Yamada K, Ishibe Y (2005) Oral administration of geranylgeranylacetone improves survival rate in a rat endotoxin shock model: administration timing and heat shock protein 70 induction. Shock 24:482–487

Noonan FP, Halliday WJ, Morton H, Clunie, GJ (1979) Early pregnancy factor is immunosuppressive. Nature 278, 649–651

Ohashi K, Burkart V, Flohe S, Kolb H (2000) Cutting edge: heat shock protein 60 is a putative endogenous ligand of the toll-like receptor-4 complex. J Immunol 164:558–561

Osterloh A, Breloer M (2008) Heat shock proteins: linking danger and pathogen recognition. Med Microbiol Immunol 197:1–8

Osterloh A, Kalinke U, Weiss S, Fleischer B, Breloer M (2007) Synergistic and differential modulation of immune responses by Hsp60 and lipopolysaccharide. J Biol Chem 282:4669–4680

Ousman SS, Tomooka BH, van Noort JM, Wawrousek EF, O'Connor KC, Hafler DA, Sobel RA, Robinson WH, Steinman L (2007) Protective and therapeutic role for alpha B-crystallin in autoimmune demyelination. Nature 448:474–477

Park KJ, Gaynor RB, Kwak YT (2003) Heat shock protein 27 association with the I kappa B kinase complex regulates tumor necrosis factor alpha-induced NF-kappa B activation. J Biol Chem 278:35272–35278

Pockley AG, Wu R, Lemne C, Kiessling R, de Faire U, Frostegard J (2000) Circulating heat shock protein 60 is associated with early cardiovascular disease. Hypertension 36:303–307

Poltorak A, He XL, Smirnova I, Liu MY, VanHuffel C, Du X, Birdwell D, Alejos E, Silva M, Galanos C, Freudenberg M, RicciardiCastagnoli P, Layton, Beutler B (1998) Defective LPS signaling in C3H/Hej and C57BL/10ScCr mice: Mutations in TLR4 gene. Science 282:2085–2088

Qureshi ST, Lariviere L, Leveque G, Clermont S, Moore KJ, Gros P, Malo D (1999) Endotoxin-tolerant mice have mutations in toll-like receptor 4 (TLR4). J Exp Med 189:615–625

Re F, Strominger JL (2003) Separate functional domains of human MD-2 mediate Toll-like receptor 4-binding and lipopolysaccharide responsiveness. J Immunol 171:5272–5276

Reyes-Del Valle J, Chávez-Salinas S, Medina F, Del Angel RM (2005) Heat shock protein 90 and heat shock protein 70 are components of dengue virus receptor complex in human cells. J Virol 79:4557–4567

Ritossa F (1962) Discovery of the heat shock response. Cell Stress Chaperones 1:97–98

Sapozhnikov AM, Ponomarev ED, Tarasenko TN, Telford WG (1999) Spontaneous apoptosis and expression of cell surface heat-shock proteins in cultured EL-4 lymphoma cells. Cell Prolif 32:363–378

Shamaei-Tousi A, D'Aiuto F, Nibali L, Steptoe A, Coates AR, Parkar M, Donos N, Henderson B (2007) Differential regulation of circulating levels of molecular chaperones in patients undergoing treatment for periodontal disease. PLoS ONE 2:e1198

Shields AM, Panayi GS, Corrigall VM (2011) Resolution-associated molecular patterns (RAMP): RAMParts defending immunological homeostasis? Clin Exp Immunol 165:292–300

Shimazu R, Akashi S, Ogata H et al (1999) MD-2, a molecule that confers lipopolysaccharide responsiveness on Toll-like receptor 4. J Exp Med 89:1777–1782

Singh-Jasuja H, Hilf N, Scherer HU, Arnold-Schild D, Rammensee HG, Toes RE, Schild H (2000) The heat shock protein gp96: a receptor-targeted cross-priming carrier and activator of dendritic cells. Cell Stress Chaperones 5:462–470

Srivastava P (2002) Interaction of heat shock proteins with peptides and antigen presenting cells: Chaperoning of the innate and adaptive immune responses. Annu Revs Immunol 20:395–425

Srivastava PK, DeLeo AB, Old LJ (1986) Tumor rejection antigens of chemically induced carcinomas of inbred mice. Proc Natl Acad Sci USA 83:3407–3411

Srivastava PK, Menoret A, Basu S, Binder RJ, McQuade KL (1998a) Heat shock proteins come of age: primitive functions acquire new roles in an adaptive world. Immunity 8:657–665

Srivastava PK, Menoret A, Basu S, Binder RJ, McQuade KL (1998b): Heat shock proteins come of age: primitive functions acquire new roles in an adaptive world. Immunity 8:657–665

Szerafin T, Hoetzenecker K, Hacker S, Horvath A, Pollreisz A, Arpád P, Mangold A, Wliszczak T, Dworschak M, Seitelberger R, Wolner E, Ankersmit HJ (2008) Heat shock proteins 27, 60, 70, 90alpha, and 20S proteasome in on-pump versus off-pump coronary artery bypass graft patients. Ann Thorac Surg 85:80–87

Tagaya Y, Maeda Y, Mitsui A, Kondo N, Matsui H, Hamuro J, Brown N, Arai K, Yokota T, Wakasugi H (1994) ATL-derived factor (ADF), an IL-2 receptor/Tac inducer homologous to thioredoxin; possible involvement of dithiol-reduction in the IL-2 receptor induction. EMBO J 13:2244

Takeuchi O, Hoshino K, Kawai T, Sanjo H, Takada H, Ogawa T, Takeda K, Akira S (1999) Differential roles of TLR2 and TLR4 in recognition of gram-negative and gram-positive bacterial cell wall components. Immunity 11:443–451

Tamura Y, Peng P, Liu K, Daou M, Srivastava PK (1997) Immunotherapy of cancers with autologous cancer-derived heat-shock protein preparations. Science 278:117

Triantafilou M, Triantafilou K (2004) Heat-shock protein 70 and heat-shock protein 90 associate with Toll-like receptor 4 in response to bacterial lipopolysaccharide. Biochem Soc Trans 32:636–639

Triantafilou K, Triantafilou M, Dedrick RL (2001) A CD14-independent LPS receptor cluster. Nat Immunol 2:338–345

Triantafilou K, Triantafilou M, Ladha S, Mackie A, Dedrick RL, Fernandez N, Cherry R (2002) Fluorescence recovery after photobleaching reveals that LPS rapidly transfers from CD14 to hsp70 and hsp90 on the cell membrane. J Cell Sci 114:2535–2545

Triantafilou M, Triantafilou K (2004): Heat-shock protein 70 and heat-shock protein 90 associate with Toll-like receptor 4 in response to bacterial lipopolysaccharide. Biochem Soc Trans 32:636–639

Triantafilou M, Lepper PM, Briault CD, Ahmed MA, Dmochowski JM, Schumann C, Triantafilou K (2008) Chemokine receptor 4 (CXCR4) is part of the lipopolysaccharide "sensing apparatus". Eur J Immunol 38:192–203

Udono H, Levey DL, Srivastava K (1994) Cellular requirements for tumor-specific immunity elicited by heat shock proteins: Tumour rejection antigen gp96 primes CD8 T cells in vivo. Proc Natl Acad Sci USA 91:3077–3081

Vabulas RM, Ahmad-Nejad P, da Costa C, Miethke T, Kirschning CJ, Hacker H, Wagner H (2001) Endocytosed HSP60s use Toll-like receptor 2 (TLR2) and TLR4 to activate the Toll/Interleukin-1 receptor signalling pathway in innate immune cells. J Biol Chem 276:31332–31339

Vabulas RM, Braedel S, Hilf N, Singh-Jasuja H, Herter S, Ahmad-Nejad P, Kirschning CJ, da Costa C, Rammensee H, Wagner H, Schild H (2002a) The endoplasmic reticulum-resident heat shock protein Gp96 activates dendritic cells via the Toll-like receptor 2/4 pathway. J Biol Chem 277, 20847–20853

Vabulas RM, Ahmad-Nejad P, Ghose S, Kirschning CJ, Issels RD, Wagner H (2002b) Hsp70 as endogenous stimulus of the Toll/Interleukin-1 receptor signal pathway. J Biol Chem 277:15107–15112

Vanags D, Williams B, Johnson B, Hall S, Nash P, Taylor A, Weiss J, Feeney D (2006) Therapeutic efficacy and safety of chaperonin 10 in patients with rheumatoid arthritis: a double-blind randomised trial. Lancet 368:855–863

Visintin A, Latz E, Monks B G, Espevik T, Golenbock DT (2003) Lysines 128 and 132 enable lipopolysaccharide binding to MD-2, leading to Toll-like receptor-4 aggregation and signal transduction. J Biol Chem 278:48313–48320

Warger T, Hilf N, Rechtsteiner G, Haselmayer P, Carrick DM, Jonuleit H, von Landenberg P, Rammensee HG, Nicchitta CV, Radsak MP, Schild H (2006) Interaction of TLR2 and TLR4 ligands with the N-terminal domain of Gp96 amplifies innate and adaptive immune responses. J Biol Chem 281:22545–22553

Wright SD, Ramos RA, Tobias PS, Ulevitch RJ, Mathison JC (1990) CD14, a receptor for complexes of lipopolysaccharide (LPS) and LPS binding protein. Science 249:1431–1433

Zhang B, Walsh MD, Nguyen KB, Hillyard NC, Cavanagh AC, McCombe PA, Morton H (2003) Early pregnancy factor treatment suppresses the inflammatory response and adhesion molecule expression in the spinal cord of SJL/J mice with experimental autoimmune encephalomyelitis and the delayed-type hypersensitivity reaction to trinitrochlorobenzene in normal BALB/c mice. J Neurol Sci 212:37–46

Zhu XT, Zhao X, Burkholder WF, Gragerov A, Ogata CM, Gottesman ME, Hendrickson WA (1996) Structural analysis of substrate binding by the molecular chaperone DnaK. Science 272:1606–1614

Chapter 11
Chaperonin 60: An Unexpected Cell Surface Receptor in Prokaryotes and Eukaryotes

Brian Henderson

Abstract The chaperonin (Cpn)60 protein is a fascinating molecule which, under different circumstances, exists on the cell surface or is secreted from the cell. In all the various compartments in which the Cpn60 protein is found in organisms ranging from bacteria, to invertebrates to vertebrates it has evolved a wide range of additional, moonlighting, functions. Among the most unusual of these functions is the ability of this protein to exist on cell surfaces and act like a conventional agonist receptor. This chapter reviews the unexpected receptor and adhesive functions of the Cpn60 pantheon.

11.1 Introduction

The aim of this chapter is to introduce the reader to the concept that the chaperonin (Cpn)60 or Hsp60 protein of prokaryotes and eukaryotes can function as a cellular receptor, in addition to acting as a cell stimulating agonist binding to a range of putative cellular receptors. The focus of this review will be on Cpn60 as a receptor on the surface of either bacteria or eukaryotic cells. The third Kingdom of Life—the Archaea—generally do not contain the type I chaperonins found in bacteria, mitochondria and chloroplasts, but contain type II chaperonins, as found in the eukaryotic cell cytosol (Horwich et al. 2007) and so will not be discussed. In addition to reviewing the literature on Cpn60, (the abbreviation of choice in this chapter) as a receptor, some attention will be given to the receptors that have been postulated to bind to Cpn60. Before describing the literature it is important to realise that major changes are occurring in our understanding of receptor biology, which impinge on the role of Cpn60 either as a receptor agonist or as a receptor for agonists.

B. Henderson (✉)
Department of Microbial Diseases, UCL-Eastman Dental Institute,
University College London, 256 Gray's Inn Road, London WC1X 8LD, UK
e-mail: brian.henderson@ucl.ac.uk

B. Henderson, A. G. Pockley (eds.), *Cellular Trafficking of Cell Stress Proteins in Health and Disease*, Heat Shock Proteins 6, DOI 10.1007/978-94-007-4740-1_11,

11.2 A Brief Review of Receptor Biology

Agonist receptors are a mental construct of pharmacological research starting with Langley's 1905, concept of a receptive substance on the surface of skeletal muscle, mediating drug action. Paul Ehrlich developed a somewhat similar concept a few years later (reviewed in Rang 2006; Prüll et al. 2009). The work of Hill, Clark and Gaddum in the early part of the twentieth century introduced drug receptor theory with the log concentration-effect curve which is now endemic in pharmacology and with it the concept of agonists and antagonists. Stephenson in the 1950s introduced the idea of the *efficacy* of a drug which is a measure of its ability to 'activate' its binding receptor rather than a measure of its affinity for the receptor. This led on to the concept of 'spare' receptors and receptor occupancy in reference to maximal receptor stimulation. Further complications in receptor theory introduced, in addition to agonists and reversible and irreversible antagonists, the concept of partial agonists, inverse agonists and so on. Much of this work was done in the absence of the actual receptors, or of the natural agonists for the receptors, that the theories were proposing, and it was only with the advent of gene cloning and protein expression that the real receptors began to be isolated and studied. In the last decade or so the idea of the receptor has begun to become more complex largely through the study of G-protein coupled receptors (GPCRs) and through use of selective cell membrane isolation and proteomic methodology. Receptors are thought to transduce but one signal. However, it is now established that different agonists binding to the same GPCR can induce different cell signalling pathways (Kenakin 2002). The same holds true for nuclear receptors (Nettles and Greene 2005). The concept of the single receptor is now giving way to the receptor complex. Many receptors obviously dimerise or trimerise when they bind to an agonist (Milligan et al. 2006). In addition, it is now recognised that receptors often form complexes with other proteins on lipid rafts on the plasma membrane (Ostrom and Insel 2004) and proteomic analyses of such complexes are revealing the complexity that they can attain (Kabbani 2008). Another change in our thinking about receptors is the finding that these proteins can have activity in the absence of ligand binding (constitutive activity) (Teitler et al. 2002). These, and other advances in our understanding of agonist/receptor interactions, are leading to a breakdown in our ability to classify agonists and antagonists (Kenakin 2011).

This rapid overview of the discipline of receptor pharmacology, which is now over 100 years old, reveals that in 2012 there have been many surprises in this field and that the original paradigm is undergoing a revolution. It is important to comprehend that we do not fully understand the interactions between agonists and their receptors even in the situation where the structure of the agonist is fully known (as most drugs are) and where there may also be a structure for the receptor. How much less do we understand the interactions between protein agonists and their known protein receptors. Endocrinology probably provides the best corpus of knowledge of protein agonist/receptor interactions. However, cytokine biology is a good example, where the affinities and efficacies of probably only a handful of cytokine/cytokine receptor pairings are known in any detail.

When we turn to understanding the receptor biology of the interactions of molecular chaperones with cells our level of ignorance seems to expand. It is 18 years since the first paper was published on the ability of a molecular chaperone (*Mycobacterium tuberculosis* Cpn60.2) to activate human monocytes to synthesise cytokines (Friedland et al. 1993). Since this time, a growing number of molecular chaperones have been shown to be able to be secreted from cells and to act as what we would assume are rationale receptor agonists (Henderson and Pockley 2010). However, with the exception of cyclophilin A, a peptidyl prolyl isomerase, for which there seems to be good, molecular, evidence for binding to CD147 (Song et al. 2011), there is substantial confusion as to the nature of the receptors for most molecular chaperones and protein-folding catalysts (see Chaps. 12 and 13 for more information about receptors for Hsp70). Three problems seem to exist in the identification of such receptors. The first is the potential contamination of recombinant proteins with bacterial contaminants. However, this criticism has been answered in recent reviews (Henderson et al. 2010; Henderson and Pockley 2010) and it is clear that if certain controls are adhered to then such contamination can be ignored. The second is the number of different receptors that have been reported to bind many of the molecular chaperones (Henderson and Pockley 2010) and this is probably related to the third problem. The third problem is the nature of the agonists that are being dealt with. The secreted molecular chaperones and protein-folding catalysts that are functioning as receptor agonists are moonlighting proteins (Jeffery 1999). Briefly, a moonlighting protein is one which has at least one additional and independent biological activity in addition to the original activity of the protein in question. In most cases it is assumed that the moonlighting activity is due to an independent 'moonlighting' site on the protein. Now, it may be that such moonlighting sites are, in themselves, a novel population of protein 'folds'. Indeed, it has been suggested that moonlighting proteins may have disordered structures in their moonlighting sites (Tompa et al. 2005) which may allow for receptor binding promiscuity. To add to this problem is the fact that the mechanism of the moonlighting process has not been identified. All glyceraldehyde-3-phosphate dehydrogenase (GAPD) proteins will utilise glyceraldehyde-3-phosphate dehydrogenase plus NAD^+ and phosphate and will generate 1,3-bisphosphoglycerate and reduced NAD^+. However, not all moonlighting protein homologues will share the same moonlighting activity. Taking the subject of this chapter, Cpn60, there are now many examples of different Cpn60 proteins having biological actions not shared by other Cpn60 proteins. For example, the *Enterobacter aerogenes* Cpn60 protein is a potent insect neurotoxin while the highly homologous *E. coli*, GroEL, has no such activity (Yoshida et al. 2001). The *Mycobacterium tuberculosis* Cpn60.1 protein inhibits the generation of the multinucleate myeloid cell known as the osteoclast. The Cpn60.2 protein from the same bacterium (over 60 % identical at the sequence level) has no inhibitory, or stimulatory effect on osteoclasts (Winrow et al. 2008). In contradistinction to these two proteins, the *E. coli* GroEL is a potent promoter of the formation of osteoclasts (Reddi et al. 1998). 'Bioinformatically', these are all the same protein. However, in terms of the rules of protein moonlighting they are very different cell signalling agonists. This has to be borne in mind when reading the literature, which may be dealing with human, mouse or bacterial Cpn60 proteins.

To conclude this section, there are many unknowns in the moonlighting actions of the Cpn60 protein as a receptor agonist. This ignorance is nothing to the lack of knowledge about Cpn60 as a receptor for agonists.

11.3 Moonlighting Receptors

The focus of this chapter is the unlikely hypothesis that the Cpn60 protein, a mitochondrial/chloroplast/hydrogenosome protein in eukaryotic cells and a cytosolic protein in bacteria, can also function as a cell surface receptor in eukaryotes and bacteria. Such a proposition looks unlikely if the reader is not aware of the literature on protein moonlighting in which a growing number of unlikely proteins are being found on the surface of various cells and show affinity for a whole array of agonist molecules. One of the earliest examples is the glycolytic enzyme, GAPD, mentioned earlier. Many bacteria express GAPD on their outer cell wall where it acts as a cellular receptor for transferrin, fibronectin (reviewed by Henderson and Martin 2011) and urokinase plasminogen activator receptor (uPAR—Jin et al. 2005). The protozoan, *Trichomonas vaginalis*, which causes the sexually trasmittted disease, Trichomonosis, binds to the vaginal epithelium through the moonlighting hydrogenosomal enzyme pyruvate:ferredoxin oxidoreductase, an iron-induced cell surface protein which acts as an epithelial cell adhesin in this organism (Moreno-Brito et al. 2005). A final example is from the human where is has recently been shown that the mitochondrial F_1F_0 ATP synthase can also function as a cell surface high affinity receptor for apoA-I, which is the main protein in high density lipoproteins (HDL) (Vantourout et al. 2010). With this unlikely group of diverse proteins apparently able to act as cell surface receptors in bacteria and eukaryotic cells it is less surprising that the Cpn60 protein can also function as a receptor. The curiosity about Cpn60 proteins is the diversity of ligands they can bind to. The reader should note that Chap. 14 is devoted to a discussion of another molecular chaperone, BiP, as a multifunctional moonlighting cell surface receptor.

11.4 Chaperonin 60 as a Receptor Agonist

With the realisation that certain cytokines, such as RANK and RANKL, function as plasma membrane bound proteins (Hanada et al. 2010) the relationship between what is a receptor and what is the receptor agonist began to become blurred. Thus to start off the discussion of Cpn60 as a receptor it is sensible to begin with the identification of the Cpn60 protein as a putative receptor agonist. This was the report from St George's Hospital Medical School that the *Mycobacterium tuberculosis* Cpn60.2 protein was an activator of human monocyte cytokine synthesis (Friedland et al. 1993). *Mycobacterium tuberculosis* is one of a number of bacteria which encode more than one Cpn60 protein (Kong et al. 1993). It is pertinent to start here, as almost 20 years later, it turns out that this self-same protein is also a mycobacterial

Table 11.1 Cell surface receptors reported to bind to chaperonin 60 proteins

Cpn60 from species	Receptor	Cell type	Reference
Homo sapiens	CD14	Monocyte	Kol et al. 2000
Homo sapiens	TLR4	Monocyte	Ohashi et al. 2000
Mus musculus	TLR2	Lymphocytes	Zanin-Zharov et al. 2003
Bacterial Cpn60 proteins	TLR4 and CD14	Mixed	Maguire et al. 2002
Homo sapiens	TREM-2	Microglia	Stefano et al. 2009
Homo sapiens	LOX-1	Macrophages	Xie et al. 2010
Homo sapiens	ATP synthase	Vascular endothelial cells	Alard et al. 2011
Mycobacterium tuberculosis Cpn60.1	DC-SIGN	Macrophage	Carroll et al. 2010
M. tuberculosis Cpn60.2	CD43	Macrophage	Hickey et al. 2009, 2010

cell surface protein functioning as an adhesin for the binding of *M. tuberculosis* to macrophages (Hickey et al. 2009, 2010). This paper by Friedland suggested that the *M. tuberculosis* Cpn60.2 protein was activating the human monocyte in the same way that Gram-negative bacterial lipopolysaccharide (LPS) or interferon-gamma would. Such macrophage activation is now termed classic macrophage activation and is the type of activation needed for macrophages to present antigens to T lymphocytes (Martinez et al. 2009). This involves the upregulation of expression of cell surface MHC class II, Fcγ receptors, co-stimulatory antigens (CD80, CD86) and increased production of oxygen-derived free radicals. Now, 1 year after Friedland's paper, a study from the late Ralph van Furth's group revealed that *M. tuberculosis* Cpn60.2, while it did stimulate macrophage cytokine synthesis, did not classically activate macrophages (Peetermans et al. 1994). A further paper from this group also revealed that the *M. tuberculosis* Cpn60.2 protein induced the expression of vascular endothelial cell adhesion proteins by a distinct mechanism to that of LPS (Verdegaal et al 1996). Further, it was shown that the *E. coli* GroEL protein could stimulate bone breakdown (due to induction of macrophage-derived osteoclasts) in C3H/HeJ mice which lack a functional TLR4 receptor (Kirby et al. 1995). Had these early papers on bacterial Cpn60 proteins been read, much of the confusion about the signalling action of the eukaryotic Cpn60 protein would have been obviated. Unfortunately, those working on eukaryotic Cpn60 proteins, mainly the human protein, have claimed that the receptor for this protein on human or rodent cells is the TLR4 receptor. This has then mired this whole field of secreted molecular chaperone signalling proteins with vague concerns over such biological activity being due to contamination with bacterial PAMPs (pathogen-associated molecular patterns) such as LPS (Tsan and Gao 2009).

So what receptors have Cpn60 proteins been claimed to bind to (Table 11.1)? As stated, initial work on the human protein suggested that CD14 (Kol et al. 2000) and or TLR4 (Ohashi et al. 2000) were the receptors for Cpn60. On T lymphocytes, TLR2 was claimed to be the receptor binding human Cpn60 (Zanin-Zhorov et al. 2003).

Others reported that some bacterial Cpn60 proteins also bound to CD14 and TLR4 (Maguire et al. 2002). Since then a number of other cell surface receptors have been suggested to bind Cpn60 proteins. These include, triggering receptor expressed in myeloid (TREM) cells 2 (Stefano et al. 2009), a cell surface protein on macrophages and dendritic cells with a number of putative activities such as being a negative regulator of immune responses and a phagocytic receptor for bacteria (see N'Dyiae et al. 2009). Little is known of the natural agonists of TREM-2. It has been shown that Cpn60 can bind to cells expressing TREM-2 and activate phagocytosis. The binding affinity was around 4 μM, which is relatively low for protein-protein interactions, and it is not clear what role this binding event plays in cell behaviour (Stefano et al. 2009). Hsp70 has been reported to bind to one of the C-type lectin superfamily, called Lox-1 (Inoue and Sawamura 2007). This same receptor has also been claimed to bind human Cpn60 (Xie et al. 2010). Indeed, the *Chlamydia pneumoniae* Cpn60.1 protein has recently been shown to induce Lox-1 expression on vascular endothelial cells (Lin et al. 2011). Another recently described Cpn60 receptor is the β-subunit of the endothelial cell surface ATPase (Alard et al. 2011). Specific binding of recombinant Hsp60 to the recombinant ATPase has been shown to occur using surface plasmon resonance. This methodology can, as will be discussed, determine the kinetics of receptor-ligand binding. However, in this paper such measurements were not reported. Another receptor, this time for the *Mycobacterium bovis* BCG Cpn60.1 protein, identified by affinity chromatography and peptide mass fingerprinting is DC-SIGN, another C-type lectin (Carroll et al. 2010). The same technology applied to the *M. tuberculosis* Cpn60.1 protein pulled out many more putative receptors (Henderson and Mesher 2007). Of course, in both of these affinity proteomic studies, the ability of the Cpn60 protein to actually bind to the identified receptors has not been studied. Finally, as reported above, the Cpn60.2 protein of *M. tuberculosis* binds to CD43 (Hickey et al. 2009, 2010). It is not known if such receptor binding is also inducing signalling. A more detailed discussion of this interaction is provided in Chapter 15.

Most agonists bind to a single receptor, or a single class of receptors. Most protein endocrine hormones have identifiable receptors with relatively high affinities. Cytokines are now speciated on the basis of the receptor family they bind to. It is therefore surprising that molecular chaperones such as the Cpn60 protein (or Cpn60 homologues) can bind to so many different receptor types. Is this because the Cpn60 protein, being a molecular chaperone, naturally binds to a range of proteins? In the case of the *M. tuberculosis* Cpn60.1 protein this cannot be the explanation as we have shown that it is not a molecular chaperone (Hu et al. 2008). Another explanation is that the Cpn60 protein is a 'super moonlighting protein' which has had over 3 billion years to evolve into a multi-binding protein with agonist and receptor binding actions. Clearly, further work is needed to explain why this set of protein homologues can have such promiscuous binding ability.

The receptors for Cpn60 are varied and new ones keep being reported, making for a confusing situation. What do we know about the agonist sites on Cpn60 that interact with the receptor. Christiane Habich's group were the first to study the kinetics of binding of mammalian Cpn60 to monocytes. Using Alexa-fluor-488-

labelled human Cpn60 and flow cytometry to assess binding to the murine monocyte cell line J774A.1, it was shown that binding was saturable and could be inhibited by unlabelled Cpn60 but not by other molecular chaperones—two measures of a specific binding process. Of interest, binding to TLR4-negative murine macrophages was the same as to wild type cells but there was no transduction of the signal in the TLR4-negative cells. Thus these data suggest that there is a cell surface receptor for human Cpn60 on murine macrophages, which is not TLR4 but is, somehow, linked to this protein for the process of cell activation (Habich et al. 2002). Competition experiments using different Cpn60 preparations revealed that labelled human Cpn60 can be competed for binding by rat and mouse Cpn60 but not by: hamster, *Escherichia coli*, *Chlamydia pneumoniae* or *Mycobacterium bovis* Cpn60 (Habich et al. 2003). Evolution seems to have been busy generating cell surface receptors for the Cpn60 homologues. Even the two highly homologous Cpn60 proteins from *Mycobacterium tuberculosis* do not compete for binding to human monocytes (Cehovin et al. 2010). Attempts have been made to identify the agonist site in human Cpn60 by using an overlapping peptide approach. This methodology has some problems as peptides may exhibit biological activity not found with the same sequence in the intact protein. One way of overcoming this problem is to generate truncation mutants lacking the sequences picked up by the peptide approach and if the recombinant protein lacking said peptide is inactive then this is good evidence that that particular stretch of amino acid sequence is responsible for the biology of the protein. Anyway, the result from using overlapping peptides suggests that the C-terminal peptide aa481–500 contains the murine monocyte cell line binding site (Habich et al. 2004). This study used the murine monocytic cell line J774A.1 which is assumed to be a reasonable facsimile of the primary murine macrophage, Curiously, a later study using primary murine macrophages suggested that regions of human Cpn60 containing residues aa241–260, aa391–410 and aa461–480 were responsible for binding to these cells. Moreover, with primary macrophages, the hamster Cpn60 competes with the human (Habich et al. 2006). The difference between these two studies is not a technical issue but reveals significant differences between receptor specificity on a macrophage cell line and a primary macrophage which is unexpected. Similar studies of the binding of human Cpn60 to adipocytes suggests another part of this protein as the ligand site for binding to the receptor. Now the binding residues are aa1–50 and aa91–110 (Märker et al. 2010). If all this data is supportable it suggests that the human Cpn60 protein has at least six sites on, presumably, its surface, for interacting with cell surface receptors on target cells. It is not surprising then that the nature of the receptor(s) for Cpn60 is unclear.

11.5 Chaperonin 60 as a Receptor

The Cpn60 protein has had a chequered history in terms of where it is found in the cell. In eukaryotic cells the Cpn60 protein is imported into the mitochondria from the cytoplasm. However, as early as 1992 it was reported that mammalian pancreatic β-cells contained Cpn60 in their secretory granules (Brudzynski et al. 1992). In

Table 11.2 Bacterial Cpn60 proteins functioning as receptors and known ligand specificity

Bacterium	Agonist	Reference
A. actinomycetemcomitans	?	Goulhen et al. 1998
Borrelia burgdorferi	Glycosphingolipid	Kaneda et al. 1997
Brucella abortis	?	Watarai et al. 2003
Chlamydia pneumoniae Cpn60.1	?	Wuppermann et al. 2008
Clostridium difficile	?	Hennequin et al. 2001
Haemophilus ducreyi	?	Frisk et al. 1998
Haemophilus ducreyi	Glycosphingolipids	Pantzar et al. 2006
Helicobacter pylori	?	Yamaguchi et al. 1996, 1997a, b
Helicobacter pylori	Lactoferrin	Amini et al. 1996
Histoplasma capsulatum[a]	CD11/CD18	Long et al. 2003
Lactobacillus johnsonii	Mucin	Bergonzelli et al. 2006
Lactococcus lactis	Yeast invertase	Katakura et al. 2010
Legionella pneumophila	?	Garduo et al. 1998a, b, Chong et al. 2009
Mycobacterium avium	$\alpha_v\beta_3$	Hayashi et al. 1997
Mycobacterium bovis BCG Cpn60.1	DC-SIGN	Carroll et al. 2010
Mycobacterium leprae Cpn60.2	?	Esaguy and Aguas 1997
Mycobacterium tuberculosis Cpn60.2	CD43	Hickey et al. 2010
Mycobacterium smegmatis	?	Esaguy and Aguas 1997
Plesiomonas shigelloides	?	Tsugawa et al. 2007
Salmonella enterica Serovar typhimurium	Mucus	Ensgraber and Loos 1992
Streptococcus agalactiae	?	Hughes et al. 2002
Streptococcus suis	?	Wu et al. 2008

[a]A fungus

addition, these early studies also found the presence of Cpn60 on the cell surface (reviewed by Gupta et al. 2008). Possibly the first paper reporting the cell surface appearance of Cpn60 was of its presence on Daudi Burkitt's lymphoma cells and of this protein being recognised by human $\gamma\delta$ T lymphocytes (Fisch et al. 1990). Since then, the Cpn60 protein has been reported to be a surface receptor on both bacteria and eukaryotic (mainly human) cells. These two populations of Cpn60 proteins will be dealt with separately and any overlaps discussed. With bacteria, it is important to recognise that to colonise any particular environment in an animal the bacterium needs to be able to bind in order to prevent it being removed by the flow of liquids such as blood, saliva and other moving body fluids. The components of bacteria that allow them to colonise are cell surface molecules, normally proteins, termed adhesins. What is surprising is the number of adhesins that are also moonlighting proteins and the number of such moonlighting proteins that are cell stress proteins.

11.5.1 Bacterial Cell Surface Cpn60

As elaborated in Table 11.2 a growing number of bacteria have been reported to express Cpn60 on the cell surface. Bacteria colonising animals have to interact with

Table 11.3 Eukaryotic Cpn60 proteins as receptors

Species and cell containing surface Cpn60	Binding partner
Histoplasma capsulatum	CR3 (CD11/CD18)
Mus musculus (sperm surface Cpn60)	?
Homo sapiens epithelial cells	*Staph. aureus* FnBPs
Homo sapiens epithelial cells	*L. monocytogenes* alcohol dehydrogenase
Homo sapiens lymphoblast cell lines	HIV Gp41
Homo sapiens VEC	?
Homo sapiens various	HDL

mucosal surfaces and binding to the mucus present on such surfaces allow them to become attached. The first report of a bacterium using Cpn60 as an adhesin was the protein from *Salmonella typhimurium* which was able to bind to mucus (Ensgraber and Loos 1992). Looking at Table 11.2 it is hard not to marvel at the creative power of the evolutionary process which has taken the one protein and shaped it into a receptor for components ranging from sphingolipids to mucus to selected proteins such as CD43, $\alpha_5\beta_3$ and lactoferrin. The *M. tuberculosis* Cpn60.2 protein has recently been found to be a cell surface protein that binds to CD43 and enables this bacterium to bind to the macrophage cell surface presumably priming it for invasion (Hickey et al. 2009, 2010). What is particularly interesting is that CD43 controls the intracellular growth of *M. tuberculosis*, potentially by generating the macrophage pro-apoptotic cytokine, TNFα (Randhawa et al. 2008). So is it possible that the binding of Cpn60.2 to CD43 provides a signal to inhibit the growth of *M. tuberculosis*? The Chlamydia, which are obligately intracellular bacteria, have three Cpn60 proteins (Karunakaran et al. 2003). Only one of these three homologues (the Cpn60.1 protein) functions as an adhesin, enabling the bacterium to bind to target cells as a prelude to invasion (Wupperman et al. 2008). Indeed, coating latex beads with recombinant Chlamydial Cpn60.1 allowed them to bind to cells. In contrast, the other two Cpn60 proteins have no adhesive function. There is divergence between Cpn60.1 and the other two Chlamydial Cpn60 proteins, with only the Cpn60.1 protein being able to complement a temperature-sensitive *E. coli groEL* mutant (Karunkaran et al. 2003). Thus, curiously, the major Cpn60 protein, and true molecular chaperon/cell stress protein, of the Chlamydia also moonlights as an adhesin, while the other two Cpn60 proteins have lost this folding function, presumably evolving some other evolutionary important functions for cell survival—as yet unknown. It is interesting how both *M. tuberculosis* and Chlamydia utilise their only protein-folding Cpn60 protein for adherence (Table 11.3).

A third intracellular bacterium, *Legionella pneumophila*, also utilises its Cpn60 protein as a cell surface modulator of cellular invasion (Garduno et al. 1998a, b). When the recombinant *L. pneumophila* Cpn60 was attached to latex beads, these beads both bound to cells (like those with attached Chlamydial Cpn60.1 above) but also internalised ending up in tight endosomes that did not fuse with other vesicles (Garduno et al. 1998a, b). Indeed, elevated surface levels of Cpn60 correlate with the virulence of *L. pneumophila* (Fernandez et al. 1996). Invasion of *L. pneumophila* causes changes in mitochondrial distribution and in the actin cytoskeleton which are thought to be related to the ability of this organism to survive within macrophages. Surprisingly, beads coated with the *L. pneumophila* Cpn60 protein also caused changes in mitochondrial distribution and well as alterations in the cytoskeleton. How this molecular chaperone induces these changes, particularly the attraction of the mitochondria, is not known, but suggests that this Cpn60 protein is 'chemotactic' for mitochondria (Chong et al. 2009). Of importance, the *E. coli* Cpn60 protein, GroEL, is unable to reproduce the actions of the *L. pneumophila* Cpn60 protein (Chong et al. 2009).

Helicobacter pylori, the causative agent of gastric ulceration and cancer also utilises a cell surface Cpn60 protein for adhesion to epithelial cells (Yamaguchi et al. 1997a). Intriguingly, the Cpn60 protein of the probiotic organism, *Lactobacillus johnsonii*, can aggregate *H. pylori*, but not other enteric bacteria and the Cpn60 protein of *H. pylori* cannot aggregate itself or other organisms (Bergonzelli et al. 2006). Thus far we have viewed the bacterial cell surface Cpn60 as a non-signalling receptor for some host component and this binding is largely to do with bacterial adhesion. It is not considered likely that bacterial cell surface Cpn60 could signal to the bacterium. However, with the *H. pylori* Cpn60 protein it has been reported that antibodies binding to the cell surface chaperone cause inhibition of bacterial growth (Yamaguchi et al. 1997b).

11.5.2 Eukaryotic Cell Surface Cpn60

Most studies of cell surface Cpn60 have focused on mammalian cells. However, the pathogenic fungus, *Histoplasma capsulatum*, has a cell surface Cpn60 protein which binds to the complement type 3 integrin receptor (CD11/CD18) on human myeloid cells (Long et al. 2003). There are a number of reports of mammalian cells (mainly mouse, rat, human) having Cpn60 proteins which bind to all manner of microbial components. For example, human Cpn60 has been reported to bind to the HIV transmembrane glycoprotein, gp41, suggesting this protein may form part of a complex receptor for HIV binding and uptake (Speth et al. 1999). In this context, the human Cpn60 protein appears to have evolved to be a target for a number of bacterial proteins. One example is the binding of *Staphylococcus aureus* fibronectin binding protein (FnBP) with cell surface Cpn60 on epithelial cells. This organism is known to bind, through FnBPA/B (two homologous proteins) to β_1 integrins on target cells surfaces as a prologue to bacterial internalisation. However, it is also established that cell surface Cpn60 can also bind to FnBPs and that antibodies to Cpn60 can inhibit internalisation (Dziewanowska et al. 2000). Human Cpn60 can also bind to LPS (Habich et al. 2005) and so could also act as a cell surface receptor for this pro-inflammatory component. This has resonance with the work of the Triantafilou's, who have shown that cell surface Hsp70 and Hsp90 on mammalian cells form part of an LPS receptor complex (Triantafilou et al. 2001). This is discussed in more detail in Chap. 9.

One of the most unexpected roles for human cell surface Cpn60 is as a receptor for the pathogenic bacterium *Listeria monocytogenes*. This bacterium binds to intestinal epithelial cells and a key cell surface adhesin for this organism is the so-called Listeria-adhesion protein (LAP) (Jaradat et al. 2003). Unexpectedly, this LAP protein was identified as the metabolic enzyme—acetaldehyde alcohol dehydrogenase (Kim et al. 2006). This is clearly a moonlighting protein. Meanwhile it had been shown that the host cell surface receptor for LAP was Cpn60 (Wampler et al. 2004). Measurement of the binding of LAP with human Cpn60, using surface plasmon resonance, revealed a Kd value in the low nanomolar range, which is a very respectable binding affinity (Kim et al. 2006). Thus, we enter into a new arena of protein moonlighting where a prokaryotic moonlighting protein binds to a eukaryotic moonlighting protein to facilitate a bacterial infection. The evolutionary background to this phenomenon would be of interest to solve. Analysis of LAP/alcohol acetaldehyde dehydrogenase binding in non-pathogenic strains of *Listeria* have found that while these strains produce this enzyme there is very little of it on the bacterial surface and so only pathogenic strains bind to target cells via LAP/Hsp60 interaction (Jagadeesan et al. 2010). As human Cpn60 is a stress protein, the role of cell stress in *Listeria* infection has been examined. Thus exposure of CaCo-2 cells, used for infection assays, to various stressors increased intracellular Hsp60 levels and enhanced the adhesion, but not invasion, of *L. monocytogenes*. Knock-down of Hsp60 with inhibitory RNA reduced the adhesion and translocation of wild-type *L. monocytogenes* but a *lap* mutant showed unchanged adhesion. Overexpression of Hsp60 enhanced wild type adhesion and cellular translocation but there was no change in the *lap* mutant. Of importance, infection with *L. monocytogenes* increased plasma membrane expression of Hsp60. Thus there is a dynamic response between these two moonlighting proteins to enhance *L. monocytogenes* infection (Burkholder and Bhunia 2010). Of interest, this ability of human Cpn60 to bind to *L. monocytogenes* is being used to develop a biochip for detecting this pathogen in body fluids (Koo et al. 2009).

There has been significant interest in the role played by human Cpn60 in the major human disease process known as atherosclerosis. This has largely focused on cross-reaction between antibodies to Cpn60 of pathogens (mainly bacterial) and the human protein (Foteinos and Xu 2009). Such cross-reactivity was assumed to be with Cpn60 on the surface of vascular endothelial cells (VECs) thus localising the pathology to the appropriate site for atherosclerosis—the blood vessels. While this hypothesis was first propounded in the 1990s it was not until 2005 that human Cpn60 was definitively shown to be present on the surface of stressed VECs (Pfister et al. 2005). It is not known if this cell surface Cpn60 has any role other than binding to antibodies to Cpn60. However, it is likely that it does. This process may also be the genesis of the circulating Cpn60 found in a proportion of the normal human population (Shamaei-Tousi et al. 2007). One role for cell surface Cpn60 of relevance to human atherosclerosis is the finding that this protein binds to high density lipoprotein with relatively high affinity (Bocharov et al. 2000).

Two final examples of the role of Cpn60 as a eukaryotic receptor will be provided. The first concerns a key process in organismal survival—the generation of functional sperm. To become functional, sperm have to undergo changes after ejaculation in a process called capacitation. In the mouse capacitation requires the participation of a cell surface located Cpn60, which undergoes tyrosine phosphorylation to generate the function sperm (Asquith et al. 2004). This is a key evolved moonlighting functional of this protein which would be assumed to be a conserved function. Surprisingly, human sperm do not have Cpn60 on their cell surface and there is no evidence of cell surface tyrosine phosphorylation (Mitchell et al. 2007). Mice and the precursors of *Homo sapiens* diverged about 75 millions years ago (Stillman and Stewart 2004). This suggests that the line resulting in *Homo sapiens* lost this particular Cpn60 moonlighting site, and its associated mechanisms, over this period, revealing a fairly rapid evolutionary dynamic in the gene(s) encoding the mammalian Cpn60 protein. At the other end of the spectrum Hsp60 has been reported to enter into the plasma membranes of apoptotic cells and be exposed on the cell surface possibly as a signal to phagocytes (Goh et al. 2011).

11.6 Conclusions

Can there be a protein with such an unexpected armamentarium of biological functions? We have barely scratched the surface of this most curious protein family. The Cpn60 protein can function as a cell signalling agonist with immune actions or the ability to cause paralysis in insects due to neurotoxic activity. Almost nothing is known about the ability of the Cpn60 protein family to signal to cells and there is tremendous uncertainty about the nature of the receptors that transduce the Cpn60 signal with CD14, TLR2, TLR4, TREM-2, LOX-1 etc being reported to bind to various Cpn60 proteins. The reasons for the richness of the receptor landscape for this protein lie in the nature of the protein. The Cpn60 protein is the moonlighting protein, *par excellence*, having at least 30 distinct biological functions at the time of writing. Moonlighting, to some extent, rewrites the rules of structure:function relationships and protein homologues, no matter how high their homology, may have completely distinct moonlighting functions (Henderson and Martin 2011). Thus the *M. tuberculosis* and *M. leprae* Cpn60.2 proteins have only four non-conservative substitutions, yet the former has no influence when administered to mice with experimental asthma and the former is a potent inhibitor of this experimental lesion (Rha et al. 2002). This is also seen in the work of Habich and co-workers who have shown that the residues involved in the binding of human Cpn60 to murine J774A.1 cells, murine bone marrow macrophages and murine adipocytes are different. This is not an expected result and suggests that we do not really understand the nature of the binding event between Cpn60 proteins and mammalian cells.

This is exemplified by the fact that in addition to acting as a soluble agonist with the most diverse of biological functions, the Cpn60 protein also functions as a receptor or adhesin. Thus many bacteria have Cpn60 on their cell walls and use it to bind to a wide variety of host molecules including glycolipids, polysaccharides and proteins. Evolutionary reciprocity has converted the human Cpn60 protein into a receptor for bacterial LPS and for other bacterial proteins allowing enhanced levels of bacterial infection. What else does cell surface mammalian Cpn60 bind to? It is likely that in the next few years many intriguing examples of Cpn60-agonist interactions will be discovered and this protein will be shown to be an essential moonlighting protein firmly ensconced in the warp-and-weave of the living organism.

References

Alard JE, Hillion S, Guillevin L, Saraux A, Pers JO, Youinou P, Jamin C (2011) Autoantibodies to endothelial cell surface ATP synthase, the endogenous receptor for hsp60, might play a pathogenic role in vasculatides. PLoS ONE 6:e14654

Amini HR, Ascencio F, Ruiz-Bustos E, Romero MJ, Wadström T (1996) Cryptic domains of a 60 kDa heat shock protein of *Helicobacter pylori* bound to bovine lactoferrin. FEMS Immunol Med Microbiol 16:247–255

Asquith KL, Baleato RM, McLaughlin EA, Nixon B, Aitken RJ (2004) Tyrosine phosphorylation activates surface chaperones facilitating sperm-zona recognition. J Cell Sci 117:3645–3657

Bergonzelli GE, Granato D, Pridmore RD, Marvin-Guy LF, Donnicola D, Corthésy-Theulaz IE (2006) GroEL of *Lactobacillus johnsonii* La1 (NCC 533) is cell surface associated: potential role in interactions with the host and the gastric pathogen *Helicobacter pylori*. Infect Immun 74:425–434

Bocharov AV, Vishnyakova TG, Baranova IN, Remaley AT, Patterson AP, Eggerman TL (2000) Heat shock protein 60 is a high-affinity high-density lipoprotein binding protein. Biochem Biophys Res Commun 277:228–235

Brudzynski K, Martinez V, Gupta RS (1992) Immunocytochemical localization of heat-shock protein 60-related protein in beta-cell secretory granules and its altered distribution in non-obese diabetic mice. Diabetologia 35:316–324

Burkholder KM, Bhunia AK (2010) *Listeria monocytogenes* uses Listeria adhesion protein (LAP) to promote bacterial transepithelial translocation and induces expression of LAP receptor Hsp60. Infect Immun 78:5062–5073

Carroll MV, Sim RB, Bigi F, Jäkel A, Antrobus R, Mitchell DA (2010) Identification of four novel DC-SIGN ligands on *Mycobacterium bovis* BCG. Protein Cell 1:859–870

Cehovin A, Coates AR, Riffo-Vasquez Y, Tormay P, Botanch C, Altare F, Henderson B (2010) Comparison of the moonlighting actions of the two highly homologous chaperonin 60 proteins of *Mycobacterium tuberculosis*. Infect Immun 78:3196–3206

Chong A, Lima CA, Allan DS, Nasrallah GK, Garduo RA (2009) The purified and recombinant Legionella pneumophila chaperonin alters mitochondrial trafficking and microfilament organization. Infect Immun 77:4724–4739

Dziewanowska K, Carson AR, Patti JM, Deobald CF, Bayles KW, Bohach GA (2000) Staphylococcal fibronectin binding protein interacts with heat shock protein 60 and integrins: role in internalization by epithelial cells. Infect Immun 68:6321–6328

Ensgraber M, Loos M (1992) A 66-kilodalton heat shock protein of *Salmonella typhimurium* is responsible for binding of the bacterium to intestinal mucus. Infect Immun 60:3072–3078

Esaguy N, Aguas AP (1997) Subcellular localization of the 65-kDa heat shock protein in mycobacteria by immunoblotting and immunogold ultracytochemistry. J Submicrosc Cytol Pathol 29:85–90

Fernandez RC, Logan SM, Lee SH, Hoffman PS (1996) Elevated levels of *Legionella pneumophila* stress protein Hsp60 early in infection of human monocytes and L929 cells correlate with virulence. Infect Immun 64:1968–1976

Fisch P, Malkovsky M, Kovats S, Sturm E, Braakman E, Klein BS, Voss SD, Morrissey LW, DeMars R, Welch WJ et al (1990) Recognition by human V gamma 9/V delta 2 T cells of a GroEL homolog on Daudi Burkitt's lymphoma cells. Science 250:1269–1273

Foteinos G, Xu Q (2009) Immune-mediated mechanisms of endothelial damage in atherosclerosis. Autoimmunity 42:627–633

Friedland JS, Shattock R, Remick DG, Griffin GE (1993) Mycobacterial 65-kD heat shock protein induces release of proinflammatory cytokines from human monocytic cells. Clin Exp Immunol 91:58–62

Frisk A, Ison CA, Lagergård T (1998) GroEL heat shock protein of *Haemophilus ducreyi*: association with cell surface and capacity to bind to eukaryotic cells. Infect Immun 66:1252–1257

Garduno RA, Faulkner G, Trevors MA, Vats N, Hoffman PS (1998a) Immunolocalization of Hsp60 in *Legionella pneumophila*. J Bacteriol 180:505–513

Garduno RA, Garduno E, Hoffman PS (1998b) Surface-associated hsp60 chaperonin of *Legionella pneumophila* mediates invasion in a HeLa cell model. Infect Immun 66:4602–4610

Goh YC, Yap CT, Huang BH, Cronshaw AD, Leung BP, Lai PB, Hart SP, Dransfield I, Ross JA (2011) Heat-shock protein 60 translocates to the surface of apoptotic cells and differentiated megakaryocytes and stimulates phagocytosis. Cell Mol Life Sci 68:1581–1592

Goulhen F, Hafezi A, Uitto VJ, Hinode D, Nakamura R, Grenier D, Mayrand D (1998) Subcellular localization and cytotoxic activity of the GroEL-like protein isolated from *Actinobacillus actinomycetemcomitans*. Infect Immun 66:5307–5313

Gupta RS, Ramachandra NB, Bowes T, Sigh B (2008) Unusual cellular disposition of the mitochondrial molecular chaperones Hsp60, Hsp70 and Hsp10. In Novartis Foundation Symposium 291:59–73

Habich C, Baumgart K, Kolb H, Burkart V (2002) The receptor for heat shock protein 60 on macrophages is saturable, specific, and distinct from receptors for other heat shock proteins. J Immunol 168:569–576

Habich C, Kempe K, van der Zee R, Burkart V, Kolb H (2003) Different heat shock protein 60 species share pro-inflammatory activity but not binding sites on macrophages. FEBS Lett 533:105–109

Habich C, Kempe K, Burkart V, van der Zee R, Lillicrap M, Gaston H, Kolb H (2004) Identification of the heat shock protein 60 epitope involved in receptor binding on macrophages. FEBS Lett 568:65–69

Habich C, Kempe K, van der Zee R, Rümenapf R, Akiyama H, Kolb H, Burkart V (2005) Heat shock protein 60: specific binding of lipopolysaccharide. J Immunol 174:1298–1305

Habich C, Kempe K, Gomez FJ, Lillicrap M, Gaston H, van der Zee R, Kolb H, Burkart V (2006) Heat shock protein 60: identification of specific epitopes for binding to primary macrophages. FEBS Lett 580:115–120

Hanada R, Hanada T, Penninger JM (2010) Physiology and pathophysiology of the RANKL/RANK system. Biol Chem 391:1365–1370

Hayashi T, Rao SP, Catanzaro A (1997) Binding of the 68-kilodalton protein of *Mycobacterium avium* to alpha(v)beta3 on human monocyte-derived macrophages enhances complement receptor type 3 expression. Infect Immun 65:1211–1216

Henderson B, Mesher J (2007) The search for the chaperonin 60 receptors. Methods 43:223–228

Henderson B, Pockley AG (2010) Molecular chaperones and protein-folding catalysts as intercellular signaling regulators in immunity and inflammation. J Leukoc Biol 88:445–462

Henderson B, Calderwood SK, Coates AR, Cohen I, van Eden W, Lehner T, Pockley AG (2010) Caught with their PAMPs down? The extracellular signalling actions of molecular chaperones are not due to microbial contaminants. Cell Stress Chaperones 15:123–141

Henderson B, Martin A (2011) Bacterial moonlighting proteins and bacterial virulence. Curr Topics Microbiol Immunol (in press)

Hennequin C, Porcheray F, Waligora-Dupriet A, Collignon A, Barc M, Bourlioux P, Karjalainen T (2001) GroEL (Hsp60) of *Clostridium difficile* is involved in cell adherence. Microbiology 147:87–96

Hickey TB, Thorson LM, Speert DP, Daffé M, Stokes RW (2009) *Mycobacterium tuberculosis* Cpn60.2 and DnaK are located on the bacterial surface, where Cpn60.2 facilitates efficient bacterial association with macrophages. Infect Immun 77:3389–3401

Hickey TB, Ziltener HJ, Speert DP, Stokes RW (2010) *Mycobacterium tuberculosis* employs Cpn60.2 as an adhesin that binds CD43 on the macrophage surface. Cell Microbiol 12:1634–1647

Horwich AL, Fenton WA, Chapman E, Farr GW (2007) Two families of chaperonin: physiology and mechanism. Annu Rev Cell Dev Biol 23:115–145

Hu Y, Henderson B, Lund PA, Tormay P, Liu HL, Gurcha SS, Besra GS, Coates ARM (2008) A *Mycobacterium tuberculosis* mutant lacking the *groEL* homologue *cpn60.*1 is viable but fails to induce an inflammatory response in animal models of infection. Infect Immun 76:1535–1546

Hughes MJ, Moore JC, Lane JD, Wilson R, Pribul PK, Younes ZN, Dobson RJ, Everest P, Reason AJ, Redfern JM, Greer FM, Paxton T, Panico M, Morris HR, Feldman RG, Santangelo JD (2002) Identification of major outer surface proteins of Streptococcus agalactiae. Infect Immun 70:1254–1259

Inoue N, Sawamura T (2007) Lectin-like oxidized LDL receptor-1 as extracellular chaperone receptor: its versatile functions and human diseases. Methods 43:218–222

Jagadeesan B, Koo OK, Kim KP, Burkholder KM, Mishra KK, Aroonnual A, Bhunia AK (2010) LAP, an alcohol acetaldehyde dehydrogenase enzyme in Listeria, promotes bacterial adhesion to enterocyte-like Caco-2 cells only in pathogenic species. Microbiology 156:2782–2795

Jaradat ZW, Wampler JW, Bhunia AW (2003) A *Listeria* adhesion protein-deficient *Listeria monocytogenes* strain shows reduced adhesion primarily to intestinal cell lines. Med Microbiol Immunol 192:85–91

Jeffery CJ (1999) Moonlighting proteins. Trends Biochem Sci 24:8–11

Jin H, Youngmia P, Boel G, Kochar J, Pancholi V (2005) Group A streptococcal surface GAPDF, SDH recognises uPAR/CD87 as its receptor on the human pharyngeal cell and mediates bacterial adherence to host cells. J Mol Biol 350:27–41

Kabbani N (2008) Proteomics of membrane receptors and signaling. Proteomics 8:4146–4155

Kaneda K, Masuzawa T, Yasugami K, Suzuki T, Suzuki Y, Yanagihara Y (1997) Glycosphingolipid-binding protein of *Borrelia burgdorferi sensu lato*. Infect Immun 65:3180–3185

Karunakaran KP, Noguchi Y, Read TD, Cherkasov A, Kwee J, Shen C, Nelson CC, Brunham RC (2003) Molecular analysis of the multiple GroEL proteins of Chlamydiae. J Bacteriol 185:1958–1966

Katakura Y, Sano R, Hashimoto T, Ninomiya K, Shioya S (2010) Lactic acid bacteria display on the cell surface cytosolic proteins that recognize yeast mannan. Appl Microbiol Biotechnol 86:319–326

Kenakin T (2002) Drug efficacy at G protein-coupled receptors. Annu Rev Pharmacol Toxicol 42:349–379

Kenakin T (2011) Functional selectivity and biased receptor signaling. J Pharmacol Exp Ther 336:296–302

Kim KP, Jagadeesan B, Burkholder KM, Jaradat ZW, Wampler JL, Lathrop AA, Morgan MT, Bhunia AK (2006) Adhesion characteristics of Listeria adhesion protein (LAP)-expressing *Escherichia coli* to Caco-2 cells and of recombinant LAP to eukaryotic receptor Hsp60 as examined in a surface plasmon resonance sensor. FEMS Microbiol Lett 256:324–332

Kirby AC, Meghji S, Nair SP, White P, Reddi K, Nishihara T, Nakashima K, Willis AC, Sim R, Wilson M, Henderson B (1995) The potent bone resorbing mediator of *Actinobacillus actinomycetemcomitans* is homologous to the molecular chaperone GroEL. J Clin Invest 96:1185–1194

Kol A, Lichtman AH, Finberg RW, Libby P, Kurt-Jones EA (2000) Cutting edge: heat shock protein (HSP) 60 activates the innate immune response: CD14 is an essential receptor for HSP60 activation of mononuclear cells. J Immunol 164:13–17

Kong TH, Coates AR, Butcher PD, Hickman CJ, Shinnick TM (1993) *Mycobacterium tuberculosis* expresses two chaperonin-60 homologs. Proc Natl Acad Sci U S A 90:2608–2612

Koo OK, Liu Y, Shuaib S, Bhattacharya S, Ladisch MR, Bashir R, Bhunia AK (2009) Targeted capture of pathogenic bacteria using a mammalian cell receptor coupled with dielectrophoresis on a biochip. Anal Chem 81:3094–3101

Lin FY, Lin YW, Huang CY, Chang YJ, Tsao NW, Chang NC, Ou KL, Chen TL, Shih CM, Chen YH (2011) GroEL1, a heat shock protein 60 of Chlamydia pneumoniae, induces lectin-like oxidized low-density lipoprotein receptor 1 expression in endothelial cells and enhances atherogenesis in hypercholesterolemic rabbits. J Immunol 186:4405–4414

Long KH, Gomez FJ, Morris RE, Newman SL (2003) Identification of heat shock protein 60 as the ligand on *Histoplasma capsulatum* that mediates binding to CD18 receptors on human macrophages. J Immunol 170:487–494

Maguire M, Coates AR, Henderson B (2002) Chaperonin 60 unfolds its secrets of cellular communication. Cell Stress Chaperones 7:317–329

Martinez FO, Helming L, Gordon S (2009) Alternative activation of macrophages: an immunologic functional perspective. Annu Rev Immunol 27:451–483

Milligan G, Canals M, Pediani JD, Ellis J, Lopez-Gimenez JF (2006) The role of GPCR dimerisation/oligomerisation in receptor signalling. Ernst Schering Found Symp Proc 2:145–161

Moreno-Brito V, Yáez-Gómez C, Meza-Cervantez P, Avila-González L, Rodríguez MA, Ortega-López J, González-Robles A, Arroyo R (2005) A *Trichomonas vaginalis* 120 kDa protein with identity to hydrogenosome pyruvate:ferredoxin oxidoreductase is a surface adhesin induced by iron. Cell Microbiol 7:245–258

Märker T, Kriebel J, Wohlrab U, Habich C (2010) Heat shock protein 60 and adipocytes: characterization of a ligand-receptor interaction. Biochem Biophys Res Commun 391:1634–1640

Mitchell LA, Nixon B, Aitken RJ (2007) Analysis of chaperone proteins associated with human spermatozoa during capacitation. Mol Hum Reprod 13:605–613

N'Diaye EN, Branda CS, Branda SS, Nevarez L, Colonna M, Lowell C, Hamerman JA, Seaman WE (2009) TREM-2 (triggering receptor expressed on myeloid cells 2) is a phagocytic receptor for bacteria. J Cell Biol 184:215–223

Nettles KW, Greene GL (2005) Ligand control of coregulator recruitment to nuclear receptors. Annu Rev Physiol 67:309–333

Ohashi K, Burkart V, Flohé S, Kolb H (2000) Cutting edge: heat shock protein 60 is a putative endogenous ligand of the toll-like receptor-4 complex. J Immunol 164:558–561

Ostrom RS, Insel PA (2004) The evolving role of lipid rafts and caveolae in G protein-coupled receptor signaling: implications for molecular pharmacology. Br J Pharmacol 143:235–245

Pantzar M, Teneberg S, Lagergård T (2006) Binding of *Haemophilus ducreyi* to carbohydrate receptors is mediated by the 58.5-kDa GroEL heat shock protein. Microbes Infect 8:2452–2458

Peetermans WE, Raats CJ, Langermans JA, van Furth R (1994) Mycobacterial heat-shock protein 65 induces proinflammatory cytokines but does not activate human mononuclear phagocytes. Scand J Immunol 39:613–617

Pfister G, Stroh CM, Perschinka H, Kind M, Knoflach M, Hinterdorfer P, Wick G (2005) Detection of HSP60 on the membrane surface of stressed human endothelial cells by atomic force and confocal microscopy. J Cell Sci 118:1587–1594

Prüll C-R, Maehle A-H, Halliwell R F (2009) A short history of the drug receptor concept. Palgrave McMillan, London

Randhawa AK, Ziltener HJ, Stokes RW (2008) CD43 controls the intracellular growth of *Mycobacterium tuberculosis* through the induction of TNF-alpha-mediated apoptosis. Cell Microbiol 10:2105–2117

Rang HP (2006). The receptor concept: pharmacology's big idea. Br J Pharmacol 147(Suppl 1):S9–S16

Reddi K, Meghji S, Nair SP, Arnett TR, Miller AD, Preuss M, Wilson M, Henderson B, Hill P (1998) The *Escherichia coli* chaperonin 60 (groEL) is a potent stimulator of osteoclast formation. J Bone Miner Res 13:1260–1266

Rha YH, Taube C, Haczku A, Joetham A, Takeda K, Duez C, Siegel M, Aydintug MK, Born WK, Dakhama A, Gelfand EW (2002) Effect of microbial heat shock proteins on airway inflammation and hyperresponsiveness. J Immunol 169:5300–5307

Shamaei-Tousi A, Steptoe A, O'Donnell K, Palmen J, Stephens JW, Hurel SJ, Marmot M, Homer K, D'Aiuto F, Coates AR, Humphries SE, Henderson B (2007) Plasma heat shock protein 60 and cardiovascular disease risk: the role of psychosocial, genetic, and biological factors. Cell Stress Chaperones 12:384–392

Song F, Zhang X, Ren XB, Zhu P, Xu J, Wang L, Li YF, Zhong N, Ru Q, Zhang DW, Jiang JL, Xia B, Chen ZN (2011) Cyclophilin A (CyPA) induces chemotaxis independent of its peptidylprolyl cis-trans isomerase activity: direct binding between CyPA and the ectodomain of CD147. J Biol Chem 286:8197–8203

Speth C, Prohászka Z, Mair M, Stöckl G, Zhu X, Jöbstl B, Füst G, Dierich MP (1999) A 60 kD heat-shock protein-like molecule interacts with the HIV transmembrane glycoprotein gp41. Mol Immunol 36:619–628

Stefano L, Racchetti G, Bianco F, Passini N, Gupta RS, Panina Bordignon P, Meldolesi J (2009) The surface-exposed chaperone, Hsp60, is an agonist of the microglial TREM2 receptor. J Neurochem 110:284–294

Stillman B, Stewart D (2004) The genome of *Homo sapiens*. Cold Spring Harbor, New York

Teitler M, Herrick-Davis K, Purohit A (2002) Constitutive activity of G-protein coupled receptors: emphasis on serotonin receptors. Curr Top Med Chem 2:529–538

Tompa P, Szász C, Buday L (2005) Structural disorder throws new light on moonlighting. Trends Biochem Sci 30:484–489

Triantafilou K, Triantafilou M, Dedrick RL (2001) A CD14-independent LPS receptor cluster. Nat Immunol 2:338–345

Tsan MF, Gao B (2009) Heat shock proteins and immune system. J Leukoc Biol 85:905–910

Tsugawa H, Ito H, Ohshima M, Okawa Y (2007) Cell adherence-promoted activity of *Plesiomonas shigelloides* groEL. J Med Microbiol 56:23–29

Vantourout P, Radojkovic C, Lichtenstein L, Pons V, Champagne E, Martinez LO (2010) Ecto-F_1-ATPase: a moonlighting protein complex and an unexpected apoA-I receptor. World J Gastroenterol 16:5925–5935

Verdegaal ME, Zegveld ST, van Furth R (1996) Heat shock protein 65 induces CD62e, CD106, and CD54 on cultured human endothelial cells and increases their adhesiveness for monocytes and granulocytes. J Immunol 157:369–376

Wampler JL, Kim KP, Jaradat Z, Bhunia AK (2004) Heat shock protein 60 acts as a receptor for the Listeria adhesion protein in Caco-2 cells. Infect Immun 72:931–936

Watarai M, Kim S, Erdenebaatar J, Makino S, Horiuchi M, Shirahata T, Sakaguchi S, Katamine S (2003) Cellular prion protein promotes Brucella infection into macrophages. J Exp Med 198:5–17

Winrow VR, Mesher J, Meghji S, Morris CJ, Fox S, Coates AR, Tormay P, Blake D, Henderson B (2008) The two homologous chaperonin 60 proteins of *Mycobacterium tuberculosis* have distinct effects on monocyte differentiation into osteoclasts. Cell Microbiol 10:2091–2104

Wu Z, Zhang W, Lu C (2008) Comparative proteome analysis of secreted proteins of *Streptococcus suis* serotype 9 isolates from diseased and healthy pigs. Microb Pathog 45:159–166

Wuppermann FN, Mölleken K, Julien M, Jantos CA, Hegemann JH (2008) *Chlamydia pneumoniae* GroEL1 protein is cell surface associated and required for infection of HEp-2 cells. J Bacteriol 190:3757–3767

Xie J, Zhu H, Guo L, Ruan Y, Wang L, Sun L, Zhou L, Wu W, Yun X, Shen A, Gu J (2010) Lectin-like oxidized low-density lipoprotein receptor-1 delivers heat shock protein 60-fused antigen into the MHC class I presentation pathway. J Immunol 185:2306–2313

Yamaguchi H, Osaki T, Taguchi H, Hanawa T, Yamamoto T, Kamiya S (1996) Flow cytometric analysis of the heat shock protein 60 expressed on the cell surface of Helicobacter pylori. J Med Microbiol 45:270–277

Yamaguchi H, Osaki T, Kurihara N, Taguchi H, Hanawa T, Yamamoto T, Kamiya S (1997a) Heat-shock protein 60 homologue of *Helicobacter pylori* is associated with adhesion of *H. pylori* to human gastric epithelial cells. J Med Microbiol 46:825–831

Yamaguchi H, Osaki T, Taguchi H, Hanawa T, Yamamoto T, Fukuda M, Kawakami H, Hirano H, Kamiya S (1997b) Growth inhibition of Helicobacter pylori by monoclonal antibody to heat-shock protein 60. Microbiol Immunol 41:909–916

Yoshida N, Oeda K, Watanabe E, Mikami T, Fukita Y, Nishimura K, Komai K, Matsuda K (2001) Protein function. Chaperonin turned insect toxin. Nature 411:44

Zanin-Zhorov A, Nussbaum G, Franitza S, Cohen IR, Lider O (2003) T cells respond to heat shock protein 60 via TLR2: activation of adhesion and inhibition of chemokine receptors. FASEB J 17:1567–1569

Chapter 12
Pathophysiological Barriers Impeding the Delivery of Heat Shock Protein (HSP)-Based Macromolecules and Nanotherapeutics to Solid Tumors

Peter Vaupel and Gabriele Multhoff

Abstract The chaotic microvasculature of solid tumors leads to significant impediment of delivery, uneven distribution and compromised penetration of macromolecules and nanotherapeutics from tumor microvessels across the interstitial compartment to cancer cells, especially to cells distant from microvessels. To reach viable tumor cells in a relevant concentration, diagnostic and therapeutic agents are confronted with several obstacles: disturbed convective transport within the chaotic vascular compartment, spatio-temporally uneven distribution within the tissue, and significant shunt flow bypassing the exchange processes between the vascular bed and the extravascular space. Extravasation and extravascular convection of macromolecules and nanoparticles are mainly impaired by high interstitial fluid pressure. Furthermore, marked gradients in concentrations of macromolecules and nanoparticles exist within the extravascular space limiting anticancer activities with increasing distance from tumor blood vessels.

12.1 Introduction

Heat shock proteins (HSPs) are highly conserved from bacteria to humans. In human tumor cells, macromolecules such as HSPs are frequently overexpressed (Schmid et al. 2007). The search for targets which are selectively located on the plasma membrane of tumor—but not normal—cells resulted in the identification of members of the 70 and 90 kDa HSP families such as Hsp70, grp78 and Hsp90 (Multhoff et al. 1995; Tamura et al. 1993; Li and Lee 2006; Lee and Hendershot 2006; Ullrich et al. 1986; Becker et al. 2004; Cid et al. 2009). These membrane-expressed

P. Vaupel (✉)
Department of Radiotherapy and Radiooncology, Klinikum rechts der Isar,
Technical University of Munich, Ismaninger Strasse 22, 81675 Munich, Germany
e-mail: vaupel@uni-mainz.de

G. Multhoff
Klinikum rechts der Isar, TU-Munchen, Department of Radiation Oncology and Helmholtz Center Munich, National Research Center for Environmental Health GmbH,
85764 Munich, Germany
e-mail: Gabriele.multhoff@lrz.tu-muenchen.de

B. Henderson, A. G. Pockley (eds.), *Cellular Trafficking of Cell Stress Proteins in Health and Disease*, Heat Shock Proteins 6, DOI 10.1007/978-94-007-4740-1_12,

HSPs qualify for the development of tumor-specific therapeutics. In this context, HSP-specific, naked antibodies for eliciting antibody dependent cellular cytotoxicity (ADCC; Stangl et al. 2011) and HSP-based antibody drug conjugates (ADC), and lipid/protein-based HSP-nanoparticles (nanoagents) are considered as examples of innovative and tumor-specific therapeutics. A major goal of this chapter is to elucidate pathophysiological barriers which might impede the delivery of these novel therapeutics. Knowledge of the bioavailability of anticancer agents is pivotal for a better understanding of their mode of action.

12.2 Vascularization is the Basic Prerequisite for the Macroscopic Growth of Solid Tumors

Most, if not all, solid tumors initiate from avascular aggregates of malignant cells. "Microscopic tumors" exchange nutrients and breakdown products of metabolism with their surroundings by simple diffusion (Brem et al. 1976; Vaupel et al. 1989, 2004, 2006; Vaupel 1994b, 2009a). The growth of avascular three-dimensional tumor aggregates to a size of 1–2 mm is therefore self-limiting. Small tumors can stay dormant for a very long time period until an angiogenic switch occurs (Schneider et al. 2009). Rapid tumor growth, tumor progression, invasion, local spread and distant metastasis to other organs or tissues following the avascular growth period are possible only if convective transport (nutrient supply and waste removal) is initiated through nutritive blood flow, i.e., flow through tumor microvessels that guarantees adequate exchange processes between the microcirculatory bed and the cancer cells. This notion has led to the dogmatic assumption that both tumor growth and tumor spread are dependent on rigorous angiogenesis (Sivridis et al. 2003). This implies that vascularization is a basic prerequisite for macroscopic tumor growth, invasion and metastasis. At the same time the tumor microcirculation is the major transport mechanism for the effective delivery of therapeutic molecules such as HSP-based vaccines. A compromised microcirculation is therefore considered as a primary obstacle in drug delivery (Jain and Stylianopoulos 2010; Vaupel 2012).

In this chapter, consequences of irregular structure and function of the tumor microcirculation are discussed for the delivery of macromolecules (e.g., HSP-antibodies, HSP-drug conjugates) and HSP nanoparticle formulations (such as liposomes, protein-based nanoparticles). Special emphasis will be given to the following three-step process: (a) delivery by the heterogeneous microcirculation, (b) extravasation through the leaky vessel walls, and (c) traversing of the special interstitial compartment. These different steps are characterized by special barriers to therapy that may not be shared by hemoblastoses.

To better understand the barriers for delivery of macromolecules and nanoparticles, basic principles of tumor angiogenesis, lymphangiogenesis, vascular functionality and peculiarities of the interstitial compartment are described first.

12.3 Basic Principles of Blood Vessel Formation in Solid Tumors

When considering the continuous and indiscriminate formation of a vascular network in rapidly growing tumors, six different mechanisms have been described (Carmeliet and Jain 2000; Ribatti et al. 2003; Vaupel et al. 2004, Vaupel 2006, 2009a, 2012):

(a) angiogenesis by endothelial sprouting from pre-existing venules,
(b) co-option of existing vessels,
(c) vasculogenesis,
(d) intussusception,
(e) vascular mimicry, and
(f) microvessel formation by bone-marrow derived myeloid cells.

12.3.1 Angiogenesis

The avascular (=prevascular) growth phase characteristic of a "dormant" tumor and the vascular phase in which "explosive" growth ensues in many solid tumors are separated by the "angiogenic switch" (Bergers and Benjamin 2003). This switch is "off" when the effect of pro-angiogenic molecules is balanced by that of anti-angiogenic molecular players. It is "on" when the net balance is tipped in favor of angiogenesis (Hanahan and Weinberg 2000). Pro- and anti-angiogenic molecules can be released from cancer cells, endothelial cells (ECs), stromal and inflammatory cells or can be mobilized from the extracellular matrix (Carmeliet and Jain 2000; Ribatti et al. 2003). The "angiogenic switch", a pivotal and early event in tumor progression greatly depends on one or more positive regulators such as growth factors, permeability regulating factors, migration stimulators, proteolytic enzymes (balanced by their inhibitors), extracellular matrix molecules and adhesion molecules (Vaupel 2012). *Vascular-specific growth factors* include vascular endothelial growth factors (VEGF) and their receptors, the angiopoietin family (Ang) and Tie receptors, and the ephrins. *Non-specific factors* comprise platelet-derived growth factor (PDGF), fibroblasts growth factors (FGFs), transforming growth factor-β (TGF-β), tumor necrosis factor-α (TNF-α), epidermal growth factor (EGF) and several others.

A central inducer of growth of new blood vessels is VEGF, which was originally described as vascular permeability factor (VPF). Expression of VEGF is regulated by hypoxia (Semenza 2000; Pugh and Ratcliffe 2003), hypoglycemia (Shweiki et al. 1995), acidosis (Carmeliet and Jain 2000), activation of oncogenes or deletion of tumor-suppressor genes that control production of angiogenesis regulators (Kerbel 2000), cytokines and hormones (M. Neeman, personal communication).

The process of angiogenesis is extremely complex and requires balanced interactions with biologic redundancy. Apart from the angiogenic switch and up-regulation of pro-angiogenic molecules and binding of the latter to specific endothelial cell receptors and ligand-receptor interaction, major steps in the "angiogenic cascade" include:

- dilation of existing vessels,
- activation of endothelial cells (ECs),
- hyperpermeability of postcapillary venules and vessel destabilization,
- localized degradation of basement membrane,
- matrix remodelling (degradation of extracellular matrix in response to activation of matrix metalloproteinases (MMPs), formation of a new provisional extravascular matrix by leaked plasma proteins),
- migration of ECs,
- cell-cell-contacts, sprout formation,
- extension of sprouts by EC proliferation,
- tube formation ("canalization"), fusion to form vascular loops,
- (non-mandatory) recruitment of pericytes and smooth muscle cells, and
- (often improper) vessel maturation.

These different steps are partly concurrent, partly in series and sequential. They occur in different parts of primary tumors and in metastases at the same time (Vaupel 2012).

12.3.2 Vascular Co-Option

Tumors often appear to co-opt vessels, i.e., they can incorporate pre-existing vessels within a vascularized host tissue to initiate vessel-dependent tumor growth as opposed to classic angiogenesis (Ribatti et al. 2003). Later on, ECs of the co-opted vessels release angiopoietin-2 (probably by autocrine action) which leads to vascular destabilization and vascular collapse. The resulting hypoxia and nutrient deprivation yields an up-regulation of VEGF and "secondary" angiogenesis (Ellis et al. 2001).

12.3.3 Vasculogenesis

For vasculogenesis, *de novo* vessel formation through incorporation of circulating endothelial precursor cells (angioblasts) from bone marrow or peripheral blood is mandatory.

12.3.4 Intussusception

In intussusception, interstitial tissue columns insert into the lumen of pre-existing vessels and lead to partition of the initial vessel lumen (Patan et al. 1996).

12.3.5 Vascular Mimicry

In central areas of melanomas *de novo* generation of pseudo-vascular channels without the participation of endothelial cells has been described. A contribution of cancer cells to the wall of tumor vessels has also been reported for tumor entities other than melanomas (Ruoslahti 2002; Ribatti et al. 2003). The concept of vascular mimicry remains controversial (McDonald et al. 2000).

12.3.6 Microvessel Formation by Myeloid Cells

Microvessel formation may also be triggered by a subset of bone-marrow derived myeloid cells infiltrating solid tumors. The tumor vasculature is characterized by vigorous proliferation leading to immature, structurally defective and, in terms of perfusion, ineffective microvessels. "Tumor vessels lack the signals to mature" and tumor vasculature is often described as an "aberrant monster" (Shchors and Evan 2007). Consequently, tumor blood flow is chaotic and heterogenous (see subchapter 5).

12.4 Tumor Lymphangiogenesis

Although the metastatic spread of tumor cells to regional lymph nodes is a common feature of many human cancers, it is not clear whether shedding tumor cells utilize existing lymphatic vessels or whether tumor dissemination requires *de novo* formation of lymphatics (Ribatti et al. 2003). The notion that tumor microcirculation may be supported by a newly formed, tumor-induced lymphatic network has so far not been convincingly confirmed. VEGF-C, VEGF-D and their corresponding receptors have been identified as specific lymphangiogenic factors in several tumors (Jussila and Alitalo 2002; Ribatti et al. 2003). It has been proposed that functional lymphatics at the tumor-normal tissue interface are sufficient for lymphatic metastasis, because the tumor center was found to contain no functional lymphatics (Padera et al. 2002). The lymphatic vessels in the tumor center are usually collapsed (compressed) due to the high interstitial pressure caused by the growing tumor mass in a confined space. In the tumor periphery VEGF-C causes lymphatics to enlarge collecting interstitial fluid and shedded cancer cells mediating metastasis (Carmeliet and Jain 2000).

12.5 Tumor Vascularity and Blood Flow

As already mentioned, the key players in tumor angiogenesis are VEGF, angiopoietins, ephrins and their corresponding receptors. Their excessive production causes the formation of structurally and functionally abnormal blood vessels. The tumor vasculature can be described as a system that is maximally stimulated, but only minimally fulfils the metabolic demands of the growing tumor (Hirst and Flitney 1997).

Microvessels in solid tumors are often dilated, tortuous, elongated, and saccular (for a recent review see Vaupel 2012). There is significant arterio-venous shunt perfusion accompanied by chaotic vascular organization which lacks any regulation matched to the metabolic demands or functional status of the tissue. Excessive branching is a common finding, often coinciding with blind vascular endings. Incomplete or even missing endothelial lining and discontinuous or even absent basement membranes (in some areas while not in others) result in an increased vascular permeability with extravasation of blood plasma and red blood cells expanding the interstitial fluid space and drastically increasing the hydrostatic pressure in the tumor interstitium. In solid tumors there is a *rise in viscous resistance to flow*, mainly due to hemoconcentration. Aberrant vascular morphology and a decrease in vessel density are responsible for an *increase in geometric resistance to flow*, which can lead to an inadequate perfusion. The average velocity of red blood cells (erythrocytes) in tumor microvessels can be an order of magnitude lower than in normal microvasculature (Jain and Stylianopoulos 2010). Substantial spatial heterogeneity in the distribution of tumor vessels and significant temporal heterogeneity in the microcirculation within a tumor (Gillies et al. 1999) may result in a considerably uneven distribution of tumor tissue oxygenation and of a number of other factors, which are usually closely linked and which define the metabolic microenvironment. Variations in these relevant parameters between tumors are often more pronounced than differences occurring between different locations or microareas within a tumor (Vaupel and Höckel 2000; Vaupel et al. 2001).

Blood flow in solid tumors can vary considerably, ranging—*grosso modo*—from 0.01 to 3.0 $ml \times g^{-1} \times min^{-1}$ (Vaupel 2012). Tumors can thus have flow rates similar to those measured in tissues with a high metabolic rate, or can exhibit perfusion rates comparable to those of tissues with low metabolic turnover. Flow data from multiple sites of measurement show marked heterogeneity within individual tumors (with even non-perfused regions). Delivery of anticancer agents is similarly compromised. When measuring the microregional distribution of perfusion within a tumor using the H_2-clearance technique, microflow rates of 0.01–3.0 μl/μl tissue/min (median microflow: 0.5 μl/μl/min) were observed with an accumulation of measured values <0.01 μl/μl/min. Considering all flow values assessed by this technique, flow data can vary by a factor of approx. 300. This flow variability in experimental animal tumors is thus significantly larger than that seen in individual tumor entities in the clinical setting (×100). This heterogeneous flow distribution clearly mirrors the chaotic microvasculature found in solid tumors (Vaupel et al. 2009a). However, tumor-to-tumor variability seems to be more pronounced than intra-tumor heterogeneity (Vaupel and Höckel 2000).

12.6 Arterio-Venous Shunt Perfusion in Tumors

First rough estimations concerning the arterio-venous shunt flow in malignant tumors showed that at least 30 % of the arterial blood can pass through experimental tumors without participating in the microcirculatory exchange processes (Vaupel et al. 1977; Endrich et al. 1982; Weiss et al. 1979). In patients receiving intra-arterial chemotherapy for head and neck cancer, shunt flow is reported to be 8–43 % of total tumor blood flow, the latter consistently exceeding normal tissue perfusion of the scalp (Wheeler et al. 1986). The mean fractional shunt perfusion of tumors was 23 ± 13 % in studies utilizing ^{99m}Tc-labeled macroaggregated albumin (diameter of the particles, 15–90 µm). The significance of this shunt flow on local, intra-tumor drug distribution (pharmacokinetics), and on relevant metabolic phenomena has not yet been systematically studied and remains speculative.

12.7 Volume and Characteristics of the Tumor Interstitial Space

The interstitial compartment of solid tumors is signficantly different from that of most normal tissues (Vaupel and Mueller-Klieser 1983, Vaupel 2009a, 2012). In general, the tumor interstitial space is characterized by

(a) a distension of its volume, which is three to five times larger than in most normal tissues,
(b) a relatively large quantity of free fluid in contrast to normal tissues where almost all of the fluid is in the gel phase,
(c) a quick diffusion of low-molecular weight, water-soluble agents (e.g., contrast agents, drugs) (Reinhold 1971),
(d) a compromised convective transport of high-molecular weight substances.

12.8 Interstitial Fluid Pressure in Tumors

As already mentioned, the growing tumor produces new, often abnormally leaky (hyper-permeable) microvessels, but is unable to form its own functional lymphatics (Fukumura and Jain 2007). As a result, there is significant bulk flow of free fluid into the interstitial space as long as a relevant pressure difference (both hydrostatic and oncotic) between the intra- and extravascular space exists. Whereas convective currents into the interstitial compartment are estimated to be about 0.5–1 % of plasma flow in normal tissue, in human cancers water efflux into the interstitium can reach 15 % of the respective plasma flow (Vaupel 1994a, b).

After seeping copiously out of the highly permeable tumor microvessels—an equilibrium is reached when the hydrostatic and oncotic pressures within the microvessels and the respective interstitial pressures become equal—fluid accumulates in the tumor extracellular space and a high interstitial fluid pressure (IFP) builds up in solid tumors (Young et al. 1988; Gutmann et al. 1992; Less et al. 1992; Milosevic et al. 2004).

Besides vessel hyper-permeability and lack of functional lymphatics, interstitial fibrosis, contraction of the interstitial space mediated by stromal fibroblasts, and high oncotic pressures within the interstitium may contribute to the development of interstitial hypertension (Heldin et al. 2004). Whereas in most normal tissues IFP is slightly subatmospheric ("negative") or just above atmospheric values (Guyton and Hall 2006), an interstitial hypertension with values up to 60–70 mmHg (Heldin et al. 2004) develops in cancers which forms a "physiologic" barrier to the delivery (via convection) of therapeutic macromolecules (e.g., HSP antibodies, HSP-drug conjugates) and HSP-nanoparticles to the cancer cells (Vaupel 2009b, 2012).

The tumor IFP seems to be rather uniform throughout the center of the tumor but drops steeply in the periphery. Fluid is squeezed out of the high- to the low-pressure regions at the tumor/normal tissue interface, carrying away antitumor drugs and diagnostic agents.

Despite increased overall leakiness, not all tumor microvessels are leaky. Vascular permeability varies from tumor to tumor and exhibits spatio-temporal heterogeneity ("4D-heterogeneity") within the same tumor as well as during tumor growth and/or regression. Furthermore, IFP in tumors fluctuates with changing microvascular pressures (Netti et al. 1995).

Transmural coupling between IFP and microvascular pressure due to the high leakiness of tumor microvessels can abolish perfusion pressure differences between up- and down-stream tumor microvessels and thus can lead to blood flow stasis in tumors without "physically" occluding (compressing) the vessels in some microregions (Fukumura and Jain 2007). The equilibration of hydrostatic pressures between the interstitial and microvascular compartments is accompanied by a similar equilibration of oncotic pressures in both spaces (20.0 mmHg in plasma vs. 20.5 mmHg in the interstitial space of solid tumors (Stohrer et al. 2000).

12.9 Role of the Disorganized, Compromised Microcirculation As an Obstacle in Delivery of Anticancer Agents

12.9.1 Blood-Borne Delivery

As already mentioned, there is a disturbed balance of pro-angiogenic and anti-angiogenic molecules which induces an unregulated angiogenesis. This leads to the development of a disorganized microvasculature and significant arterio-venous shunt perfusion and thus to an *inefficient delivery of therapeutic molecules* (e.g., HSP antibodies, HSP antibody-drug conjugates and nanotherapeutics) and nutrients (e.g., oxygen and glucose) through the vascular system of the tumor. The situation is further aggravated by flow-dependent spatio-temporal *heterogeneities in the distribution of plasma-borne agents* (and their metabolites). These "4D-heterogeneities" are not static, but instead are quite dynamic, and therefore more complex than has been previously assumed.

The considerable impediment of fluctuating (intermittent) perfusion to successful cancer therapy has been comprehensively reviewed by Durand (2001) and Durand and Aquino-Parsons (2001a, b).

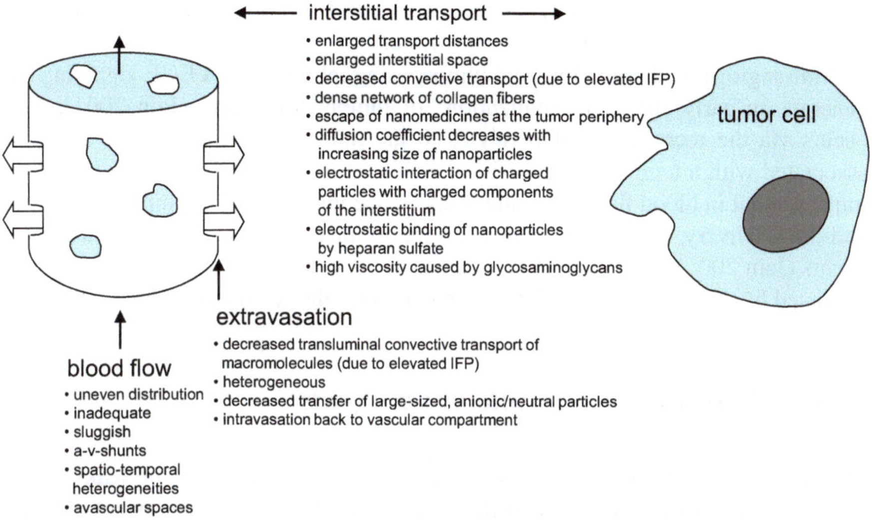

Fig. 12.1 Pathophysiological barriers impeding the delivery of HSP-based macromolecules and nanotherapeutics to solid tumors. Obstacles in blood born delivery are shown on the *left*, hindrances in the extravasation are listed in the *center*, and barriers to interstitial transport are depicted on the *right* hand side

The mean vascular density in most tumor areas is generally lower than that in normal tissues, and thus *diffusion distances are enlarged* (see Fig. 12.1). Penetration of anticancer agents from tumor capillaries to tumor cells is therefore compromised. As shown by Primeau et al. (2005), the concentration of low-molecular drugs including antibodies decreases exponentially with distance from tumor blood vessels, decreasing to half its perivascular concentration at a distance of about 40–50 μm. (Note: Intervessel distances in solid tumors can vary between 10 and 1,000 μm with a median value of approx. 100 μm (Folarin et al. 2010). Thus, many viable tumor cells are not exposed to detectable concentrations of low-molecular weight drugs following a single injection. In these tumor regions distant to patent microvessels some agents (i.e., preferentially drugs with a short half-life within the circulation) cannot achieve sufficient concentrations to exert lethal toxicity for all of the viable cells further away from the tumor microvasculature system (Primeau et al. 2005; Minchinton and Tannock 2006; Di and Bocci 2007). In addition, in these tumor regions, concentrations of key nutrients are also low. This results in marked gradients with higher cellular turnover rates very close to blood vessels and lower cell proliferation (and cell cycle arrest) in regions which are further away from microvessels before treatment. After/between different treatment cycles this might enhance a repopulation of surviving tumor cells (Tannock 1968, 2001; Hirst and Denekamp 1979).

Cells dividing at reduced rate would be protected from effects of cytotoxic therapies whose activity is "selective" for rapidly dividing cell populations with a short cell cycle. This specifically comprises cells in the S-phase (Hall and Giaccia 2006;

Tredan et al. 2007). There is evidence that growth fraction decreases as tumor size increases, at least in experimental tumor systems.

Anti-angiogenic therapy for solid tumors using inhibition of VEGF-signaling can generate an early-phase "normalization" of tumor vasculature (Jain 2001). This occurs via the recruitment of pericytes to the tumor microvasculature, an effect associated with a temporary, short-lived stabilization of vessels and a hypothetical improvement in blood flow. The latter may be accompanied by an improved oxygen and drug delivery, creating a window of opportunity for the delivery of anticancer agents (Jain 2005). The postulated increase in pericyte recruitment is thought to be mediated by angiopoietin-1 and matrix metalloproteinases (Lin and Sessa 2004).

12.9.2 Extravasation of Anticancer Agents

Therapeutic (and diagnostic) molecules and nanomedicines cross the leaky vessel walls by two major mechanisms: diffusion and convection. Large pore sizes of tumor microvessels facilitate these transport processes. Diffusion is the prevailing molecular transport of small-size molecules driven by concentration gradients. Convection is driven by hydrostatic pressure gradients and is the dominant mode of transport for large molecules, liposomes and other nanoparticles (Kuszyk et al. 2001). Due to the increased interstitial pressure (interstitial hypertension) significant hindering of the transport of macromolecules and nanoparticles into the extravascular space by convection has to be considered (see Fig. 12.1). For this reason, the main mechanism of mass transport across vessel walls is diffusion. This process is significantly slower than convection, especially for macromolecules and nanoparticles (Jain and Stylianopoulos 2010). Vessel wall hyperpermeability (enhanced porosity) is thus counteracted by elevated IFP in tumors (and by the large size of nanoparticles).

Vascular permeability decreases with increasing size of the transported nanoparticles and is higher for cationic compounds than for their anionic or neutral counterparts (Jain and Stylianopoulos 2010).

12.10 Barriers to Interstitial Transport

As already outlined, the interstitial compartment of tumors differs significantly from that of normal tissues. As a result of (a) vessel leakiness, (b) lack of functional lymphatics, (c) interstitial fibrosis, and (d) contraction of the interstitial matrix mediated by stromal fibroblasts, most solid tumors have an *increased interstitial (hydrostatic) fluid pressure* (Jain 1987, 1990; Heldin et al. 2004; Milosevic et al. 2004; Cairns et al. 2006).

As already mentioned above, increased interstitial fluid pressure (IFP) within solid tumors decreases extravasation and inhibits interstitial transport of larger particles (e.g., antibodies, antibody drug conjugates) by convection (see Fig. 12.1). Macromolecules rely more heavily on convection as opposed to simple diffusional transport of low-molecular weight drugs. Compounds larger than 60 nm in diameter are not able to effectively diffuse through the extracellular matrix of highly fibrotic tumors. Interstitial transport of macromolecules is further impaired by a much *denser network of interconnected collagen fibers* in the extracellular matrix of tumors as compared to normal tissues leaving them in higher concentrations in perivascular areas only (Jain and Stylianopoulos 2010). The transport of compounds with sizes of up to 1,000 nm is further hindered by highly negative charged heparan sulfate in the matrix.

Heterogeneous mobility and distribution of nanomedicines is additionally caused by two phases in the matrix: a more aqueous phase is found in regions with low fiber content (fast component with relatively high diffusivity), and a more viscous phase is due to a high concentration of collagen fibers in a dense matrix (slow component with high retention of compounds). Collagen content in tumors is much higher and collagen fibers are much thicker than in normal tissues leading to an increased mechanical stiffness of the tissue (Netti et al. 2000; Heldin et al. 2004). The interstitium also contains stromal cells and enzymes that can affect the activity and delivery of agents to the tumor cells (Kuszyk et al. 2001).

IFP is almost uniform throughout a tumor and drops precipitously at the tumor/normal tissue interface. For this reason, the *interstitial fluid oozes out of the tumor* into the surrounding normal tissue, carrying away anticancer agents, growth factors or actively released heat shock proteins and cancer cells with it (see Fig. 12.1) (Fukumura and Jain 2007). Shedded cancer cells may mediate metastases. As another consequence of this drop in IFP, blood may be diverted away from the tumor center toward the periphery where anticancer agents may be lost from larger vessels.

Transmural coupling between IFP and microvascular pressure can critically reduce perfusion pressure between up- and downstream tumor blood vessels leading to blood flow stasis and thus, inadequate delivery of anticancer agents, in addition to the mechanisms impairing blood flow already mentioned above.

Interactions between cancer cells and the extracellular matrix can affect their response to therapy. The basic mechanisms involved in the so-called *adhesion-mediated drug resistance* are rather complex and still under investigation. Agents that can modulate cell adhesion might enhance the effects of anticancer therapy (Tredan et al. 2007).

Since increased interstitial fluid pressure (IFP) in malignant tumors can impair the delivery of therapeutic agents, interventions that can reduce IFP may improve drug delivery. Several types of treatment have been shown to decrease tumor IFP in patients. These compounds include VEGF-antagonists, corticosteroids, hyaluronidase, and prostaglandin E1 (Cairns et al. 2006; Heldin et al. 2004; Lee et al. 2000; Willett et al. 2004; Willett et al. 2005; Batchelor et al. 2007).

12.11 Conclusions

Besides "classical" drug resistance which is based on the molecular background of tumor cells, (e.g., mutations, alterations of gene expression, epigenetic changes), substantial experimental and clinical evidence suggests that the irregular and heterogeneous structure and function of the microcirculation and the self-perpetuating hostile pathophysiological microenvironment of solid tumors may—to a large extent—mediate resistance of solid tumors to anticancer drugs. Therefore, distribution of tumor blood flow, extravasation and interstitial transport are increasingly receiving attention in the clinical setting. One of the goals of translational cancer research is to obtain a better understanding of the compromised delivery and distribution of anticancer compounds in solid tumors (intratumor pharmacokinetics) in order to improve patients' outcomes.

Acknowledgments The authors want to thank Anett Lange for excellent editorial assistance and for art work.

References

Batchelor TT, Sorensen AG, di TE, Zhang WT, Duda DG, Cohen KS, Kozak KR, Cahil DP, Chen PJ, Zhu M, Ancukiewicz M, Mrugala MM, Plotkin S, Drappatz J, Louis DN, Ivy P, Scadden DT, Benner T, Loeffler JS, Wen PY, Jain RK (2007) AZD2171, a pan-VEGF receptor tyrosine kinase inhibitor, normalizes tumor vasculature and alleviates edema in glioblastoma patients. Cancer Cell 11:83–95

Becker B, Multhoff G, Farkas B, Wild PJ, Landthaler M, Stolz W, Vogt T (2004) Induction of Hsp90 protein expression in malignant melanomas and melanoma metastases. Exp Dermatol 13:27–32

Bergers G, Benjamin LE (2003) Tumorigenesis and the angiogenic switch. Nat Rev Cancer 3:401–410

Brem S, Brem H, Folkman J, Finkelstein D, Patz A (1976) Prolonged tumor dormancy by prevention of neovascularization in the vitreous. Cancer Res 36:2807–2812

Cairns R, Papandreou I, Denko N (2006) Overcoming physiologic barriers to cancer treatment by molecularly targeting the tumor microenvironment. Mol Cancer Res 4:61–70

Carmeliet P, Jain RK (2000) Angiogenesis in cancer and other diseases. Nature 407:249–257

Cid C, Regidor I, Poveda PD, Alcazar A (2009) Expression of heat shock protein 90 at the cell surface in human neuroblastoma cells. Cell Stress Chaperones 14:321–327

Di PA, Bocci G (2007) Drug distribution in tumors: mechanisms, role in drug resistance, and methods for modification. Curr Oncol Rep 9:109–114

Durand RE (2001) Intermittent blood flow in solid tumours—an under-appreciated source of 'drug resistance'. Cancer Metastasis Rev 20:57–61

Durand RE, Aquino-Parsons C (2001a) Clinical relevance of intermittent tumour blood flow. Acta Oncol 40:929–936

Durand RE, Aquino-Parsons C (2001b) Non-constant tumour blood flow—implications for therapy. Acta Oncol 40:862–869

Ellis LM, Liu W, Ahmad SA, Fan F, Jung YD, Shaheen RM, Reinmuth N (2001) Overview of angiogenesis: biologic implications for antiangiogenic therapy. Semin Oncol 28:94–104

Endrich B, Hammersen F, Götz A, Messmer K (1982) Microcirculatory blood flow, capillary morphology and local oxygen pressure of the hamster amelanotic melanoma A-Mel-3. J Natl Cancer Inst 68:475–485

Folarin AA, Konerding MA, Timonen J, Nagl S, Pedley RB (2010) Three-dimensional analysis of tumour vascular corrosion casts using stereoimaging and micro-computed tomography. Microvasc Res 80:89–98

Fukumura D, Jain RK (2007) Tumor microenvironment abnormalities: causes, consequences, and strategies to normalize. J Cell Biochem 101:937–949

Gillies RJ, Schornack PA, Secomb TW, Raghunand N (1999) Causes and effects of heterogeneous perfusion in tumors. Neoplasia 1:197–207

Gutmann R, Leunig M, Feyh J, Goetz AE, Messmer K, Kastenbauer E, Jain RK (1992) Interstitial hypertension in head and neck tumors in patients: correlation with tumor size. Cancer Res 52:1993–1995

Guyton AC, Hall JE (2006) Textbook of medical physiology, 11th edn. Elsevier, Philadelphia

Hall EJ, Giacca AJ (2006) Radiobiology for the radiologist, 6th edn. Lippincott Williams & Wilkins, Baltimore

Hanahan D, Weinberg RA (2000) The hallmarks of cancer. Cell 100:57–70

Heldin CH, Rubin K, Pietras K, Ostman A (2004) High interstitial fluid pressure—an obstacle in cancer therapy. Nat Rev Cancer 4:806–813

Hirst DG, Denekamp J (1979) Tumour cell proliferation in relation to the vasculature. Cell Tissue Kinet 12:31–42

Hirst DG, Flitney FW (1997) The physiological importance and therapeutic potential of nitric oxide in the tumour-associated vasculature. In: Bicknel R, Lewis CE, Ferrara N (eds) Tumour angiogenesis. Oxford University Press, Oxford, pp 153–167

Jain RK (1987) Transport of molecules across tumor vasculature. Cancer Metastasis Rev 6:559–593

Jain RK (1990) Physiological barriers to delivery of monoclonal antibodies and other macromolecules in tumors. Cancer Res 50:814s–819s

Jain RK (2001) Normalizing tumor vasculature with anti-angiogenic therapy: a new paradigm for combination therapy. Nat Med 7:987–989

Jain RK (2005) Normalization of tumor vasculature: an emerging concept in antiangiogenic therapy. Science 307:58–62

Jain RK, Stylianopoulos T (2010) Delivering nanomedicine to solid tumors. Nat Rev Clin Oncol 7:653–664

Jussila L, Alitalo K (2002) Vascular growth factors and lymphangiogenesis. Physiol Rev 82:673–700

Kerbel RS (2000) Tumor angiogenesis: past, present and the near future. Carcinogenesis 21:505–515

Kuszyk BS, Corl FM, Franano FN, Bluemke DA, Hofman LV, Fortman BJ, Fishman EK (2001) Tumor transport physiology: implications for imaging and imaging-guided therapy. AJR Am J Roentgenol 177:747–753

Lee AS, Hendershot LM (2006) ER stress and cancer. Cancer Biol Ther 5:721–722

Lee CG, Heijn M, di TE, Griffon-Etienne G, Ancukiewicz M, Koike C, Park KR, Ferrara N, Jain RK, Suit HD, Boucher Y (2000) Anti-Vascular endothelial growth factor treatment augments tumor radiation response under normoxic or hypoxic conditions. Cancer Res 60:5565–5570

Less JR, Posner MC, Boucher Y, Borochovitz D, Wolmark N, Jain RK (1992) Interstitial hypertension in human breast and colorectal tumors. Cancer Res 52:6371–6374

Li J, Lee AS (2006) Stress induction of GRP78/BiP and its role in cancer. Curr Mol Med 6:45–54

Lin MI, Sessa WC (2004) Antiangiogenic therapy: creating a unique "window" of opportunity. Cancer Cell 6:529–531

McDonald DM, Mun L, Jain RK (2000) Vasculogenic mimicry: how convincing, how novel, and how significant? Am J Pathol 156:383–388

Milosevic M, Fyles A, Hedley D, Hil R (2004) The human tumor microenvironment: invasive (needle) measurement of oxygen and interstitial fluid pressure. Semin Radiat Oncol 14:249–258

Minchinton AI, Tannock IF (2006) Drug penetration in solid tumours. Nat Rev Cancer 6:583–592

Multhoff G, Botzler C, Wiesnet M, Müller E, Meier T, Wilmanns W, Issels RD (1995) A stress-inducible 72-kDa heat-shock protein (HSP72) is expressed on the surface of human tumor cells, but not on normal cells. Int J Cancer 61:272–279

Netti PA, Baxter LT, Boucher Y, Skalak R, Jain RK (1995) Time-dependent behavior of interstitial fluid pressure in solid tumors: implications for drug delivery. Cancer Res 55:5451–5458

Netti PA, Berk DA, Swartz MA, Grodzinsky AJ, Jain RK (2000) Role of extracellular matrix assembly in interstitial transport in solid tumors. Cancer Res 60:2497–2503

Padera TP, Kadambi A, di TE, Carreira CM, Brown EB, Boucher Y, Choi NC, Mathisen D, Wain J, Mark EJ, Mun LL, Jain RK (2002) Lymphatic metastasis in the absence of functional intratumor lymphatics. Science 296:1883–1886

Patan S, Mun LL, Jain RK (1996) Intussusceptive microvascular growth in a human colon adenocarcinoma xenograft: a novel mechanism of tumor angiogenesis. Microvasc Res 51:260–272

Primeau AJ, Rendon A, Hedley D, Lilge L, Tannock IF (2005) The distribution of the anticancer drug Doxorubicin in relation to blood vessels in solid tumors. Clin Cancer Res 11:8782–8788
Pugh CW, Ratcliffe PJ (2003) Regulation of angiogenesis by hypoxia: role of the HIF system. Nat Med 9:677–684
Reinhold HS (1971) Improved microcirculation in irradiated tumors. Eur J Cancer 7:273–280
Ribatti D, Vacca A, Dammacco F (2003) New non-angiogenesis dependent pathways for tumour growth. Eur J Cancer 39:1835–1841
Ruoslahti E (2002) Specialization of tumour vasculature. Nat Rev Cancer 2:83–90
Schmid E, Gehrmann M, Brunet M, Multhoff G, Garrido C (2007) Intracellular and extracellular functions of heat shock proteins: repercussions in cancer therapy. J Leukoc Biol 81:15–27
Schneider V, Rischke HC, Drevs J (2009) Tumor angiogenesis. In: Molls M, Vaupel P, Nieder C, Anscher MS (eds) The impact of tumor biology on cancer treatment and multidisciplinary strategies. Springer, Berlin, pp 39–50
Semenza GL (2000) HIF-1: using two hands to flip the angiogenic switch. Cancer Metastasis Rev 19:59–65
Shchors K, Evan G (2007) Tumor angiogenesis: cause or consequence of cancer? Cancer Res 67:7059–7061
Shweiki D, Neeman M, Itin A, Keshet E (1995) Induction of vascular endothelial growth factor expression by hypoxia and by glucose deficiency in multicell spheroids: implications for tumor angiogenesis. Proc Natl Acad Sci U S A 92:768–772
Sivridis E, Giatromanolaki A, Koukourakis MI (2003) The vascular network of tumours—what is it not for? J Pathol 201:173–180
Stangl S, Gehrmann M, Riegger J, Kuhs K, Riederer I, Sievert W, Hube K, Mocikat R, Dressel R, Kremmer E, Pockley AG, Friedrich L, Vigh L, Skerra A, Multhoff G (2011) Targeting membrane heat-shock protein 70 (Hsp70) on tumors by cmHsp70.1 antibody. Proc Natl Acad Sci U S A 108:733–738
Stohrer M, Boucher Y, Stangassinger M, Jain RK (2000) Oncotic pressure in solid tumors is elevated. Cancer Res 60:4251–4255
Tamura Y, Tsuboi N, Sato N, Kikuchi K (1993) 70 kDa heat shock cognate protein is a transformation-associated antigen and a possible target for the host's anti-tumor immunity. J Immunol 151:5516–5524
Tannock IF (1968) The relation between cell proliferation and the vascular system in a transplanted mouse mammary tumour. Br J Cancer 22:258–273
Tannock IF (2001) Tumor physiology and drug resistance. Cancer Metastasis Rev 20:123–132
Tredan O, Galmarini CM, Patel K, Tannock IF (2007) Drug resistance and the solid tumor microenvironment. J Natl Cancer Inst 99:1441–1454
Ullrich SJ, Robinson EA, Law LW, Willingham M, Appella E (1986) A mouse tumor-specific transplantation antigen is a heat shock-related protein. Proc Natl Acad Sci U S A 83:3121–3125
Vaupel P (1994a) Blood flow and metabolic microenvironment of brain tumors. J Neurooncol 22:261–267
Vaupel P (1994b) Blood flow, oxygenation, tissue pH distribution, and bio-energetic status of tumors. Lecture 23, Ernst Schering Research Foundation
Vaupel P (2006) Abnormal microvasculature and defective microcirculatory function in solid tumors. In: Siemann DW (ed) Vascular-targeted therapies in oncology. Wiley, Chichester, pp 9–29
Vaupel P (2009a) Pathophysiology of solid tumors. In: Molls M, Vaupel P, Nieder C, Anscher MS (eds) The impact of tumor biology on cancer treatment and multidisciplinary strategies. Springer, Berlin, pp 51–92
Vaupel P (2009b) Physiological mechanisms of treatment resistance. In: Molls M, Vaupel P, Nieder C, Anscher MS (eds) The impact of tumor biology on cancer treatment and multidisciplinary strategies. Springer, Berlin, pp 273–290
Vaupel P (2012) Pathophysiological and vascular characteristics of solid tumors in relation to drug delivery. In: Kratz F, Senter P, Steinhagen H (eds) Drug delivery in oncology, vol. 1. Wiley-VCH, Weinheim, pp 33–64

Vaupel P, Mueller-Klieser W (1983) Interstistieller Raum und Mikromilieu in malignen Tumoren. Progr Appl Microcirc 1:78–90

Vaupel P, Höckel M (2000) Blood supply, oxygenation status and metabolic micromilieu of breast cancers: characterization and therapeutic relevance. Int J Oncol 17:869–879

Vaupel P, Grunewald WA, Manz R, Sowa W (1977) Intracapillary HbO_2 saturation in tumor tissue of DS-carcinosarcoma during normoxia. Adv Exp Med Biol 94:367–375

Vaupel P, Kallinowski F, Okunief P (1989) Blood flow, oxygen and nutrient supply, and metabolic microenvironment of human tumors: a review. Cancer Res 49:6449–6465

Vaupel P, Thews O, Hoeckel M (2001) Treatment resistance of solid tumors: role of hypoxia and anemia. Med Oncol 18:243–259

Vaupel P, Mayer A, Höckel M (2004) Tumor hypoxia and malignant progression. Methods Enzymol 381:335–354

Vaupel P, Mayer A, Höckel M (2006) Impact of hemoglobin levels on tumor oxygenation: the higher, the better? Strahlenther Onkol 182:63–71

Weiss L, Hultborn R, Tveit E (1979) Blood flow characteristics in induced rat mammary neoplasia. Microvasc Res 17:S119

Wheeler RH, Ziessman HA, Medvec BR, Juni JE, Thral JH, Keyes JW, Pit SR, Baker SR (1986) Tumor blood flow and systemic shunting in patients receiving intraarterial chemotherapy for head and neck cancer. Cancer Res 46:4200–4204

Willett CG, Boucher Y, di TE, Duda DG, Mun LL, Tong RT, Chung DC, Sahani DV, Kalva SP, Kozin SV, Mino M, Cohen KS, Scadden DT, Hartford AC, Fischman AJ, Clark JW, Ryan DP, Zhu AX, Blaszkowsky LS, Chen HX, Shellito PC, Lauwers GY, Jain RK (2004) Direct evidence that the VEGF-specific antibody bevacizumab has antivascular effects in human rectal cancer. Nat Med 10:145–147

Willett CG, Boucher Y, Duda DG, di TE, Mun LL, Tong RT, Kozin SV, Petit L, Jain RK, Chung DC, Sahani DV, Kalva SP, Cohen KS, Scadden DT, Fischman AJ, Clark JW, Ryan DP, Zhu AX, Blaszkowsky LS, Shellito PC, Mino-Kenudson M, Lauwers GY (2005) Surrogate markers for antiangiogenic therapy and dose-limiting toxicities for bevacizumab with radiation and chemotherapy: continued experience of a phase I trial in rectal cancer patients. J Clin Oncol 23:8136–8139

Young SD, Marshal RS, Hil RP (1988) Hypoxia induces DNA overreplication and enhances metastatic potential of murine tumor cells. Proc Natl Acad Sci U S A 85:9533–9537

Chapter 13
The Chaperokine Activity of HSPA1A

Punit Kaur and Alexzander Asea

Abstract There is a dichotomy between the intracellular and extracellular effector functions of HSPA1A. The enhanced expression of intracellular HSPA1A primarily promotes protein chaperoning, transport and folding of naïve, aberrantly folded, or mutated proteins, resulting in cytoprotection when cells are exposed to a variety of stressful stimuli. In contrast, exposure of immunocompetent cells to extracellular HSPA1A activates antigen presenting cell-mediated effectors functions; including enhanced pro-inflammatory and anti-inflammatory responses, chemokine and co-stimulatory molecule expression and in anti-tumor surveillance. In addition, extracellular HSPA1A has been shown to play a role in situations of acute psychological stress and exercise. This chapter covers recent advances in understanding the complex nature of extracellular HSPA1A and briefly discusses the biological significance of circulating serum HSPA1A to host physiology and includes recent application of HSPA1A-based immunotherapies.

Abbreviation

APC	antigen presenting cells
CTL	cytotoxic T lymphocytes
HspA8	constitutive form of the 73 kDa heat shock protein
HSPA1A	inducible form of the 72 kDa heat shock protein
IFN-λ	interferon-gamma
IL	interleukin
TLR	Toll-Like Receptors

A. Asea (✉) · P. Kaur
Department of Microbiology, Biochemistry and Immunology (MBI),
Morehouse School of Medicine, 720 Westview Ave., SW, Atlanta, GA 30310, USA
e-mail: aasea@msm.edu

B. Henderson, A. G. Pockley (eds.), *Cellular Trafficking of Cell Stress Proteins in Health and Disease*, Heat Shock Proteins 6, DOI 10.1007/978-94-007-4740-1_13,

13.1 Introduction

Many of the effects of extracellular stress proteins are mediated through cell-surface receptors. Such receptors include TLR2, TLR4, CD40, CD91, CCR5, and members of the scavenger receptor family, such as LOX-1 and SREC-1 (see Chaps. 10, 11, 14 and 15 for more detail). This wide range of receptors for the HSP and Grp family permits binding to a diverse range of cells and the performance of complex functions in multicellular organisms, particularly in immune cells and neurons (Calderwood et al. 2007; Murshid et al. 2011). Although the exact nature of the HSPA1A surface receptor is still under investigation (see Chap. 14 for details), a variety of cells have been studied for their ability to bind HSPA1A. In particular, natural killer (NK) cells (Multhoff et al. 1995, 1997; Gross et al. 2003) and APC including DC (Reed and Nicchitta 2000; Asea et al. 2002), macrophages, peripheral blood monocytes (Asea et al. 2000; Sondermann et al. 2000) and B cells (Arnold-Schild et al. 1999) have been shown to efficiently bind HSPA1A. In contrast, there is no evidence for the binding of HSPA1A to T lymphocytes (Arnold-Schild et al. 1999).

The authors have previously demonstrated that the binding of extracellular HSPA1A to human monocytes and dendritic cells results in production of the proinflammatory cytokines: TNF-α, IL-1β, IL-6 and IL-12 (Asea et al. 2000, 2002) and IFN-γ (Fig. 13.1b). Pre-treatment of cells with 1 μg/ml *Rhodopseudomonas sphaeroides* LPS (RSLP), an endotoxic LPS antagonist, did not significantly affect HSPA1A-induced IFN-γ expression, thereby negating the possibility that endotoxin contamination might have resulted in enhanced IFN-γ expression (Fig. 13.1c). However, heat denaturation at 100 °C for 1 h completely abrogated HSPA1A-induced IFN-γ expression (data not shown). Pre-treatment of cells with RSLP completely inhibited LPS-induced IFN-γ expression (Fig. 13.1d). Controls were cells pre-treated with control protein OVA (Fig. 13.1a). All HSPA1A preparations were tested for LPS content by *Limulus* amebocyte lysate assay and found to have LPS levels below the sensitivity limits of the assay (<0.01 endotoxin U/ml). Exposure of human dendritic cells for 3–5 days with extracellular HSPA1A resulted in the significant increase in proliferation of immature dentritic cells and augmentation of co-stimulatory molecules, MHC class II and CD86 (Binder et al. 2001; Somersan et al. 2001; Asea et al. 2002; Vabulas et al. 2002). The mechanism by which peptides within the HSPA1A-peptide binding groove get incorporated into the MHC was addressed by Binder et al. (2001). These authors showed that free peptides introduced into the cytosol become ligands of MHC I molecules, and that treatment of cells with deoxyspergualin, a drug known to specifically bind HSPA1A and Hsp90, abrogates the ability of cells to present antigenic peptides through MHC I molecules (Binder et al. 2001). Recently, an article comprehensively and concisely addressed the question of endotoxin contamination and concluded that sufficient evidence exists to support fully the hypothesis that molecular chaperones have cell-cell signaling actions that are likely to be part of the homeostatic mechanism of the vertebrate (Henderson et al. 2010).

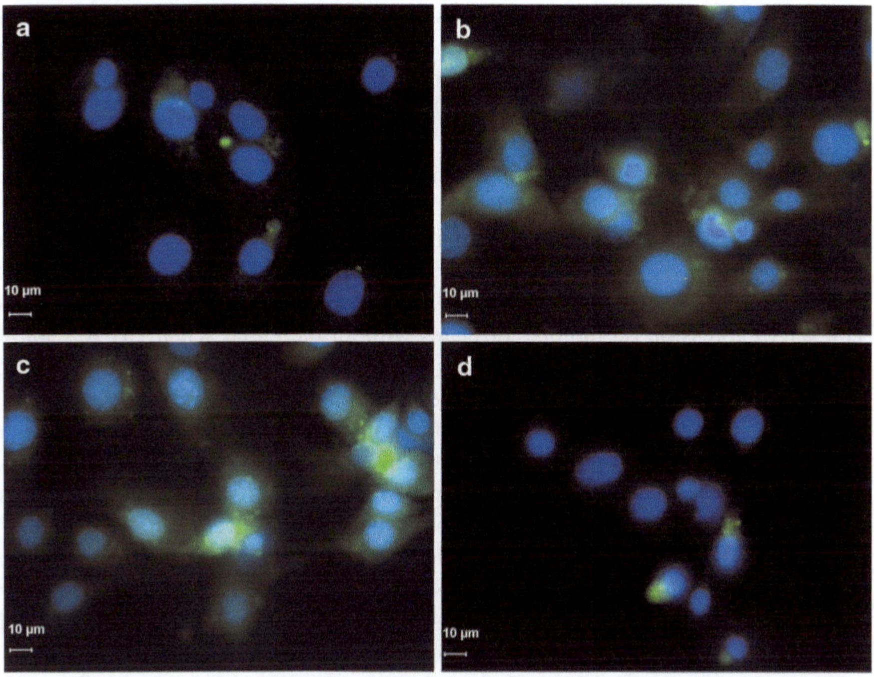

Fig. 13.1 Chaperokine-induced IFN-γ expression. Human monocytic cells THP1 were grown on Falcon culture slides (BD Labware, Franklin Lakes, NJ) overnight and cells were stimulated with 100 ng/ml control protein OVA (**a**), or 100 ng/ml HSPA1A alone (**b**), or pre-treated with 1 µg/ml *Rhodopseudomonas spheroids* (RSLP), a LPS antagonist, for 1 h prior to stimulation with 100 ng/ml HSPA1 A (**c**), or pre-treated with 1 µg/ml RSLP for 1 h prior to stimulation with 100 ng/ml LPS (**d**). Cells were incubated at 37 °C for 4 h in the presence of 10 µM Brefeldin A (Sigma), and simultaneously fixed and permeabilized using Cytofix/CytopermTM kit (BD Biosciences, San Diego, CA) according to the manufacturer's instructions and counter stained with anti-human IFN-γ-FITC (BD Biosciences). One drop of mounting media containing DAPI stain to visualize nuclear staining (Oncogene, Boston, MA) was placed onto the glass slide before the cover slip was sealed with nail polish. Results show fluorescence microscope pictograms of overlays of DAPI stain that reveal the cells nuclear morphology (*blue*) overlaid onto FITC stained cells which reveals the intensity of intracellular IFN-γ expression (*green*). Results are a representative experiment from three independently performed experiments with similar results

Stimulation of murine and human macrophages with HSPA1A and gp96 results in induction of inducible nitric oxide (iNO) synthase and the release of NO (Panjwani et al. 2002). HSPA1A-induced NO release was unaffected by pre-treatment of cells with the LPS antagonist, RSLP, negating the possibility of endotoxin contamination. However, HSPA1A-induced NO release was completely abrogated by pre-treatment of cells with NO inhibitor L-NMMA, an analog of arginine. Co-incubation of macrophages with IFN-γ and HSPA1A resulted in the synergistic production of NO (Panjwani et al. 2002). *Mycobacterium tuberculosis*-derived HSPA1A (Hsp70/DnaK) stimulated the production of the CC chemokines RANTES, MIP-1α and MIP-1β from three leukocyte subsets: PBMC, $CD8^+$ and $CD4^+$ T lymphocytes and $CD14^+$ monocytes from macaques. This is significant since $CD8^+$ produce human immunodeficiency virus (HIV) and simian immunodeficiency virus (SIV) suppressor factors (Walker et al. 1986). Protection against SIV infection is associated with CD8-suppresor factors and the three CC chemokines: RANTES, MIP-1α and MIP-1β in a number of non-human primate experiments. See reviews by (Lehner and Anton 2002; Lehner 2003).

In comparison with immunocompetent cells, malignant tumor cells express high levels of surface bound HSPA1A (Botzler et al. 1996; Multhoff and Hightower 1996; Multhoff et al. 1997; Botzler et al. 1998; Hantschel et al. 2000). This HSPA1A expression on tumors correlates with an increased sensitivity to natural killer (NK)-mediated cytolysis following cytokine stimulation (Botzler et al. 1998; Multhoff et al. 1999, 2001). Recent studies have shown that the cytolytic activity of HSPA1A can also be transduced by specific fragments of the HSPA1A protein. Both the full-length HSPA1A protein and the C-terminal domain of HSPA1A stimulates the cytolytic activity of naive NK cells against HSPA1A-positive tumor target cells (Gross et al. 2003). In addition, tumor growth in mice with severe combined immunodeficiency was shown to be inhibited by HSPA1A-peptide-activated, CD94-positive natural killer cells (Moser et al. 2002). Recent work from the Multhoff laboratory demonstrates that a 14 amino acid peptide sequence of the HSPA1A protein, termed TKD (TKDNNLLGRFELSG, aa450–463) is the extracellular recognition site for NK cells (Multhoff et al. 2001). These authors demonstrate that granzyme B specifically binds to portions of the HSPA1A expressed on the plasma surface of tumors but not normal cells (Gross et al. 2003); thus, demonstrating a hitherto unknown mechanism by which cytolytic effector cells eliminate HSPA1A-expressing tumors in a perforin-independent, granzyme B-dependent manner and demonstrating that a $CD8^+$ CTL-independent mechanism, associated with HSPs, is an important mechanism by which tumor surveillance is achieved. The surface expression of HSPA1A in metastatic melanoma (Farkas et al. 2003), acute myeloid leukemia (Gehrmann et al. 2003), and head and neck cancer (Kleinjung et al. 2003) stimulates specific NK cell-mediated cytolytic functions are some recent examples. The development of a HSPA1A peptide which stimulates NK cell-mediated killing of leukemic blast cells (Gross et al. 2008; Stangl et al. 2008), and the demonstration that NK cell-mediated targeting of membrane HSPA1A on tumors can be greatly enhanced after treatment with the cmHsp70.1 monoclonal antibody (Stangl et al. 2011), are additional $CD8^+$ CTL-independent mechanisms.

13.2 Chaperokine-Induced Signal Transduction Pathways

To date, several cell surface proteins have been described as the receptor for HSPA1A including Toll-like receptors 2 and 4 with their cofactor CD14 (Asea et al. 2002), the scavenger receptor, CD36 (Delneste et al. 2002; Nakamura et al. 2002), and the co-stimulatory molecule, CD40 (Becker et al. 2002). Further discussion of the receptors for HSPA1A is provided in Chap. 14. Our group addressed the various steps involved in the HSPA1A-induced signal transduction cascade and revealed that HSPA1A binds with high affinity to the plasma membrane of APC and within 10 s elicits a rapid intracellular Ca^{2+} ($[Ca^{2+}]I$) flux (Asea et al. 2000). This is an important signalling step that distinguishes HSPA1A- from LPS-induced signaling, since treatment of APC with LPS does not result in $[Ca^{2+}]i$ flux (McLeish et al. 1989). The possibility that endotoxin contamination might confound our results was addressed by using Polymyxin B and Lipid IVa (LPS inhibitor) which abrogates LPS-induced, but not HSPA1A-induced, cytokine expression. Boiling the proteins at 100 °C for 1 h abrogates HSPA1A-induced, but not LPS-induced, cytokine expression. We noted that rapid HSPA1A-induced $[Ca^{2+}]i$ flux is followed by the phosphorylation of I-κBα (Asea et al. 2000). Activation of NF-κB is regulated by its cytoplasmic inhibitor, I-κBα, *via* phosphorylation at Serine 32 (Ser-32) and 36 (Ser-36) which targets it for degradation by the proteosome and releases NF-κB to migrate to the nucleus and activate the promoter of target genes (Baeuerle and Baltimore 1988). As early as 30 min post exposure to extracellular HSPA1A, I-κBα was phosphorylated at Serine 32 (Ser-32) and 36 (Ser-36) resulting in the release and nuclear translocation of NF-κB (Asea et al. 2000). Mechanistic studies using the HEK293 model system revealed that HSPA1A-induced NF-κB promoter activity is MyD88-dependent, CD14-dependent and transduced *via* both TLR2 and TLR4 (Asea et al. 2002). Further, the presence of both TLR2 and TLR4 synergistically stimulates HSPA1A-induced cytokine production (Asea et al. 2002). Interestingly, we found that the synergistic activation of NF-κB promoter by co-expression of both TLR2 and TLR4 is MyD88-independent, suggesting an alternative pathway by which extracellular HSPA1A stimulates cells of the immune system. Independent studies by Vabulas et al. (2002) confirmed these findings, thereby classifying extracellular HSPA1A as an intracellular natural adjuvant.

CD40 is a co-stimulatory molecule expressed on APC and found to play an important role in B lymphocyte function and autoimmunity (Bodmer et al. 2002). CD40 has been demonstrated to bind HSPA1A-peptide complexes *via* its exoplasmic domain (Becker et al. 2002). The HSPA1A-CD40 interaction was shown to be mediated by the NH_2-terminal ATPase domain of HSPA1A in its ADP-bound state and is strongly augmented by the presence of substrate peptides in the COOH-terminal domain of HSPA1A. The HSPA1A-CD40 interaction was suppressed by Hip, a co-chaperone that is known to stabilize the HSPA1A ATPase domain in the ADP state (Becker et al. 2002). Using the HEK293 cell model system, these authors demonstrated that specific HSPA1A-CD40 binding stimulates signal transduction *via* the phosphorylation of p38 (previously shown to induce the release of TNF-α and secretion of IFN-γ (Pullen et al. 1999), which results in the activation of NF-κB and uptake of peptide (Becker et al. 2002).

HSPA1A has been discovered to also bind to two classes of receptor: c-type lectin receptors (CLR) and scavenger receptors (SR) (See Chap. 14). Although the receptor-ligand interactions is incompletely understood, HSPA1A binds LOX-1 (a member of both the CLR and SR), with the c-type lectin binding domain (CTLD) as well as the SR family members SREC-I and FEEL-1/CLEVER-1/STABILIN-1, which by contrast have arrays of EGF-like repeats in their extracellular domains (Theriault et al. 2005; Calderwood et al. 2007; Murshid et al. 2011). Theriault and colleagues demonstrated that three members of the SR family (lectin-like oxidized low density lipoprotein receptor 1; fasciclin, epidermal growth factor-like, laminin-type epidermal growth factor-like, and link domain-containing scavenger receptor-1; and SR expressed by endothelial cells-1) bind Hsp70-peptide complexes and mediate its efficient internalization (Theriault et al. 2006). The scavenger receptor, CD36 is another protein that has been shown to bind HSPA1A (Delneste et al. 2002). Specifically, LOX-1, on human dendritic cells was shown to bind HSPA1A, and incubation of cells with a neutralizing anti-LOX-1 mAb abrogated HSPA1A binding to dendritic cells and suppressed HSPA1A-induced antigen cross-presentation (Delneste et al. 2002; Wang et al. 2004; Tang et al. 2005; Theriault et al. 2005, 2006; Calderwood et al. 2007). In addition, GST pull-down assays and immunoprecipitation analyses showed that HSPA1A, HSP90 and glyceraldehyde 3-phosphate dehydrogenase (GAPDH) bind to the cytoplasmic domain of scavenger receptors (Nakamura et al. 2002; Wang et al. 2004; Tang et al. 2005; Theriault et al. 2005, 2006; Calderwood et al. 2007). *In vivo* studies show that pre-treatment of animals with anti-LOX-1 mAb induces antitumor immunity (Delneste et al. 2002; Murshid et al. 2011).

13.3 Biological Significance of Circulating Serum HAPA1A: Role in Host Physiology

It has been proposed that the host releases endogenous "danger signals" capable of stimulating immunity (Matzinger 1994, 1998—see Chap. 10). Briefly, the danger theory postulates that immune activation involves danger/non-danger molecular recognition schemas. The danger theory suggests that innate immune cells are activated by danger signals that are derived from stressed or damaged self-proteins. Our working hypothesis is that indeed extracellular HSPA1A is such a danger signal. This hypothesis is further reinforced by recent data that demonstrates acute psychological stress results in the release of extracellular HSPA1A (Noble et al. 1999; Campisi and Fleshner 2003; Campisi et al. 2003). Rats exposed to inescapable tail-shock stress (IS; 100 1.6 mA tail shocks, 5 s duration, 60 s intertrial interval) or exhaustive exercise stress (EXS; treadmill running to exhaustion), but not sedentary rats, had elevated Hsp72 in dorsal vagal complex, frontal cortex, hippocampus, pituitary, adrenal, liver, spleen, mesenteric lymph nodes, and heart, as judged by blood, brain, and peripheral tissue samples collected 2 h post stress termination (Campisi et al. 2003). In an *in vivo* model, adult, male Fischer 344 rats were allowed to be physically active or sedentary using running wheels. Six weeks later, animals were exposed to IS or no stress and immediately injected subcutaneously with live *Escherichia coli*. Rats exposed to IS resolved their inflammation faster than non-stressed rats. In addition, splenocytes from physically active rats responded to extracellular HSPA1A stimulation *ex vivo* with a greater nitric oxide as compared to sedentary rats.

Studies in humans showed that physical exercise results in marked increase in circulating serum HSPA1A levels (Walsh et al. 2001). A comparison of arteriovenous differences showed the source of HSPA1A is not from the contracting skeletal muscle (Febbraio et al. 2002), but from the hepatosplanchnic tissues (Febbraio et al. 2002). The exact role of the released HSPA1A is thus far unknown, but it is hypothesized that the released circulating HSPA1A serves as a chaperokine and enhances the host defense against pathogens during exercise. Indeed, we demonstrated that HSPA8 is released from neoplastic, but not normal, cells in response to cytokines including IFN-γ and IL-10, cytokines normally found in high concentrations within the tumor microenvironment, and in response to heat shock (Barreto et al. 2003), indicating a hitherto unknown mechanism by which the host immune system gets access to tumor-derived antigenic peptides without the need for necrotic cell death. Using a model for human glia and neurons, this point is reinforced in studies that demonstrate human glioma cells export HSPA1A into culture medium under normal and stressed conditions (Guzhova et al. 2001). In addition, neuroblastoma cells took up biotinylated HSPA1A and exhibited a thermotolerance phenotype when exposed to lethal heat shock (44 °C) and to staurosporine-induced apoptosis, suggesting a mechanism by which extracellular HSPA1A might affect neuronal function (Guzhova et al. 2001). Further discussion of the role of circulating HSPA1A can be found in Chaps. 3 and 18.

13.4 Application of Extracellular HSPA1A: HSPA12A-Based Immunotherapies

Due to the unique ability to: chaperone antigenic peptides, interact with APC, mediate release of chemokine and cytokines from APC and induce the maturation of dendritic cells, makes heat shock proteins an important new generation of prophylactic and therapeutic vaccine against cancers and infectious diseases. For reviews on this concept see (Srivastava 2000, 2004). To this end, phase I/II studies in melanoma and colorectal cancer patients using personalized HSP vaccines have now been completed and show great promise (Castelli et al. 2003). Oncophage®, HSP/peptide-complex (HSPPC) derived from autologous liver metastases of colorectal carcinoma induces cancer-specific T-cell responses in patients rendered disease-free by surgery (Mazzaferro et al. 2003). The studies show that autologous tumor-derived HSPPC-96 vaccination after resection of colorectal liver metastases is safe and elicits a significant increase in CTL $CD8^+$ T-cell response against colon cancer. Further, the study demonstrates that in this limited number of patients, 2-year overall survival and disease-free survival were significantly improved in subjects with post-vaccination anti-tumor immune response, independently from other clinical prognostic factors (Mazzaferro et al. 2003; Pilla et al. 2006; Maki et al. 2007; Jonasch et al. 2008).

Malaria is one of the most devastating health problems in large parts of the world. It is currently estimated that between 300–500 million people are infected annually and 1.5–2.7 million people die from the disease each year. However, the development of an effective vaccine is marred because there are significant antigenic changes associated with the malaria parasite during its various life cycles, thus making development of an effective malaria vaccine that provides protective immunity at the different stages of parasite development particularly challenging. The merozoite surface protein 1 (MSP1) fused to heat-shock cognate protein 70 (HSPA8) was constructed and its efficacy tested against the malaria parasite (Kawabata et al. 2002). Merozoite surface protein 1 (MSP1) is expressed during blood-stage in the exoerythrocytic forms of malaria. Immunization of C57BL/6 mice with the fusion protein prior to challenge infection with *Plasmodium yoelii* sporozoites, resulted in the significant delay in or complete inhibition of parasitemia. Interestingly, the vaccine-induced protection in A/J, C3H, and BALB/c mice suggested that MSP1-specific protective immunity at the exoerythrocytic stage can be induced in animals over a wide range of genetic backgrounds (Kawabata et al. 2002). Additional ways to combat malaria have been to use inhibitors of HSP that are currently developed for cancer therapy as anti-malarials. The hypothesis is that the joint disruption of the parasite antioxidant enzymes/HSP system would interfere with parasite growth and open new perspectives for anti-malaria therapy. Akide-Ndunge and colleagues reported that mRNA expression of parasite antioxidant enzymes and HSPs was co-ordinated and stage-dependent. In addition, both systems were redox-responsive and showed remarkably increased and co-ordinated expression in oxidatively-stressed parasites and in parasites growing in antioxidant blunted G6PD-deficient RBCs (Akide-Ndunge et al. 2009).

In order to control the human immunodeficiency virus (HIV) pandemic, it is imperative that investigators develop an effective vaccine. It has been suggested that the major difficulties in developing a preventive vaccine is because of the rapid rate of mutation of HIV and the lack of induction of neutralizing antibodies to the virus (Lehner and Anton 2002; Lehner 2003; Morner et al. 2011; Wang and Lehner 2011; Whittall et al. 2011). Although current chemotherapeutic regimes have stopped the high mortality from HIV infection, the high cost of these drugs makes it of little or no use to developing countries. In addition, the anti-HIV drug treatment is associated with severe side effects and the development of drug-resistant strains of HIV. Therefore, new therapies are needed to address this important question. Investigators have looked to using HSPA1A constructs in HIV immunotherapy. Immunization of nonhuman primates with HSPA1A359-610 induced the production of RANTES and IL-12, and acted as an adjuvant when loaded with CC5-peptide (Wang et al. 2002), suggesting a possible alternative vaccine strategy for combating HIV infection (Lehner and Anton 2002; Lehner and Shearer 2002; Bergmeier et al. 2010).

13.5 Conclusion

This chapter has covered the most salient features of the extracellular HSPA1A (Fig. 13.2). Although the biological significance of extracellular HSPA1A on effector functions of immunocompetent cells is now well characterized, and has been termed the chaperokine activity of HSPA1A to better describe its dual role as both chaperone

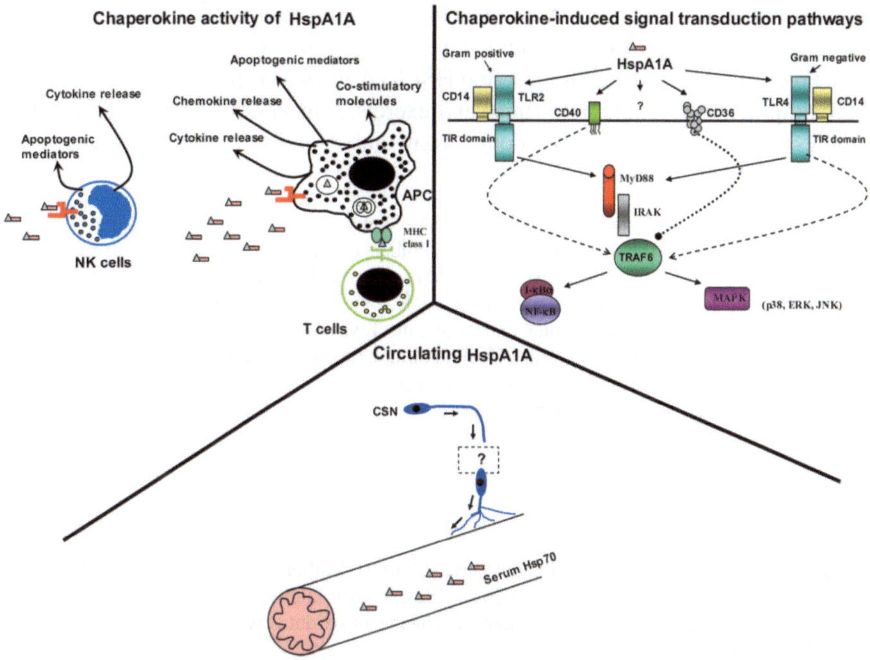

Fig. 13.2 Schematic representation of the spectrum of events mediated by exogenous HSPA1A. Chaperokine activity of HSPA1A; exogenous HSPA1A-peptide complexes composed of the HSPA1A protein and the infected cell/tumor-derived peptide when found in the zone of inflammation activates NK cell-mediated cytolytic events and cytokine release. The HSPA1A-peptide complex binds to specific receptors on APC and stimulates co-stimulatory molecule expression and the release of pro-inflammatory cytokines, chemokines and apoptogenic mediators. The HSPA1A-peptide complex is internalized and the infected cell/tumor-derived peptide is degraded within phagosomes and presented in the context of MHC class I on the surface of the APC where it will activate specific $CD8^+$ CTL responses. Chaperokine-induced signal transduction pathways; exogenous HSPA1A binds to a yet unknown receptor (?) and/or TLR4 and TLR2 in a CD14-dependent fashion and activates the signal cascade that results in the phosphorylation of adaptor protein MyD88 → IRAK → TRAF6 at which the signal bifurcates to stimulate either NF-κB or MAPK signaling pathways. Exogenous HSPA1A also binds surface bound CD40 and CD36 and activates NF-κB or MAPK signaling pathways in a MyD88-independent fashion (*dotted lines*). Circulating HSPA1A signals originating from the CSN stimulates the release of HSPA1A into the blood circulation either directly or through a yet unknown organ (?) during acute psychological stress and exercise. *APC* antigen presenting cell, *CSN* central nervous system, *CTL* cytotoxic T lymphocytes, *HSPA1A* the inducible 70 kDa heat shock protein, *IRAK* IL-1 receptor-associated kinase, *MAPK* mitogen activated protein kinase, *MHC* major histocompatability complex, *MyD88* myeloid differentiation factor 88, *NF-κB* nuclear factor-κB, *TLR* toll-like receptor, *TRAF6* TNF receptor associated factor 6

and cytokine, the exact nature of the HSPA1A receptor is still under investigation. However, this has not deterred investigators from developing unique HSPA1A-based immunotherapy for diseases including cancer, malaria and HIV. Studies on the signal transduction pathway activated by extracellular HSPA1A has begun to shed some light on the nature of the HSPA1A-receptor and seems to suggest that there might be more than one receptor for this unique protein. The next frontier is to elucidate the exact role of elevated circulating serum HSPA1A in various physiological situations including acute psychological stress and exercise. Successful completion of these studies promises to greatly further our current understanding of the role of heat shock proteins and help decipher how the host communicates when it perceives danger.

Acknowledgments We thank Princess Bempong, Shahrum Lillard and Viraj Mehta for expert technical assistance, and all the faculty and staff members of the Division of Investigative Pathology for helpful discussions. This work was supported in part a Research Advancement Award from Scott & White Memorial Hospital and Clinic (to P.K.), the US National Institutes of Health grant RO1CA91889, Scott & White Hospital and Clinic, the Texas A&M Health Science Center College of Medicine, the Central Texas Veterans Health Administration and an Endowment from the Cain Foundation (to A.A.). The funders had no role in study design, data collection and analysis, decision to publish or preparation of the chapter.

References

Akide-Ndunge OB, Tambini E, Giribaldi G, McMillan PJ, Muller S, Arese P, Turrini F (2009) Co-ordinated stage-dependent enhancement of Plasmodium falciparum antioxidant enzymes and heat shock protein expression in parasites growing in oxidatively stressed or G6PD-deficient red blood cells. Malar J 8:113

Arnold-Schild D, Hanau D, Spehner D, Schmid C, Rammensee H-G, de la Salle H, Schild H (1999) Receptor-mediated endocytosis of heat shock proteins by professional antigen-presenting cells. J Immunol 162:3757–3760

Asea A, Kraeft SK, Kurt-Jones EA, Stevenson MA, Chen LB, Finberg RW, Koo GC, Calderwood SK (2000) HSP70 stimulates cytokine production through a CD14-dependant pathway, demonstrating its dual role as a chaperone and cytokine. Nat Med 6:435–442

Asea A, Rehli M, Kabingu E, Boch JA, Bare O, Auron PE, Stevenson MA, Calderwood SK (2002) Novel signal transduction pathway utilized by extracellular HSP70: role of toll-like receptor (TLR) 2 and TLR4. J Biol Chem 277:15028–15034

Baeuerle PA, Baltimore D (1988) I kappa B: a specific inhibitor of the NF-kappa B transcription factor. Science 242:540–546

Barreto A, Gonzalez JM, Kabingu E, Asea A, Fiorentino S (2003) Stress-induced release of HSC70 from human tumors. Cell Immunol 222:97–104

Becker T, Hartl FU, Wieland F (2002) CD40, an extracellular receptor for binding and uptake of Hsp70-peptide complexes. J Cell Biol 158:1277–1285

Bergmeier LA, Babaahmady K, Pido-Lopez J, Heesom KJ, Kelly CG, Lehner T (2010) Cytoskeletal proteins bound to heat-shock protein 70 may elicit resistance to simian immunodeficiency virus infection of CD4(+) T cells. Immunology 129:506–515

Binder RJ, Blachere NE, Srivastava PK (2001) Heat shock protein-chaperoned peptides but not free peptides introduced into the cytosol are presented efficiently by major histocompatibility complex I molecules. J Biol Chem 276:17163–17171

Bodmer JL, Schneider P, Tschopp J (2002) The molecular architecture of the TNF superfamily. Trends Biochem Sci 27:19–26

Botzler C, Issels R, Multhoff G (1996) Heat-shock protein 72 cell-surface expression on human lung carcinoma cells in associated with an increased sensitivity to lysis mediated by adherent natural killer cells. Cancer Immunol Immunotherap 43:226–230

Botzler C, Schmidt J, Luz A, Jennen L, Issels R, Multhoff G (1998) Differential Hsp70 plasma-membrane expression on primary human tumors and metastases in mice with severe combined immunodeficiency. Int J Cancer 77:942–948

Calderwood SK, Theriault J, Gray PJ, Gong J (2007) Cell surface receptors for molecular chaperones. Methods 43:199–206

Campisi J, Fleshner M (2003) Role of extracellular HSP72 in acute stress-induced potentiation of innate immunity in active rats. J Appl Physiol 94:43–52

Campisi J, Leem TH, Greenwood BN, Hansen MK, Moraska A, Higgins K, Smith TP, Fleshner M (2003) Habitual physical activity facilitates stress-induced HSP72 induction in brain, peripheral, and immune tissues. Am J Physiol Regul Integr Comp Physiol 284:R520–R530

Castelli C, Rivoltini L, Rini F, Belli F, Testori A, Maio M, Mazzaferro V, Coppa J, Srivastava PK, Parmiani G (2003) Heat shock proteins: biological functions and clinical application as personalized vaccines for human cancer. Cancer Immunol Immunother 53:227–233

Delneste Y, Magistrelli G, Gauchat J, Haeuw J, Aubry J, Nakamura K, Kawakami-Honda N, Goetsch L, Sawamura T, Bonnefoy J, Jeannin P (2002) Involvement of LOX-1 in dendritic cell-mediated antigen cross-presentation. Immunity 17:353–362

Farkas B, Hantschel M, Magyarlaki M, Becker B, Scherer K, Landthaler M, Pfister K, Gehrmann M, Gross C, Mackensen A, Multhoff G (2003) Heat shock protein 70 membrane expression and melanoma-associated marker phenotype in primary and metastatic melanoma. Melanoma Res 13:147–152

Febbraio MA, Ott P, Nielsen HB, Steensberg A, Keller C, Krustrup P, Secher NH, Pedersen BK (2002) Exercise induces hepatosplanchnic release of heat shock protein 72 in humans. J Physiol 544:957–962

Gehrmann M, Schmetzer H, Eissner G, Haferlach T, Hiddemann W, Multhoff G (2003) Membrane-bound heat shock protein 70 (Hsp70) in acute myeloid leukemia: a tumor specific recognition structure for the cytolytic activity of autologous NK cells. Haematologica 88:474–476

Gross C, Hansch D, Gastpar R, Multhoff G (2003) Interaction of heat shock protein 70 peptide with NK cells involves the NK receptor CD94. Biol Chem 384:267–279

Gross C, Holler E, Stangl S, Dickinson A, Pockley AG, Asea AA, Mallappa N, Multhoff G (2008) An Hsp70 peptide initiates NK cell killing of leukemic blasts after stem cell transplantation. Leuk Res 32:527–534

Guzhova I, Kislyakova K, Moskaliova O, Fridlanskaya I, Tytell M, Cheetham M, Margulis B (2001) In vitro studies show that Hsp70 can be released by glia and that exogenous Hsp70 can enhance neuronal stress tolerance. Brain Res 914:66–73

Hantschel M, Pfister K, Jordan A, Scholz R, Andreesen R, Schmitz G, Schmetzer H, Hiddemann W, Multhoff G (2000) Hsp70 plasma membrane expression on primary tumor biopsy material and bone marrow of leukemic patients. Cell Stress Chaperones 5:438–442

Henderson B, Calderwood SK, Coates AR, Cohen I, van Eden W, Lehner T, Pockley AG (2010) Caught with their PAMPs down? The extracellular signalling actions of molecular chaperones are not due to microbial contaminants. Cell Stress Chaperones 15:123–141

Jonasch E, Wood C, Tamboli P, Pagliaro LC, Tu SM, Kim J, Srivastava P, Perez C, Isakov L, Tannir N (2008) Vaccination of metastatic renal cell carcinoma patients with autologous tumour-derived vitespen vaccine: clinical findings. Br J Cancer 98:1336–1341

Kawabata Y, Udono H, Honma K, Ueda M, Mukae H, Kadota J, Kohno S, Yui K (2002) Merozoite surface protein 1-specific immune response is protective against exoerythrocytic forms of Plasmodium yoelii. Infect Immun 70:6075–6082

Kleinjung T, Arndt O, Feldmann HJ, Bockmuhl U, Gehrmann M, Zilch T, Pfister K, Schonberger J, Marienhagen J, Eilles C, Rossbacher L, Multhoff G (2003) Heat shock protein 70 (Hsp70) membrane expression on head-and-neck cancer biopsy-a target for natural killer (NK) cells. Int J Radiat Oncol Biol Phys 57:820–826

Lehner T (2003) Innate and adaptive mucosal immunity in protection against HIV infection. Vaccine 21(Suppl 2):S68–S76

Lehner T, Anton PA (2002) Mucosal immunity and vaccination against HIV. AIDS 16(Suppl 4):S125–S132

Lehner T, Shearer GM (2002) Alternative HIV vaccine strategies. Science 297:1276–1277

Maki RG, Livingston PO, Lewis JJ, Janetzki S, Klimstra D, Desantis D, Srivastava PK, Brennan MF (2007) A phase I pilot study of autologous heat shock protein vaccine HSPPC-96 in patients with resected pancreatic adenocarcinoma. Dig Dis Sci 52:1964–1972

Matzinger P (1994) Tolerance, danger, and the extended family. Annu Rev Immunol 12:991–1045

Matzinger P (1998) An innate sense of danger. Semin Immunol 10:399–415

Mazzaferro V, Coppa J, Carrabba MG, Rivoltini L, Schiavo M, Regalia E, Mariani L, Camerini T, Marchiano A, Andreola S, Camerini R, Corsi M, Lewis JJ, Srivastava PK, Parmiani G (2003) Vaccination with autologous tumor-derived heat-shock protein gp96 after liver resection for metastatic colorectal cancer. Clin Cancer Res 9:3235–3245

McLeish KR, Dean WL, Wellhausen SR, Stelzer GT (1989) Role of intracellular calcium in priming of human peripheral blood monocytes by bacterial lipopolysaccharide. Inflammation 13:681–692

Morner A, Jansson M, Bunnik EM, Scholler J, Vaughan R, Wang Y, Montefiori DC, Otting N, Bontrop R, Bergmeier, LA, Singh M, Wyatt RT, Schuitemaker H, Biberfeld G, Thorstensson R, Lehner T (2011) Immunization with recombinant HLA classes I and II, HIV-1 gp140, and SIV p27 elicits protection against heterologous SHIV infection in rhesus macaques. J Virol 85:6442–6452

Moser C, Schmidbauer C, Gurtler U, Gross C, Gehrmann M, Thonigs G, Pfister K, Multhoff G (2002) Inhibition of tumor growth in mice with severe combined immunodeficiency is mediated by heat shock protein 70 (Hsp70)-peptide-activated, CD94 positive natural killer cells. Cell Stress Chaperones 7:365–373

Multhoff G, Hightower LE (1996) Cell surface expression of heat shock proteins and the immune response. Cell Stress Chaperones 1:167–176

Multhoff G, Botzler C, Wiesnet M, Eissner G, Issels R (1995) CD3-large granular lymphocytes recognize a heat-inducible immunogenic determinant associated with the 72-kD heat shock protein on human sarcoma cells. Blood 86:1374–1382

Multhoff G, Botzler C, Jennen L, Schmidt J, Ellwart J, Issels R (1997) Heat shock protein 72 on tumor cells: a recognition structure for natural killer cells. J Immunol 158:4341–4350

Multhoff G, Mizzen L, Winchester CC, Milner CM, Wenk S, Eissner G, Kampinga HH, Laumbacher B, Johnson J (1999) Heat shock protein 70 (Hsp70) stimulates proliferation and cytolytic activity of natural killer cells. Exp Hematol 27:1627–1636

Multhoff G, Pfister K, Gehrmann M, Hantschel M, Gross C, Hafner M, Hiddemann W (2001) A 14-mer Hsp70 peptide stimulates natural killer (NK) cell activity. Cell Stress Chaperones 6:337–344

Murshid A, Theriault J, Gong J, Calderwood SK (2011) Investigating receptors for extracellular heat shock proteins. Methods Mol Biol 787:289–302

Nakamura T, Hinagata J, Tanaka T, Imanishi T, Wada Y, Kodama T, Doi T (2002) HSP90, HSP70, and GAPDH directly interact with the cytoplasmic domain of macrophage scavenger receptors. Biochem Biophys Res Commun 290:858–864

Noble EG, Moraska A, Mazzeo RS, Roth DA, Olsson MC, Moore RL, Fleshner M (1999) Differential expression of stress proteins in rat myocardium after free wheel or treadmill run training. J Appl Physiol 86:1696–1701

Panjwani NN, Popova L, Srivastava PK (2002) Heat shock proteins gp96 and hsp70 activate the release of nitric oxide by APCs. J Immunol 168:2997–3003

Pilla L, Patuzzo R, Rivoltini L, Maio M, Pennacchioli E, Lamaj E, Maurichi A, Massarut S, Marchiano A, Santantonio C, Tosi D, Arienti F, Cova A, Sovena G, Piris A, Nonaka D, Bersani I, Di Florio A, Luigi M, Srivastava PK, Hoos A, Santinami M, Parmiani G (2006) A phase II trial of vaccination with autologous, tumor-derived heat-shock protein peptide complexes Gp96, in combination with GM-CSF and interferon-alpha in metastatic melanoma patients. Cancer Immunol Immunother 55:958–968

Pullen SS, Dang TT, Crute JJ, Kehry MR (1999) CD40 signaling through tumor necrosis factor receptor-associated factors (TRAFs). Binding site specificity and activation of downstream pathways by distinct TRAFs. J Biol Chem 274:14246–14254

Reed RC, Nicchitta CV (2000) Chaperone-mediated cross-priming: a hitchhiker's guide to vesicle transport (review). Int J Mol Med 6:259–264

Somersan S, Larsson M, Fonteneau JF, Basu S, Srivastava P, Bhardwaj N (2001) Primary tumor tissue lysates are enriched in heat shock proteins and induce the maturation of human dendritic cells. J Immunol 167:4844–4852

Sondermann H, Becker T, Mayhew M, Wieland F, Hartl FU (2000) Characterization of a receptor for heat shock protein 70 on macrophages and monocytes. Biol Chem 381:1165–1174

Srivastava P (2004) Heat shock proteins and immune response: methods to madness. Methods 32:1–2

Srivastava PK (2000) Heat shock protein-based novel immunotherapies. Drug News Perspect 13:517–522

Stangl S, Gross C, Pockley AG, Asea AA, Multhoff G (2008) Influence of Hsp70 and HLA-E on the killing of leukemic blasts by cytokine/Hsp70 peptide-activated human natural killer (NK) cells. Cell Stress Chaperones 13:221–230

Stangl S, Gehrmann M, Dressel R, Alves F, Dullin C, Themelis G, Ntziachristos V, Staeblein E, Walch A, Winkelmann I, Multhoff G (2011) In vivo imaging of CT26 mouse tumours by using cmHsp70.1 monoclonal antibody. J Cell Mol Med 15:874–887

Tang D, Khaleque MA, Jones EL, Theriault JR, Li C, Wong WH, Stevenson MA, Calderwood SK (2005) Expression of heat shock proteins and heat shock protein messenger ribonucleic acid in human prostate carcinoma in vitro and in tumors in vivo. Cell Stress Chaperones 10:46–58

Theriault JR, Mambula SS, Sawamura T, Stevenson MA, Calderwood SK (2005) Extracellular HSP70 binding to surface receptors present on antigen presenting cells and endothelial/epithelial cells. FEBS Lett 579:1951–1960

Theriault JR, Adachi H, Calderwood SK (2006) Role of scavenger receptors in the binding and internalization of heat shock protein 70. J Immunol 177:8604–8611

Vabulas RM, Ahmad-Nejad P, Ghose S, Kirschning CJ, Issels RD, Wagner H (2002) HSP70 as endogenous stimulus of the Toll/interleukin-1 receptor signal pathway. J Biol Chem 277:15107–15112

Walker CM, Moody DJ, Stites DP, Levy JA (1986) $CD8^+$ lymphocytes can control HIV infection in vitro by suppressing virus replication. Science 234:1563–1566

Walsh RC, Koukoulas I, Garnham A, Moseley PL, Hargreaves M, Febbraio MA (2001) Exercise increases serum Hsp72 in humans. Cell Stress Chaperones 6:386–393

Wang Y, Lehner T (2011) Induction of innate immunity in control of mucosal transmission of HIV. Curr Opin HIV AIDS 6:398–404

Wang Y, Kelly CG, Singh M, McGowan EG, Carrara AS, Bergmeier LA, Lehner T (2002) Stimulation of Th1-polarizing cytokines, C-C chemokines, maturation of dendritic cells, and adjuvant function by the peptide binding fragment of heat shock protein 70. J Immunol 169:2422–2429

Wang Y, Theriault JR, He H, Gong J, Calderwood SK (2004) Expression of a dominant negative heat shock factor-1 construct inhibits aneuploidy in prostate carcinoma cells. J Biol Chem 279:32651–32659

Whittall T, Peters B, Rahman D, Kingsley CI, Vaughan R, Lehner T (2011) Immunogenic and tolerogenic signatures in human immunodeficiency virus (HIV)-infected controllers compared with progressors and a conversion strategy of virus control. Clin Exp Immunol 166:208–217

Chapter 14
Molecular Chaperones and Scavenger Receptors: Binding and Trafficking of Molecular Chaperones by Class F and Class H Scavenger Receptors

Ayesha Murshid, Jianlin Gong and Stuart K. Calderwood

Abstract Extracellular molecular chaperones bind to a number of receptors including pattern recognition receptors, c-type lectins and scavenger receptor families. We have discussed the properties of two such receptors, the class F scavenger receptor SRECI and class H scavenger receptor stabilin-1. These receptors have, in common, the possession of extracellular domains containing multiple repeats of the EGF-like motif and the ability to bind HSPs. However, their regulatory properties are quite distinct. SRECI is an important receptor in the function of HSP vaccines and is able to mediate binding and endocytosis of HSP-peptide antigen complexes and to facilitate antigen cross presentation to T lymphocytes. Uptake of SRECI involves recruitment of the receptor to lipid raft domains and endocytosis, via the newly characterized GEEC compartment, prior to entry into early endosomes. Stabilin-1, by contrast, is localized largely to intracellular endosomes and the trans Golgi network (TGN). Although Stabilin-1 mediates internalization of HSPs, it does not appear to play an important role in antigen cross presentation and appears to be involved in other processes in response to extracellular molecular chaperones. Stabilin-1 is an unusual receptor in that it can mediate at least three processes in macrophages, including: (1) receptor-mediated endocytosis; (2) trafficking between the TGN and endosomal compartments and; (3) secretion through the lysosomal pathway. Possession of a range of HSP receptors may provide modulated responses to the molecular chaperones and is likely to impact on use of HSP based immunotherapy.

14.1 Introduction

The evidence now seems overwhelming that molecular chaperones, such as Hsp70 (HSPA1A), can exit from cells under conditions ranging from the physiological to the outright pathological (Mambula and Calderwood 2006a, b; Pockley et al. 1998 see

S. K. Calderwood (✉) · A. Murshid
Molecular and Cellular Radiation Oncology, Beth Israel Deaconess Medical Center, Harvard Medical School, 99 Brookline Avenue, Boston, MA 02215, USA
e-mail: scalderw@bidmc.harvard.edu

J. Gong
Boston University Medical School, 650 Albany Street, Boston, MA 02118, USA

B. Henderson, A. G. Pockley (eds.), *Cellular Trafficking of Cell Stress Proteins in Health and Disease*, Heat Shock Proteins 6, DOI 10.1007/978-94-007-4740-1_14,

Chap. 3). Extracellular HSPs appear to have many functions, including immune modulation of monocytes, macrophages, dendritic cells, B cells and T cells and neuronal cells (Brown 2007; Delneste et al. 2002; Srivastava 2002). Uncertainty remains however, as to how cells sense the presence of external HSPs, engage in HSP-mediated signaling and take up the proteins into the cytosol. Considerable evidence points to the existence of surface HSP receptors (Calderwood et al. 2007). However, no dedicated receptors for HSPs have been discovered. Instead HSP binding properties are shared with other ligands and belong to families such as pattern recognition receptors (PRR) and scavenger receptors (SR) (Calderwood et al. 2007). The reader should also look at Chap. 11 for a detailed discussion of receptors for chaperonin 60. Despite the rather generalized specificity of HSP receptors, there does appear to be selectivity involved, as other ligands such as BSA are not internalized through these receptors. The first receptor shown to interact with and take up HSP was the LDL receptor-related protein, LRP, also known as LRP1 and CD91 (Basu et al. 2001; Binder et al. 2000). LRP1 is a large multidomain protein that binds at least 30 distinct extracellular ligands, in addition to Hsp70 (Calderwood et al. 2007). An important function of this receptor appears to involve uptake of extracellular ligands and their transport to the lysosomes (Calderwood et al. 2007). We and others have recently shown that the class F scavenger receptor (SR) SRECI is an important receptor for HSP70, 90 and calreticulin and mediates binding, internalization and antigen cross-presentation (Berwin et al. 2004; Gong et al. 2009; Murshid et al. 2010; Theriault et al. 2006). SREC1 was originally characterized as *scavenger receptor expressed by endothelial cells* and is encoded by the SCARF1 gene (Adachi and Tsujimoto 2002a; Adachi et al. 1997; Ishii and Arai 2001). Aliases include: SREC-I, MGC47738 and the acetyl LDL receptor. Other receptors that bind and internalize HSPs include the class E scavenger family member LOX-1 and class H scavenger receptor FEEL-I or stabilin-1 (Calderwood et al. 2007; Delneste et al. 2002; Theriault et al. 2006). In this chapter we will discuss the shared and distinct molecular properties of SRECI and FEEL-1/Stabilin-1 and how they might be involved in responses to extracellular molecular chaperones.

14.2 SREC-1 and Genes Involved in Apoptotic Cell Clearance

Although SREC-I is a class F scavenger receptor, this information alone casts little light on its function. Scavenger receptors are a heterogeneous group of proteins grouped along functional lines, sharing the ability to bind and internalize damaged/modified proteins (as epitomized by oxidized low density lipoprotein) in the extracellular domain (Krieger 1997; Pluddemann et al. 2007). However, SREC-I is also structurally related to the *C. elegans* surface receptor shown to mediate the uptake of apoptotic cell corpses (May 2001; Zhou et al. 2001). Both proteins have a similar overall structure, with an N-terminal extracellular domain (ED), a single transmembrane domain and a C-terminal intracellular domain (ID) (Fig. 14.1). The area of greatest homology is the ED, which in each case contains repeats of

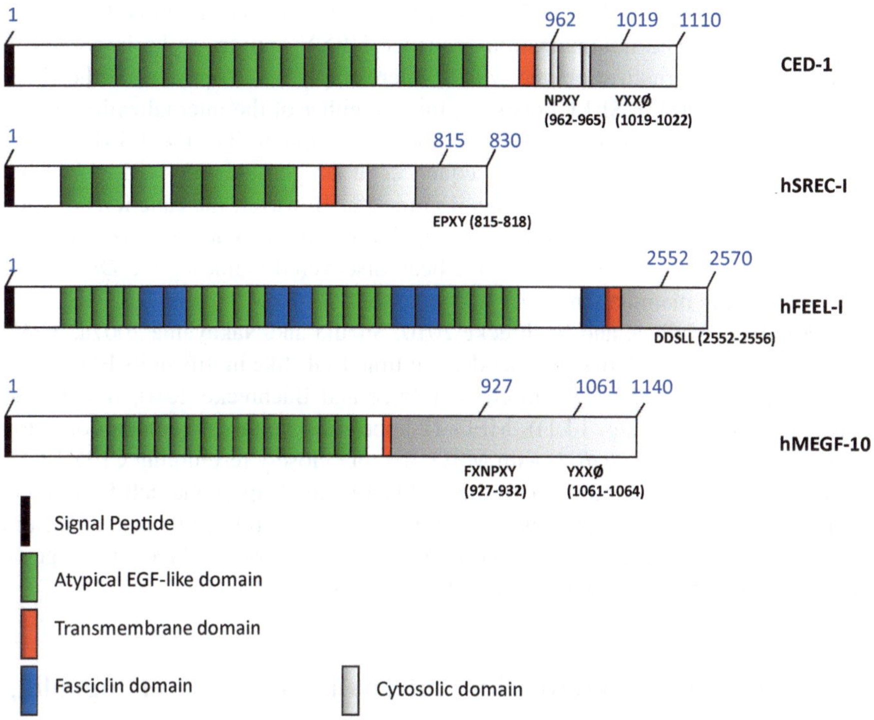

Fig. 14.1 The domain structures of CED-1 and related proteins. The locations of atypical EGF-like domains are indicated in all four proteins. CED-1, hSREC-I, hFEEL-I/Stabilin-1 and MEGF10 share EGF-like consensus repeats in extracellular domains. Tyrosine-based sorting signals are known to interact with the phosphor-tyrosine domain of clathrin adaptors (NPXY for CED-1, FXNPXY and YXXØ for hMEGF10) are shown in the figure. A dileucine based (DXXLL for hFEEL-1) sorting signal is present in the cytosolic tails of hFEEL-1 and can also be found in mannose 6-phosphate receptors that mediate sorting between trans-Golgi network (TGN) and endosomes. The hSREC-I has somewhat similar tyrosine based sorting signal (EPXY)

the EGF-like domain, a 40–45 amino acid sequence with eight conserved cysteines and four conserved glycine residues (May 2001; Zhou et al. 2001). In addition, both proteins contain multiple serine and threonine residues in the ID that may be involved in regulation. CED-1 contains an NPXY (phosphotyrosine binding site) and an YXXØ (YXXL) motifs (src homology domain involved in endocytosis). The carboxy terminal of the β-sandwich domain of the medium subunit (μ-) of AP-2 (adaptor protein-2) binds directly to the YXXØ internalization sequence (Owen et al. 2004). AP also recognizes a dileucine based motif [DE]XXXL[IL] (Owen et al. 2004). These adaptors have the ability to recognize and bind both clathrin and cargo molecules synchronously. Other internalization sequences include FXNPXY used by tyrosine kinase receptors, LDL receptor and ubiquitin (Traub 2003). The adaptor protein AP1 can also bind an acidic cluster sorting signal found in the cytosolic domain of the cation-dependent and cation-independent mannose

6-phosphate receptors (CD-M6PRs and CI-M6PRs) (Stöckli et al. 2004; Ghosh and Kornfeld 2004). The internalization sequence NPXY appears to be important in CED-6/GULP internalization and which, interestingly is also present in the ID of LRP1 (Su et al. 2002). SRECI does not include either of the internalization motifs in its ID and is relatively deficient in tyrosine in this domain. The ID of SREC1 does however contain VPXY and EPXY sequences at amino acids 534 and 818 which somewhat resemble NPXY internalization motifs. However, the functions of such sequences, or their potential ability to mimic NPXY, are not known. Recently, more closely-related paralogs of CED-1 have been discovered including the *Drosophila* gene *draper* and mammalian MEGF10 (multiple EGF-like-domains 10), MEGF-11 and MEGF12 (McPhee and Baehrecke 2010; Suzuki and Nakayama 2007a, b; Wu et al. 2009). The MEGF10 gene encodes multiple EGF-like motifs in its ED as well as internal NPXY and YXXL motifs (McPhee and Baehrecke 2010; Suzuki and Nakayama 2007a, b (Fig. 14.1)). MEG-12 appears to be the mammalian ortholog of the *Drosophila* gene *Jedi-1*, with MEG-10 more closely resembling CED-1 (Wu et al. 2009). These receptors share the ability to engulf apoptotic cell bodies and decrease inflammation by preventing accumulation of necrotic cells and, at least in mammalian tissues, after cellular digestion of the cell corpses, inducing tolerogenic responses in professional scavenging cells such as macrophages.

14.3 SRECI: Ligand Binding, Internalization and Cell Signaling

Thus SRECI contains within its ED, multiple EGF-like motifs that are associated with engulfment of apoptotic cell bodies (Fig. 14.1). However, although a number of such EGF-like domain proteins have been identified, the specific ligands expressed by cell corpses and recognized by these proteins are not known. Engulfing proteins containing multiple EGF-like repeats likely respond to "eat me" signals on the apoptotic cell surface such as externalized, oxidized phosphatidylserine, that resemble low density lipoprotein particles and the thrombospondin-1 binding site (Oka et al. 1998; Savill et al. 1992). Soluble ligands such as oxidized LDL or HSPs may thus be recognized and taken up by endocytosis by a similar process to the phagocytosis of apoptotic cell. However little is known regarding this process and the structure/function characteristics of these domains are not well characterized. Solving the solution structures of SRECI and other EGF-like repeat proteins, particularly when bound to their ligands, may thus offer considerable information regarding the roles of the proteins in cell regulation.

14.4 SRECI, the GEEC Pathway, Lipid Rafts and Cell Signaling

We have carried out a careful study of the pathway of Hsp90 internalization after binding to SRECI in macrophages and dendritic cells. Hsp90 with or without association with antigens, binds to SRECI either when ectopically expressed on

CHO or HeLa cells or in wild type dendritic cells and macrophages (Murshid et al. 2010). Hsp90-SRECI complexes are then internalized and rapidly accumulate in early endosomes. Somewhat surprisingly, internalization does not involve the classical clathrin-coated pit pathway, but instead appeared to use the newly characterized GPI-AP (glycophosphatidylinositol-anchored proteins)-enriched early endosome compartment (GEEC) (Mayor and Pagano 2007; Mayor and Riezman 2004). The GPI moiety is the membrane anchor for a large and heterogenous group of cell surface proteins with various functions. These proteins are usually parasite antigens, immune antigens and receptors (Ferguson et al. 1999). Unlike other surface receptors they do not concentrate in caveolae or CCPs and are localized in caveolae only after cross linking induced by multivalent antibodies (Fujimoto 1996; Mayor and Maxfield 1995; Mayor et al. 1994; Parton et al. 1994). These proteins are internalized through a dynamin and clathrin independent pathway (Sabharanjak et al. 2002). We showed that a range of inhibitors of the clathrin- and caveolin-dependent endocytosis pathways failed to inhibit uptake of ligand-associated SRECI and that antagonism of the Rho GTPase, CDC42, blocked uptake (Murshid et al. 2010). Cdc42 is known to regulate uptake of GPI-AP into the GEEC domain. Interestingly although the GEEC pathway is required for uptake of GPI-AP family proteins, SRECI does not belong to this family but is instead a classical transmembrane protein. The GEEC pathway is the major pathway for uptake of proteins sorted into lipid raft domains and these include mostly GPI-AP (Mayor and Pagano 2007; Mayor and Riezman 2004). However, internalized (Hsp90 bound) SRECI was shown to rapidly co-localize with the GPI-AP protein CD59 in the GEEC compartment (Murshid et al. 2010) (Fig. 14.2). Like endocytosis of GPI-anchored proteins, HSP90/SRECI uptake depends on Cdc42 and actin, can be blocked by depletion of sphingomyelin, and is not inhibited by dominant-negative mutants of dynamin, Rac or Rho. These are the hallmarks of endocytosis into GEECs (Sabharanjak et al. 2002), and consistent with this, after short pulses of Hsp90 (10 μg/ml), we find endocytosed Hsp90-SRECI colocalized with fluorescently-labeled folate in endosomes lacking conventional markers for early endosomes such as EEA1 (data not shown). Based on these data, we hypothesize that Hsp90/SRECI is endocytosed into GEECs at 10 μg/ml dose of Hsp90-Ova or at least into a compartment containing GPI-anchored proteins that is distinct from early endosomes.

Another protein modification is known to target transmembrane proteins to lipid rafts—the S-acylation of cysteine residues—closely adjacent to the transmembrane domain, usually with highly saturated palmitate residues that can dissolve in the raft environment of high cholesterol and sphingolipids (Levental et al. 2010a, b; Rajendran and Simons 2005). SRECI would seem a particularly good candidate for such a modification with five cysteine residues (amino acids 440, 441, 443, 444, 445) adjacent to the transmembrane domain (amino acids 422–442). In addition, proteins with bulky side chains are excluded from coated pits (Nichols 2009). SRECI contains a number of potential N-glycosylation sites in the ED, which, if avidly glycosylated, would create such a bulky conformation (Adachi et al. 1997). Raft targeting signals provided by acylation and prenylation of proteins are heterogeneous in terms of the general types of lipid modification. Targeting to lipid raft also depends on the

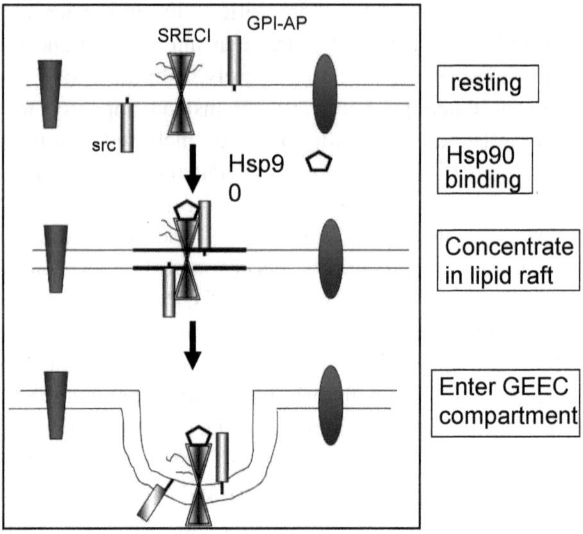

Fig. 14.2 Hsp90-triggered sorting of SRECI into lipid rafts and the GEEC internalization pathway Under resting conditions, SRECI is shown in the bulk membrane domain containing a range of surface proteins which are either transmembrane proteins such as SRECI, GPI-AP proteins or proteins anchored to the inside of the membrane such as c-src. Upon Hsp90 binding, SRECI becomes localized into lipid raft domains and co-localized with c-src. Within 5 min of ligand binding, Hsp90-SRECI complexes enter the GEEC compartment and are internalized

number and length of the lipid groups of proteins. Other protein-based modifications are also responsible for targeting to lipid raft in addition to this lipid modification such as palmitoylation. Ras proteins which are both prenylated and palmitoylated are found to be localized to lipid raft whereas only prenylated protein has very little affinity for lipid rafts due to the presence of its bulky branched structure. McCabe and Berthiaume (2001) suggested that the lipid modifications might cause targeting of the protein to the membrane but protein-protein interactions may be necessary and important for directing to the lipid raft. The transmembrane domain of proteins can also target proteins to rafts, depending on the length of transmembrane domain. Again, Yamabhai and Anderson (2002) used mutagenesis to prove that the EGF receptor has a cysteine-rich region in its membrane proximal region which targeted it to rafts. Both protein-protein interactions and protein-lipid interactions can help partitioning of proteins into lipid rafts. We observed the presence of SRECI in lipid raft domains as evidenced by co-localization with binding sites for cholera toxin (gangliosides) and such localization was disrupted by cholesterol sequestering drug, cyclodextrin (Murshid et al. 2010). Lipid rafts appear to be important regions for clustering of signaling molecules that can occur after antibody cross-linking or ligand binding (Harder et al. 1998; Harder and Simons 1999; Levental et al. 2010a). Members of the Src family of non-receptor tyrosine kinases are frequently recruited to lipid rafts under these conditions and may form signaling platforms within the raft regions. Interestingly, Src family kinases, as with SRECI are not GPI-AP but appear to associate with lipid rafts after S-acylation (Arni et al. 1996). We have found that Hsp90-activated SRECI becomes associated with c-src and leads to trans-phosphorylation and activation of the kinase (Murshid et al. 2010) (Fig. 14.2). Inhibition of one Src family member, c-src in dendritic cells led to a block in SRECI-mediated cross presentation of antigens associated with Hsp90 to T lymphocytes. Although the exact mechanism for this is not completely understood, SREC1 contains a number of intracellular tyrosine residues that might mediate effects of c-src on its endocytosis. Therefore, although we are in the early stages of characterizing this protein, our evidence suggests that SRECI can mediate Hsp90-induced transmembrane signaling from the cell surface as well as endocytosis of Hsp90-peptide complexes. SRECI also contains multiple proline, serine and threonine residues in the ID. This might suggest a largely unstructured sequence for the ID, with the potential for multiple phosphorylation events. Ishii et al. observed binding of the ID of SRECI to protein phosphatase 2α in L cells induced to form neurite outgrowth suggesting non-scavenger roles for SRECI mediated through phosphorylation (Ishii et al. 2007).

There is also evidence that SRECI can lead to cell signalling through indirect pathways, by recruitment of other molecules. It was shown that the outer membrane protein (OmpA) from *Klebsiella pneumoniae* can bind to SRECI (as well as LOX-1) and give rise to transcription of cytokines. This process appears to involve the recruitment of Toll-like receptor 2 (TLR2) to these receptors and downstream signaling, leading to cytokine gene expression (Jeannin et al. 2005). In addition, SRECI as well as scavenger receptor 1 (SR-1) can cooperate with TLR2 in the recognition of *hepatitis virus non-structural protein 3* by dendritic cells (Beauvillain et al. 2010). There have been considerable interesting potential interactions between HSP and TLR family members and their immune consequences, although most studies indicate that these effects may be indirectly mediated through primary HSP receptors such as SRECI or LOX-1 (Delneste et al. 2002; Shi et al. 2006). Our *in vivo* studies indicated that knockout of TLR2 and TLR4 and knockdown of SRECI completely abrogate the effects of HSP70 based anticancer vaccines (Gong et al. 2009).

14.5 Stabilin-1/FEEL1 the Endosomal HSP Receptor-Apoptotic Cells

HSPs also bind to class H scavenger receptor stabilin-1 (also known as FEEL-1 (fascilin, epidermal growth factor-like, laminin-type EGF-like and link domain containing scavenger receptor1) and CLEVER-1 (the common lymphatic endothelial and vascular endothelial receptor) (Irjala et al. 2003; Politz et al. 2002). Stabilin-1 is mostly restricted to the Golgi and endosomal compartments, although a small fraction is found on the cell surface (Kzhyshkowska et al. 2004; Pluddemann et al. 2007; Tamura et al. 2003). The receptor has been discovered in cells involved in scavenging of exogenous proteins and cells such as sinusoidal endothelial cells and is also expressed inducibly in monocytes and tissue macrophages (Kzhyshkowska and Krusell 2009; Pluddemann et al. 2007). We have shown that, when overexpressed in CHO cells, stabilin-1 can bind and internalize extracellular Hsp70 and Hsp90 (Theriault et al. 2006; Murshid and Calderwood, unpublished). Stabilin-1 is a high molecular weight (2570 amino acid) protein with a large ED (amino acids 1–2479), a single membrane spanning domain and a relatively small ID (amino acids 2499–2570) (Fig. 14.1). The ED contains 16 EGF-like domains, 7 fascilin domains, 2 laminin EGF-like domains and 1 link domain. Stabilin-1 has been shown to act as a scavenger receptor for acetylated LDL (but not oxidized LDL), Gram positive and Gram negative bacteria (Adachi and Tsujimoto 2002b; Kzhyshkowska 2010; Tamura et al. 2003). In addition, recent studies have shown that stabilin-1 is a receptor for externalized phosphatidylserine (PS) and can bind and lead to the engulfment of PS-expressing apoptotic bodies in a similar manner to other receptors with multiple EGF-like repeats in the extracellular domain (Park et al. 2009). Although the ED of stabilin-1 contains other structural domains, it was shown that competition with an excess of purified single EGF-like domains blocked apoptotic body engulfment by stabilin-1 (Park et al. 2009). Thus stabilin-1 is likely to mediate the tolerogenic effects of apoptotic body removal and digestion in macrophages. Stabilin-1 also binds to two other proteins when expressed in macrophages. These include SPARC—the universal regulator of tissue turnover and the hormone placental lactogen (Kzhyshkowska and Krusell 2009). It may be significant that different motifs in the large ED of Stabilin-1 are involved in binding to these individual ligands. SPARC binds to the EGF-like domain located between fascilin four motifs and five, while placental lactogen binds to fascilin domain 7 (Kzhyshkowska et al. 2006). Thus the remarkably large ED of stabilin-1 may be required for discrete binding to a range of ligands including HSPs, SPARC and placental lactogen.

The role of stabilin-1 in mediating immune responses to extracellular HSP is however, not clear. When overexpressed in CHO cells, stabilin-1 leads to efficient Hsp70 binding to the cell surface and rapid internalization into early endosomes (Calderwood et al. 2007; Theriault et al. 2006). Stabilin-1 is known to be involved in at least two major intracellular trafficking pathways that may be significant for HSP metabolism, including: (1) plasma membrane receptor-mediated endocytosis and, (2) shuttling of ligands between the endosomal compartment and the trans-Golgi network (TGN) (Zhang et al. 2009). Stabilin-1 appears to be able to move rapidly between internal endosomal compartments and the cell surface (Prevo et al. 2004). However despite these powerful molecular trafficking properties, we did not observe a role for overexpressed stabilin-1 in antigen cross-presentation *in vitro* (unlike SRECI) (Murshid and Calderwood unpublished). In addition, while in mice treated with Hsp70-based anticancer vaccines we observed migration of SRECI-expressing DC to T-lymphocyte rich regions of efferent lymph nodes, we failed to observe significant levels of accumulation of stabilin-1-expressing DC in the lymph nodes: in fact dendritic cells appeared to express very low levels of stabilin-1 either *in vitro* or *in vivo* (Gong and Calderwood, unpublished). Endocytosis of cell surface stabilin-1 appears to involve different mechanisms compared to SRECI, utilizing the clathrin coated pit pathway in association with clathrin adaptor family proteins (Hansen et al. 2005; Kzhyshkowska et al. 2004). In addition, stabilin-1 endocytosis possesses regulatory requirements common to CED-1 and LRP1, interacting with the adaptor protein CED-6/GULP that mediates endocytosis through an NPXY-like domain in the ID (Park et al. 2009) (Fig. 14.1). Shuttling of stabilin-1 between the TGN and endosomes appears to involve different motifs in the ID of stabilin-1 in addition to the NPXY-like domain. TGN-endosomal transport appears to be mediated by GGAs (Golgi-localized, γ-ear-containing, Arf-[ADP-ribosylation factor]-binding proteins (Bonifacino 2004; Kzhyshkowska et al. 2004). The well-studied mannose-6-phosphate receptor can traffic from the Golgi via clathrin-coated vesicles (CCVs) as tubular intermediates and fuse with endosomes with the help of GGA (Puertollano et al. 2001; Puertollano et al. 2003). AP1 containing intermediates can also emerge from the TGN and merge with endosomes (Waguri et al. 2003). Stabilin-1 contains two motifs within its rather meager ID that mediate this interaction with GGAs, including a dileucine (DXXLL) motif that is responsible for anterograde trafficking of receptors from TGN to endosomes (Bonifacino 2004; Zhang et al. 2009) (Fig. 14.1).

Stabilin-1 thus has the potential to facilitate at least two types of interaction involving the intracellular trafficking of HSPs, based on its known properties. These include (1) the endocytosis of HSPs associated with cargo and we have shown that Hsp70 and Hsp90 can be bound and endocytosed by cell surface stabilin-1 (Theriault et al. 2006). Hsp90-peptide-stabilin-1 complexes do not appear to feed the antigen cross presentation pathway suggesting that, in contrast to SRECI, stabilin-1 does not enter the compartments where it can mediate antigen processing and presentation to MHC class I molecules. However, stabilin-1 also has the potential to mediate trafficking between intracellular organelles. For example, the chitinase-like protein SI-CLIP is secreted through a pathway involving stabilin-1 mediated transport of this protein from the TGN to the endosomal/lysosomal system. SI-CLIP can there dissociate from stabilin-1 before transport to secretary lysosomes and secretion (Kzhyshkowska et al. 2006). Indeed, stabilin-1 appear to mimic the mannose-6-phosphate receptor in its ability mediate the secretion of molecules through the non-conventional lysosomal pathway (Kzhyshkowska and Krusell 2009). The TGN -lysosomal secretion pathway is important in release of hydrolytic enzymes into the extracellular milieu and may play both inflammatory and anti-inflammatory roles. It is notable that our studies have indicated a role for secretary lysosomes in release of Hsp70 (along with cathepsin D) from macrophages and tumor cells stimulated by a range of conditions (Mambula and Calderwood 2006b) (SS Mambula and SK Calderwood, in preparation). It is also likely that stabilin-1 may mediate other interactions between internalized HSP and molecules in intracellular organelles and our laboratory is pursuing these possibilities. One recent study shows that stabilin-1 is involved in trans-endothelial migration of T cells in inflamed liver. Stabilin-1 levels increase at sites of inflammation and selectively stimulate trans-endothelial migration of CD4 FoxP3(+) regulatory T cells into the sites of inflammation (Shetty et al. 2011). Stabilin-1 may thus mediate immunoregulatory processes under some conditions.

It appears therefore that cells possess a range of scavenger receptors that can regulate the trafficking and sorting of extracellular HSP and cargo into different compartments and mediate cell signaling. Extracellular HSP-peptide complexes can be internalized by SRECI, Lox-1 and stabilin-1 through different endocytic pathways and likely to different destinations. In addition, these receptors are differentially expressed with both SRECI and LOX-1 found in DC where they can mediate antigen cross presentation and stabilin-1 inducible expressed in macrophages These receptors may also mediate intracellular trafficking and secretion of HSP into the extracellular microenvironment.

Acknowledgments We thank the department of Radiation Oncology, Beth Israel Deaconess Medical Center for support and encouragement. SKC is supported by NIH research grants R01CA047407; R01CA094397 and R01CA119045.

References

Adachi H, Tsujimoto M (2002a) Characterization of the human gene encoding the scavenger receptor expressed by endothelial cell and its regulation by a novel transcription factor, endothelial zinc finger protein-2. J Biol Chem 277:24014–24021

Adachi H, Tsujimoto M (2002b) FEEL-1, a novel scavenger receptor with in vitro bacteria-binding and angiogenesis-modulating activities. J Biol Chem 277:34264–34270

Adachi H, Tsujimoto M, Arai H, Inoue K (1997) Expression cloning of a novel scavenger receptor from human endothelial cells. J Biol Chem 272:31217–31220

Arni S, Ilangumaran S, van Echten-Deckert G, Sandhoff K, Poincelet M, Briol A, Rungger-Brandle E, Hoessli DC (1996) Differential regulation of Src-family protein tyrosine kinases in GPI domains of T lymphocyte plasma membranes. Biochem Biophys Res Commun 225:801–807

Basu S, Binder RJ, Ramalingam T, Srivastava PK (2001) CD91 is a common receptor for heat shock proteins gp96, hsp90, hsp70, and calreticulin. Immunity 14:303–313

Beauvillain C, Meloni F, Sirard JC, Blanchard S, Jarry U, Scotet M, Magistrelli G, Delneste Y, Barnaba V, Jeannin P (2010) The scavenger receptors SRA-1 and SREC-I cooperate with TLR2 in the recognition of the hepatitis C virus non-structural protein 3 by dendritic cells. J Hepatol 52:644–651

Berwin B, Delneste Y, Lovingood RV, Post SR, Pizzo SV (2004) SREC-I, a type F scavenger receptor, is an endocytic receptor for calreticulin. J Biol Chem 279:51250–51257

Binder RJ, Han DK, Srivastava PK (2000) CD91: a receptor for heat shock protein gp96. Nat Immunol 1:151–155

Bonifacino JS (2004) The GGA proteins: adaptors on the move. Nat Rev Mol Cell Biol 5:23–32

Brown IR (2007) Heat shock proteins and protection of the nervous system. Ann N Y Acad Sci 1113:147–158

Calderwood SK, Theriault J, Gray PJ, Gong J (2007) Cell surface receptors for molecular chaperones. Methods 43:199–206

Delneste Y, Magistrelli G, Gauchat J, Haeuw J, Aubry J, Nakamura K, Kawakami-Honda N, Goetsch L, Sawamura T, Bonnefoy J, Jeannin P (2002) Involvement of LOX-1 in dendritic cell-mediated antigen cross-presentation. Immunity 17:353–362

Ferguson MA, Brimacombe JS, Brown JR et al (1999) The GPI biosynthetic pathway as a therapeutic target for African sleeping sickness. Biochim Biophys Acta 1455:327–340

Fujimoto T (1996) GPI-anchored proteins, glycosphingolipids, and sphingomyelin are sequestered to caveolae only after crosslinking. J. Histochem. Cytochem 44:929–941

Ghosh P, Kornfeld S (2004) The GGA proteins: key players in protein sorting at the trans-Golgi network. Eur J Cell Biol 83:257

Gong J, Zhu B, Murshid A, Adachi H, Song B, Lee A, Liu C, Calderwood SK (2009) T Cell Activation by Heat Shock Protein 70 Vaccine Requires TLR Signaling and Scavenger Receptor Expressed by Endothelial Cells-1. J Immunol 183:3092–3098

Hansen B, Longati P, Elvevold K, Nedredal GI, Schledzewski K, Olsen R, Falkowski M, Kzhyshkowska J, Carlsson F, Johansson S, Smedsrod B, Goerdt S, Johansson S, McCourt P (2005) Stabilin-1 and stabilin-2 are both directed into the early endocytic pathway in hepatic sinusoidal endothelium via interactions with clathrin/AP-2, independent of ligand binding. Exp Cell Res 303:160–173

Harder T, Simons K (1999) Clusters of glycolipid and glycosylphosphatidylinositol-anchored proteins in lymphoid cells: accumulation of actin regulated by local tyrosine phosphorylation. Eur J Immunol 29:556–562

Harder T, Scheiffele P, Verkade P, Simons K (1998) Lipid domain structure of the plasma membrane revealed by patching of membrane components. J Cell Biol 141:929–942

Irjala H, Elima K, Johansson EL, Merinen M, Kontula K, Alanen K, Grenman R, Salmi M, Jalkanen S (2003) The same endothelial receptor controls lymphocyte traffic both in vascular and lymphatic vessels. Eur J Immunol 33:815–824

Ishii J, Arai H (2001) Scavenger receptor expressed by endothelial cells (SREC). Nippon Rinsho 59 Suppl 2:379–383

Ishii J, Adachi H, Shibata N, Arai H, Tsujimoto M (2007) Scavenger receptor expressed by endothelial cells (SREC)-I interacts with protein phosphatase 1alpha in L cells to induce neurite-like outgrowth. Biochem Biophys Res Commun 360:269–274

Jeannin P, Bottazzi B, Sironi M, Doni A, Rusnati M, Presta M, Maina V, Magistrelli G, Haeuw JF, Hoeffel G, Thieblemont N, Corvaia N, Garlanda C, Delneste Y, Mantovani A (2005) Complexity and complementarity of outer membrane protein A recognition by cellular and humoral innate immunity receptors. Immunity 22:551–560

Krieger M (1997) The other side of scavenger receptors: pattern recognition for host defense. Curr Opin Lipidol 8:275–280

Kzhyshkowska J (2010) Multifunctional receptor stabilin-1 in homeostasis and disease. ScientificWorldJournal 10:2039–2053

Kzhyshkowska J, Krusell L (2009) Cross-talk between endocytic clearance and secretion in macrophages. Immunobiology 214:576–593

Kzhyshkowska J, Gratchev A, Martens JH, Pervushina O, Mamidi S, Johansson S, Schledzewski K, Hansen B, He X, Tang J, Nakayama K, Goerdt S (2004) Stabilin-1 localizes to endosomes and the trans-Golgi network in human macrophages and interacts with GGA adaptors. J Leukoc Biol 76:1151–1161

Kzhyshkowska J, Gratchev A, Goerdt S (2006) Stabilin-1, a homeostatic scavenger receptor with multiple functions. J Cell Mol Med 10:635–649

Levental I, Grzybek M, Simons K (2010a) Greasing their way: lipid modifications determine protein association with membrane rafts. Biochemistry 49:6305–6316

Levental I, Lingwood D, Grzybek M, Coskun U, Simons K (2010b) Palmitoylation regulates raft affinity for the majority of integral raft proteins. Proc Natl Acad Sci USA 107:22050–22054

Mambula SS, Calderwood SK (2006a) Heat induced release of Hsp70 from prostate carcinoma cells involves both active secretion and passive release from necrotic cells. Int J Hyperthermia 22:575–585

Mambula SS, Calderwood SK (2006b) Heat shock protein 70 is secreted from tumor cells by a nonclassical pathway involving lysosomal endosomes. J Immunol 177:7849–7857

May RC (2001) Phagocytosis in C. elegans: CED-1 reveals its secrets. Trends Cell Biol 11:150

Mayor S, Maxfield FR (1995) Insolubility and redistribution of GPI-anchored proteins at the cell surface after detergent treatment. Mol Biol Cell 6:929–944

Mayor S, Riezman H (2004) Sorting GPI-anchored proteins. Nat Rev Mol Cell Biol 5:110–120

Mayor S, Pagano RE (2007) Pathways of clathrin-independent endocytosis. Nat Rev Mol Cell Biol 8:603–612

Mayor S, Rothberg KG, Maxfield FR (1994) Sequestration of GPI-anchored proteins in caveolae triggered by cross-linking. Science 264:1948–1951

McCabe JB, Berthiaume LG (2001) N-terminal protein acylation confers localization to cholesterol, sphingolipid-enriched membranesbut not to lipid rafts/caveolae. Mol Biol Cell 12:3601–3617

McPhee CK, Baehrecke EH (2010) The engulfment receptor Draper is required for autophagy during cell death. Autophagy 6:1192–1193

Murshid A, Gong J, Calderwood SK (2010) Heat shock protein 90 mediates efficient antigen cross presentation through the scavenger receptor expressed by endothelial cells-I. J Immunol 185:2903–2917

Nichols B (2009) Endocytosis of lipid-anchored proteins: excluding GEECs from the crowd. J Cell Biol 186:457–459

Oka K, Sawamura T, Kikuta K, Itokawa S, Kume N, Kita T, Masaki T (1998) Lectin-like oxidized low-density lipoprotein receptor 1 mediates phagocytosis of aged/apoptotic cells in endothelial cells. Proc Natl Acad Sci USA 95:9535–9540

Owen DJ, Collins BM, Evans PR (2004) Adaptors for clathrin coats: structure and function. Annu Rev Cell Dev Biol 20:153–191

Park SY, Jung MY, Lee SJ, Kang KB, Gratchev A, Riabov V, Kzhyshkowska J, Kim IS (2009) Stabilin-1 mediates phosphatidylserine-dependent clearance of cell corpses in alternatively activated macrophages. J Cell Sci 122:3365–3373

Parton RG, Joggerst B, Simons K (1994) Regulated internalization of caveolae. J Cell Biol 127:1199–1215

Pluddemann A, Neyen C, Gordon S (2007) Macrophage scavenger receptors and host-derived ligands. Methods 43:207–217

Pockley AG, Shepherd J, Corton JM (1998) Detection of heat shock protein 70 (Hsp70) and anti-Hsp70 antibodies in the serum of normal individuals. Immunol Invest 27:367–377

Politz O, Gratchev A, McCourt PA, Schledzewski K, Guillot P, Johansson S, Svineng G, Franke P, Kannicht C, Kzhyshkowska J, Longati P, Velten FW, Johansson S, Goerdt S (2002) Stabilin-1 and −2 constitute a novel family of fasciclin-like hyaluronan receptor homologues. Biochem J 362:155–164

Prevo R, Banerji S, Ni J, Jackson DG (2004) Rapid plasma membrane-endosomal trafficking of the lymph node sinus and high endothelial venule scavenger receptor/homing receptor stabilin-1 (FEEL-1/CLEVER-1). J Biol Chem 279:52580–52592

Puertollano R, Randazzo PA, Presley JF, Hartnell LM, Bonifacino JS (2001) The GGAs promote ARF-dependent recruitment of clathrin to the TGN. Cell 105:93–102

Puertollano R, van der Wel NN et al (2003) Morphology and dynamics of clathrin/GGA1-coated carriers budding from the trans-Golgi network. Mol Bio Cell 14:1545–1557

Rajendran L, Simons K (2005) Lipid rafts and membrane dynamics. J Cell Sci 118:1099–1102

Sabharanjak S, Sharma P, Parton RG, Mayor S (2002) GPI-anchored proteins are delivered to recycling endosomes via a distinct cdc42-regulated, clathrin-independent pinocytic pathway. Dev Cell 2:411–423

Savill J, Hogg N, Ren Y, Haslett C (1992) Thrombospondin cooperates with CD36 and the vitronectin receptor in macrophage recognition of neutrophils undergoing apoptosis. J Clin Invest 90:1513–1522

Shetty S, Weston CJ, Oo YH, Westerlund N, Stamataki Z, Youster J, Hubscher SG, Salmi M, Jalkanen S, Lalor PF, Adams DH (2011) Common lymphatic endothelial and vascular endothelial receptor-1 mediates the transmigration of regulatory T cells across human hepatic sinusoidal endothelium. J Immunol 186:4147–4155

Shi Y, Tu Z, Tang D, Zhang H, Liu M, Wang K, Calderwood SK, Xiao X (2006) The inhibition of LPS-induced production of inflammatory cytokines by HSP70 involves inactivation of the NF-kappaB pathway but not the MAPK pathways. Shock 26:277–284

Srivastava P (2002) Interaction of heat shock proteins with peptides and antigen presenting cells: chaperoning of the innate and adaptive immune responses. Annu Rev Immunol 20:395–425

Stöckli J, Höning S, Rohrer J (2004) The acidic cluster of the CK2 site of the cation-dependent mannose 6-phosphate receptor (CD-MPR) but not its phosphorylation is required for GGA1 and AP-1 binding. J Biol Chem 279:23542–23549

Su HP, Nakada-Tsukui K, Tosello-Trampont AC, Li Y, Bu G, Henson PM, Ravichandran KS (2002) Interaction of CED-6/GULP, an adapter protein involved in engulfment of apoptotic cells with CED-1 and CD91/low density lipoprotein receptor-related protein (LRP). J Biol Chem 277:11772–11779

Suzuki E, Nakayama M (2007a) MEGF10 is a mammalian ortholog of CED-1 that interacts with clathrin assembly protein complex 2 medium chain and induces large vacuole formation. Exp Cell Res 313:3729–3742

Suzuki E, Nakayama M (2007b) The mammalian Ced-1 ortholog MEGF10/KIAA1780 displays a novel adhesion pattern. Exp Cell Res 313:2451–2464

Tamura Y, Adachi H, Osuga J, Ohashi K, Yahagi N, Sekiya M, Okazaki H, Tomita S, Iizuka Y, Shimano H, Nagai R, Kimura S, Tsujimoto M, Ishibashi S (2003) FEEL-1 and FEEL-2 are endocytic receptors for advanced glycation end products. J Biol Chem 278:12613–12617

Theriault JR, Adachi H, Calderwood SK (2006) Role of scavenger receptors in the binding and internalization of heat shock protein 70. J Immunol 177:8604–8611

Traub LM (2003) Sorting it out: AP-2 and alternate clathrin adaptors in endocytic cargo selection. J Cell Biol 163:203–208

Waguri S, Dewitte F, Le Borgne R et al (2003) Visualization of TGN to endosome traficking through fluorescently labeled MPR and AP-1 in living cells. Mol Biol Cell 14:142–155

Wu HH, Bellmunt E, Scheib JL, Venegas V, Burkert C, Reichardt LF, Zhou Z, Farinas I, Carter BD (2009) Glial precursors clear sensory neuron corpses during development via Jedi-1, an engulfment receptor. Nat Neurosci 12:1534–1541

Yamabhai M, Anderson RGW (2002) Second cysteine-rich region of EGFR contains targeting information for caveolae/rafts. J Biol Chem 277:24843–24846

Zhang J, Gratchev A, Riabov V, Mamidi S, Schmuttermaier C, Krusell L, Kremmer E, Workman G, Sage EH, Jalkanen S, Goerdt S, Kzhyshkowska J (2009) A novel GGA-binding site is required for intracellular sorting mediated by stabilin-1. Mol Cell Biol 29:6097–6105

Zhou Z, Hartwieg E, Horvitz HR (2001) CED-1 is a transmembrane receptor that mediates cell corpse engulfment in C. elegans. Cell 104:43–56

Chapter 15
GRP78 (BiP): A Multifunctional Cell Surface Receptor

Mario Gonzalez-Gronow, Salvatore V. Pizzo and Uma K. Misra

Abstract The endoplasmic reticulum (ER) 78 kDa glucose-regulated protein (GRP78), also known as BiP, plays a central role in a variety of physiological processes in human cells, including protein biogenesis, signal transduction, and calcium homeostasis. When expressed in plasma cell membranes, GRP78 functions as a receptor which recognizes extracellular ligands that stimulate cell proliferation, and may also behave as an autoantigen. GRP78 is a signaling receptor for activated α_2-macroglobulin, plasminogen kringle 5, and microplasminogen, and serves as a co-receptor, associated with MHC-I, in viral entry of cocksackie B. It is also a receptor for entry of dengue fever and Borna disease viruses. Furthermore, it regulates tissue factor procoagulant activity, it functions as a receptor for the angiogenic peptides RoY and ADAM15, and is also a partner of the teratocarcinoma-derived growth factor 1 (Cripto), T-cadherin, Par-4, and the DnaJ-like protein MTJ-1. These associations suggest a unique cell surface GRP78 topography which is compartmentalized to respond differently to agonists that bind to either its NH_2- or COOH-terminal domains. In this chapter, we discuss the physiological characteristics of these interactions, and the possible mechanisms involved in transportation of GRP78 from the ER to the cell surface.

15.1 Introduction

The glucose-regulated protein of 78-kDa (GRP78, also known as binding immunoglobulin protein (BiP) and, more recently HSPA5) was initially reported in 1974 (Stone et al. 1974) and then confirmed in 1977 (Pouyssegur et al. 1977) as one of the major proteins overproduced in normal chicken fibroblasts when the cells are starved of glucose, or exposed to inhibitors of glycosylation such as tunicamycin (Olden et al. 1979). As a member of the heat-shock protein 70 kDa (Hsp 70/HSPA) family, GRP78 synthesis and structure is highly conserved in animal cells (Lee 1987), showing about 60 % structural homology with Hsp70/HSPA1A (Munro and Pelham 1986); however, the rates of expression of these two proteins under stress, such

M. Gonzalez-Gronow (✉) · S. V. Pizzo · U. K. Misra
Department of Pathology, Duke University Medical Center, Box 3712,
Durham, NC 27710, USA
e-mail: gonza002@mc.duke.edu

B. Henderson, A. G. Pockley (eds.), *Cellular Trafficking of Cell Stress Proteins in Health and Disease*, Heat Shock Proteins 6, DOI 10.1007/978-94-007-4740-1_15,

as hypoxia or low extracellular pH, are significantly different (Scandra et al 1984; Whelan and Hightower 1985).

GRP78 is found primarily in the endoplasmic reticulum (ER) where it plays important roles in the quality control of proteins processed in the ER and the regulation of ER signaling in response to ER stress (Zhang and Zhang 2010) or the unfolded protein response (UPR) (Ma and Hendershot 2004). Expression of GRP78 on the cell surface was first reported in hamster fibroblasts in 1984 (Lee et al. 1984). Later, it was found in cells from the prostate (Mintz et al. 2003), breast (Wang et al. 2009), stomach (Zheng et al. 2008), malignant melanomas (Zhuang et al. 2009), ovary (Chinni et al. 1997) and olfactory neuroblastoma cancer patient tissues (Weinreb et al. 2009). GRP78 is also expressed on the surface of proliferating endothelial cells and monocytic cells (Bhattacharjee et al. 2005; Davidson et al. 2005).

Cell surface GRP78 functions as a cell surface signaling receptor for activated α_2-macroglobulin (α_2M*) (Misra et al. 2002, 2004). It is also a receptor for coxsackie A9 virus (Triantafilou et al. 2002), dengue virus serotype 2 (Jindadamrongwech et al. 2004), plasminogen kringle 5 (K5) and microplasminogen (Gonzalez-Gronow et al. 2007). In plasma membranes, it also associates with the major histocompatibility complex class I (MHC-I) (Triantafilou et al. 2001), the cell surface voltage-dependent anion channel (VDAC) (Gonzalez-Gronow et al. 2007), tissue factor (TF) (Bhattacharjee et al. 2005), and the teratocarcinoma-derived growth factor 1 (Cripto) (Shani et al. 2008). It is also involved the regulation of tissue factor (TF) procoagulant activity (Watson et al. 2003) and angiogenesis on endothelial cells (Hardy et al. 2008; Raiter et al. 2010).

Cell-surface GRP78 is a major autoantigen in prostate (Mintz et al. 2003), ovarian (Chinni et al. 1997), and gastric cancer patients (Rauschert et al. 2008). Autoantibodies against GRP78 isolated from cancer patient sera show different affinities and functions. For example, an IgG antibody from the serum of prostate cancer patients, which binds to an N-terminal region of GRP78, enhances cell survival and proliferation Gonzalez-Gronow et al (2006). In contrast, a serum IgM antibody isolated from gastric cancer patients, recognizing a C-terminal region of GRP78, induces apoptosis in tumor cells (Rauschert et al. 2008).

The crucial roles that GRP78 plays on the cell surface are finely regulated and depend on interactions with a large group of functional partners, including co-chaperones, nucleotide exchange factors, and signaling molecules Dudek et al. (2009). Such a diversity of functions implies that plasma membrane GRP78 must be conditioned to respond accurately and efficiently. In the following sections, we will describe some of the possible mechanisms explaining how GRP78 overproduced in the ER may be transported to the plasma membrane and performs such functions.

15.2 Expression and Function of GRP78 on the Cell Surface Under Stress Conditions

In response to several kinds of stress, including ER Ca^{2+}-depletion and accumulation of unglycosylated proteins, the cell increases expression of GRP78 which is at the core of multiple functions essential to maintain cell viability (Gonzalez-Gronow et al. 2009). Under certain conditions, a fraction of the GRP78 synthesized during ER stress is translocated to the cell surface. ER stress induced by agents that promote Ca^{2+} depletion include ionomycin (IM) (Miyake et al. 2000) and thapsigargin (TG) (Yoshida et al. 2006). ER stress resulting from accumulation of unglycosylated proteins may be induced by tunicamycin™ (Miyake et al. 2000). Both mechanisms stimulate cell surface expression of GRP78, thereby suggesting the existence of at least two different pathways mediating the GRP78 stress response.

On the cell surface, the agonists may induce signaling pathways that depend on whether they bind to the N- or C-termini of GRP78. For example, binding of α_2M^* or autoantibodies, from serum of prostate cancer patients, to a GRP78 amino acid segment in the N-terminus (Leu^{98}-Leu^{115}) (NT-Ab) induce cell proliferation and promote cell survival (Gonzalez-Gronow et al. 2006). Conversely, a commercial polyclonal antibody against the GRP78 COOH-terminal domain (CT-Ab) induces apoptosis in prostate cancer cells (Misra et al. 2009). The mechanisms by which cell surface GRP78 N-terminus respond to α_2M^* or the patient autoantibodies inducing cell proliferation (Misra et al. 2006) differ from those induced by ligation of the GRP78 C-terminus which promote cell apoptosis (Misra et al. 2010; Misra and Pizzo 2010a).

Binding of α_2M^* to cell surface GRP78 induces signaling cascades that resemble those observed in cells undergoing ER stress (Misra and Pizzo 2004a) such as increases of intracellular Ca^{2+}, and augmented expressions of the transmembrane protein kinase/endoribonuclease IRE1α, the activating transcription factor 6 (ATF6), and the endoplasmic reticulum resident kinase (PERK) (Misra et al. 2006). Also, the translational factor eIF2α is phosphorylated producing inhibition of protein synthesis as observed during the UPR (Misra et al. 2006). The activating transcription factor 4 (ATF4) is also up-regulated, leading to an increased expression of the growth arrest and DNA-damage inducible protein (GADD34) which later restores protein synthesis through inhibition of eIF2α (Misra et al. 2006). In contrast to what is observed in the UPR, α_2M^*-induced up-regulation of IRE1α, the apoptosis signal-regulating kinase 1 (ASK1), and the TNF receptor associated factor 2 (TRAF2) do not promote an increase in the c-Jun N-terminal kinase (JNK) or caspase-12 activation which result in apoptosis (Misra et al. 2006). The induction of Akt, which phosphorylates and inactivates ASK1 and NF-κB, up-regulates GADD45β and inhibits JNK, thereby blocking the pro-apoptotic signaling expected from activation of UPR proteins (Misra et al. 2006). In the ER, GRP78 and its partners IRE1α, ATF6, and PERK span the ER membrane to facilitate signal transduction, affecting also phosphatidylinositol-3-kinase (PI3-kinase) or downstream Akt phosphorylation and NF-κB induction. However, cell surface GRP78 forms a complex with the ER membrane Dna J-like protein MTJ-1 and the G-protein-alpha-11 (Gαq11) (Misra and Pizzo 2008) which facilitates signaling through PI3-kinase/Akt as well as NF-κB-dependent pathways (Xie et al. 2000). Therefore, cell surface GRP78 signaling resembles that of ER GRP78 with the addition that it induces a rise in cytosolic Ca^{2+} and increases in IP3, Akt phosphorylation, and NF-κB activities. These functions suggest that GRP78, in association with α_2M^*, functions like a growth factor receptor in cancer cells, where expression of GRP78 on the cell surface appears to enhance carcinogenesis and tumor progression (Zhang and Zhang 2010).

GRP78 on the surface of vascular endothelial cells forms a complex with Glycosylphosphatidylinositol (GPI)-anchored T-cadherin and influences cell survival (Philippova et al. 2008), via PI3-kinase/Akt signaling pathways which facilitate cell growth and motility (Joshi et al. 2005). Cancer cell proliferation is also stimulated by GRP78 in association with the GPI-anchored oncogene Cripto, which inhibits transforming growth factor β (TGF-β) signaling (Shani et al. 2008). Cripto is normally absent in healthy tissues, but is expressed at high levels in tumors, where it upregulates mitogen-activated MAPK/ERK and PI3K/Akt signaling pathways, and inhibits activin signaling (Strizzi et al. 2005; Gray et al. 2006). This mechanism operates only when both GRP78 and Cripto are colocalized on the cell surface (Shani et al. 2008).

GRP78 on the cell surface also associates with the prostate apoptosis response-4 protein (Par-4) near its NH_2-terminal domain (Burikhanov et al. 2009). The Par-4 gene was initially identified as an upregulated immediate early apoptotic gene in response to elevated intracellular Ca^{2+} in AT-3 rat prostate tumor cells, but is also found in renal-cell carcinoma, neuroblastoma, acute lymphoblastic leukemia, and chronic lymphocytic leukemia (Zhao and Rangnekar 2008). Par-4 was previously regarded as a cytosolic and nuclear pro-apoptotic protein; however, it is secreted by both normal and tumor cells via conventional secretory pathways (Burikhanov et al. 2009).

Cell surface GRP78 NH_2-terminal domain also binds K5, a potent angiogenic inhibitor, causing apoptosis of proliferating human microvascular endothelial cells (Davidson et al. 2005). Association of the N-terminal domain of GRP78 with either K5 or secreted Par-4 induces cell death. However; Par-4 elicits the extrinsic apoptotic pathway by activation of caspases-3 and -8, whereas GRP78/K5-mediated apoptosis involves the extrinsic pathway by activation of caspase-7 (Davidson et al. 2005; Gonzalez-Gronow et al. 2007).

Cell surface GRP78 NH_2-terminal domain on endothelial also functions as a receptor for a novel 12 amino acid peptide, RoY, which induces *in vitro* angiogenesis under hypoxic conditions (Hardy et al. 2008). Furthermore, another peptide derived from the disintegrin and metalloproteinase 15 (ADAM15) also activates cell surface GRP78 on endothelial cells under hypoxic conditions, inducing VEGF-independent angiogenesis (Raiter et al. 2010). Furthermore, the GRP78 N-terminal region also serves as a receptor for entry of Borna disease virus, GP1 (Honda et al. 2009).

The GRP78 COOH-terminal domain also functions as a receptor independently of its chaperone activity. For example, it inhibits TF procoagulant activity (Watson et al. 2003) via physical interaction between the TF extracellular domain and a region localized in the COOH-terminal domain of GRP78, distant from either its ATP or peptide binding domain (Bhattacharjee et al. 2005). The COOH-terminal region of cell surface GRP78 is also a receptor for microplasminogen (Gonzalez-Gronow et al. 2007) and the dengue virus serotype 2 (Jindadamrongwech et al. 2004). A region in microplasminogen, including plasminogen amino acid residues Ser^{759}-Phe^{778} is homologous to amino acid sequence Gly^{100}-Phe^{119} of the viral protein coat (Markoff et al. 1991). The binding of both these region to GRP78 C-terminus was further confirmed in experiments using the homologous peptide CTVALPGGYVRVC which binds to cell surface GRP78 and is rapidly internalized (Kim et al. 2006). The GRP78

Table 15.1 Comparison of the cellular responses induced by ligation of growth factor receptors or GRP78 with their agonists on the cell surface

Response to agonist	Growth factor receptors		GRP78	
	Growth factors	α_2M^*	T-cadherin	Cripto
$[Ca^{2+}]_i$ increase	+++ [55]	+++ [60]	+++ [68]	+++ [69]
Act. PIP_2 hydrolisis	+++ [55]	+++ [60]	N.D.	+++ [69]
PKC activation	+++ [55]	+++ [61]	+++ [43]	+++ [69]
PI-3 kinase activation	+++ [56]	+++ [61]	+++ [43]	+++ [70]
PLCγTyr-Phosp.	+++ [56]	+++ [62]	+++ [44]	+++ [71]
$[pH]_i$ alkalinization	+++ [56]	+++ [62]	N.D.	N.D.
Enhanced protein synthesis	+++ [56]	+++ [60]	+++ [44]	+++ [71]
Enhanced DNA synthesis	+++ [56]	+++ [60]	+++ [44]	+++ [72]
PLD activation	+++ [57]	+++ [63]	N.D.	+++ [73]
PLA_2 activation	+++ [57]	+++ [64]	N.D.	N.D.
Receptor oligomerization	+++ [57]	+++ [61]	N.D.	N.D.
Receptor kinase activity	+++ [58]	+++ [61]	+++ [43]	+++ [71]
Tyr-P of effector proteins, containing SH2 and SH3 domains	+++ [58]	+++ [61]	+++ [44]	+++ [71]
Ras and G protein activation	+++ [56]	+++ [65]	+++ [68]	+++ [74]
Ser/threo Tyr-kinase activation	+++ [58]	+++ [66]	+++ [44]	+++ [45]
Expression of genes	+++ [59]	+++ [67]	N.D.	+++ [45]
Single membrane spanning domain	+++ [59]	+++ [40]	N.D.	N.D.

N.D. non-determined

COOH-terminal region involved in these interactions is located between amino acids 467–530 (Limjindaporn et al. 2009). In addition, GRP78 COOH-terminal region also serves as a receptor for Coxsackie virus A9, which interacts only with GRP78, followed by internalization via MHC-I-associated endocytosis (Triantafilou et al. 2002).

A comparison of the cellular responses given by growth factor receptors after ligation with their agonists (Bickel 2002; Guderman et al. 2000; Neves et al. 2002; Poppleton et al. 1996; Patel 2004) and by cell surface GRP78 ligation to α_2M^* (Misra and Pizzo 1998a, b, 1999, 2004b, c, 2008; Misra et al. 1993, 1995, 2002), T-cadherin (Philippova et al. 2008; Joshi et al. 2005; Angst et al. 2001), and Cripto (Strizzi et al. 2005; De Santis et al. 1997; Ebert et al. 1999; Kannan et al. 1997; Kelber et al. 2009; Saloman et al. 2000; Watanabe et al. 2007) is shown in Table 15.1. There are several common pathways suggesting that, in addition to the multiple functions that GRP78 plays in the ER, it may also function as a growth factor receptor on the cell surface.

15.3 Potential Mechanisms Involved in Trafficking of GRP78 from the ER to the Cell Surface

Most of the mechanisms involved in translocation of GRP78 from the ER to the plasma membrane have been studied in cancer cells. Due to poor vascularization and the resulting hypoxia and glucose starvation, tumor cells are prone to ER stress

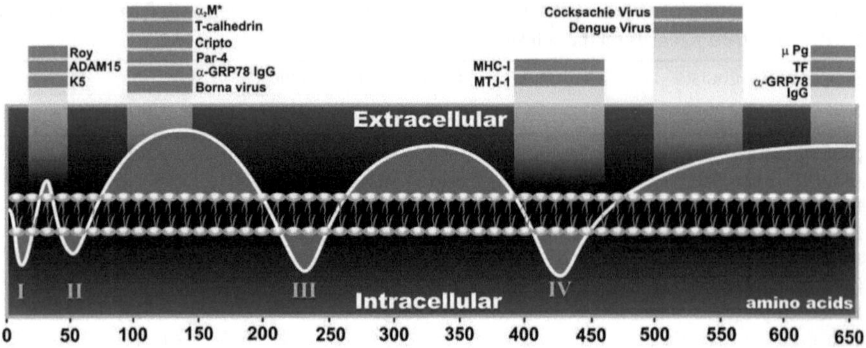

Fig. 15.1 Transmembrane model of GRP78. Four putative intracellular domains (I–IV) are identified. The GRP78 extracellular domains with their ligands and antibodies show the amino acid patches involved in these interactions

and UPR (Macario et al. 2007) (see also Chap. 12 for a fuller description of this problem). GRP78 plays a central role in protecting cancer cells against ER-stress induced apoptosis (Fu et al. 2007). For these reasons, understanding the pathways of how GRP78 is expressed and then internalized from the cell surface is central to cancer biology.

MTJ-1, a DnaJ-like transmembrane protein, is essential for cell surface localization of GRP78 (Chevalier et al. 2000). How a complex between these two chaperones appears at the cell surface can be inferred from studies using TG (Ying et al. 2002). TG induces depletion of Ca^{2+} stores in normal cells, abolishing the retention of the KDEL-containing GRP78 in the ER, and promotes its translocation to the cell surface without inducing any new synthesis (Ying et al. 2002). GRP78 is normally associated with the DnaJ protein P58 (IPK) in the ER (Rutkowski et al. 2007). In TG-treated cells, GRP78 is transported from the Golgi to p58-containing pre-Golgi intermediate compartments (IC), stopping any further transportation to the ER. Similarly, p58 that normally cycles between the ER, IC, and cis-Golgi, was largely depleted from the cell periphery and arrested in large-sized IC elements and numerous vesicles in the Golgi region, demonstrating that TG selectively blocks p58 recycling from the IC back to the ER (Ying et al. 2002). Therefore, depletion of ER- Ca^{2+} induced by stress signaling facilitates dissociation of GRP78 from p58. At this point, GRP78 possibly forms a complex with MTJ-1 which binds to a site upstream from its KDEL sequence in the COOH-terminal domain Chevalier et al. 2000). This complex is then ready to be exported by vesicular transport from the ER to the plasma membrane. It is known that GRP78 exhibits the properties of a transmembrane protein in CHO cell microsomes (Reddy et al. 2003). Using these properties, we constructed a structural model showing GRP78 transmembrane four hydrophobic domains (I through IV) with NH_2- or COOH-terminal domains localized outside of the membrane (Fig. 15.1). In this figure, we also show the GRP78 regions possibly involved in interactions with several of its ligands and autoantibodies.

Studies performed in our laboratory (Misra and Pizzo 2008) show that cell surface GRP78 localized in lipid raft/caveolae pits forms a ternary complex with MTJ-1 and Gαq11 protein that initiates signaling induced by α_2M*. GPI-anchored T-cadherin also co-localizes with caveolin-rich membrane domains and G proteins on the cell surface (Philippova et al. 1998). The GPI-anchored oncogene Cripto is also associated with caveolin (Bianco et al. 2008) and G proteins (Bianco et al. 2008). Association of GRP78 with lipid rafts may compartmentalize its partners, thereby facilitating their endocytosis via clathrin-dependent or -independent pathways (Simons and Toomre 2000). Mechanisms like this are the ones functioning in internalization of GPI-anchored proteins (Kirkham and Parton 2005; Peklmans 2005; Sharma et al. 2004; Sabharanjak et al. 2002). In a situation where the NH_2-terminal region of GRP78 acts as a receptor for several different agonists using similar transducing pathways, their responses must be individually coordinated within each pathway. Such a coordination suggests a phenomenon similar to that observed with the TGF-beta receptor which responds to different members of the TGF-beta superfamily via association with both lipid raft and non-raft membrane domains (Di Guglielmo et al. 2003). Whether such a mechanism operates with GRP78 is further supported by the fact that internalization of agonists that recognize its C-terminus are done via clathrin-mediated endocytosis (Kim et al. 2006), thereby suggesting that segregation of GRP78 receptors into distinct endocytic compartments regulates GRP78 receptor physiology. Although the evidence is indirect, the presence of lipid rafts/caveolae and GPI-anchored proteins demonstrates that GRP78 can function as a link in signalling platforms in microdomains distinct from clathrin coated pits. Although the molecular mechanisms involved in lipid raft-mediated endocytosis have not been fully elucidated, future studies of GRP78 receptor signaling and trafficking should provide important insights into its physiology.

15.4 Cell Surface GRP78 as an Autoantigen

Autoantibodies to GRP78 have been identified in serum from rheumatoid arthritis (Bodman-Smith et al. 2004), prostate cancer (Mintz et al. 2003), ovarian (Chinni et al. 1997), gastric (Zheng et al. 2008), and malignant melanoma cancer patients (Selim et al. 2011). The frequently opposite responses of GRP78 to NT-Ab from prostate (Gonzalez-Gronow et al. 2006) and malignant melanoma cancer patients (Selim et al. 2011), or against CT-Ab from gastric cancer patients (Pohle et al. 2004) or commercial polyclonal antibodies (Raiter et al. 2010), suggest that the topography of GRP78 is an important factor in the modulation of the autoimmune response. Indeed, the autoantibodies from prostate or malignant melanoma cancer patients, specific against an epitope in the NH_2-terminal region of GRP78, induce proliferation and protect human prostate and malignant melanoma cancer cells from apoptosis via a mechanism which mimics the function of α_2M* (Gonzalez-Gronow et al. 2006; Selim et al. 2011). In a recent study, we demonstrated a causal link between the humoral response to the NH_2-terminal region of GRP78 and the progression of cancer in a murine malignant melanoma model (deRidder et al. 2011).

15.5 Cell Surface Expression of GRP78 and MHC-I Down-Regulates Autoimmunity

GRP78 on the cell surface also associates with MHC-I (Triantafilou et al. 2001). Expression of MHC-I in cancer cells is down-regulated as a widespread mechanism to evade immune surveillance (Marincola et al. 2000; Seliger et al. 2003). It is possible that over-expression of GRP78 compensates for the loss of MHC-I molecules, thereby acting as an alternative antigen-presenting structure (Ciupitu et al. 1998). It is known that association of heat-shock proteins with MHC-I regulate the endogenous pathway of antigen presentation by MHC-I molecules (Li et al. 2002). In this context, over-expression of GRP78 may down-regulate MHC-I expression on the cell surface, thereby limiting the capacity of the immune system to control tumor progression.

15.6 Modulation of Signaling Cascades Transduced by the GRP78 N-terminal Domain Compartment on the Cell Surface

As discussed above, interaction of GRP78 with α_2M^* induces activation of PAK-2, which together with LIMK1 and cofilin phosphorylation, increases cell motility (Misra et al. 2005). At the same time, activations of ERK1/2, p38 MAPK and PI3K promote cell proliferation, followed by activation of the Akt and NF-κB signaling cascades that promote cell survival (Misra et al. 2006). Both the proliferative and antiapoptotic signaling mechanisms induced by α_2M^* are part of an autoregulatory feedback loop in prostate cancer, where α_2M^* stimulates synthesis and secretion of the prostate-specific antigen (PSA) forming a complex with α_2M^*, which in turn serves as a ligand for GRP78 causing a further increase in the signaling cascades described above (Misra et al. 2011a). Recently, we demonstrated that α_2M^* also upregulates TFII-I which binds to the GRP78 promoter enhancing GRP78 synthesis and regulating Ca^{2+} entry (Misra et al. 2011b). All these studies conclusively show that cell surface GRP78 functions as a protector from apoptosis, thereby promoting a more aggressive behavior of human prostate cancer cells.

As described above, GPI-anchored T-cadherin and Cripto also promote proliferation and survival of cancer cells; however, unlike α_2M^*, the end result of the association of these partners to GRP78 remains to be determined.

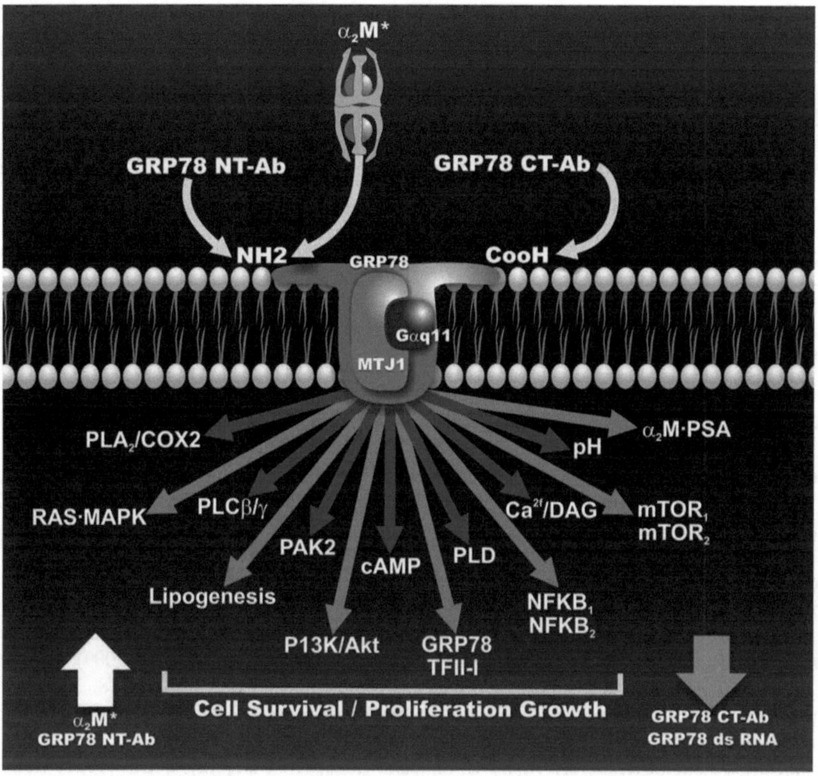

Fig. 15.2 Cell survival and proliferation signaling pathways induced by α_2M^* upon activation of cell surface GRP78. Antibodies against the GRP78 NH_2-terminal region mimic α_2M^*, whereas antibodies to the GRP78 COOH-terminal region inhibit α_2M^* induced downstream signaling. Details are described in the text

15.7 Modulation of Signaling Cascades Transduced by the GRP78 COOH-terminal Domain Expression on the Cell Surface

The different signaling pathways induced by ligation of GRP78 NH_2- or COOH-terminal domains were determined with specific antibodies raised in sheep which showed that NT-Abs induced a sharp rise in cytosolic Ca^{2+} rapidly decreasing after 200s of exposure, whereas CT-Abs induced a fast rise in cytosolic Ca^{2+} which remained as a constant plateau after 200s of exposure (Gonzalez-Gronow et al. 2009). Both antibodies were also used to stain sections of human prostatic intraepithelial neoplasma, thereby demonstrating the availability of both GRP78 regions on the cell surface (Gonzalez-Gronow et al. 2009). Using a commercial polyclonal GRP78 CT-Ab, we showed that this IgG induced apoptosis in prostate and melanoma cancer cells (Misra et al. 2009), via mechanisms involving up-regulation of p53, inhibition of NF-κB1 and NF-κB2 activation, and suppression of Ras/MAPK and PI3K/Akt signaling (Misra et al. 2009, 2010; Misra and Pizzo, 2010a), and down-regulation of IRE1-α, PERK, and ATF6α-dependente signaling (Misra and PIzzo 2010b). Up-regulation of p53 by this antibody is central in promoting activation of caspases-3, -7, and 9 (Misra and Pizzo 2010c). The opposite signaling cascades induced by ligation of GRP78 NH_2- or COOH-terminal domains, suggest the latter as a useful target to suppress proliferative signaling in cancer cells. These responses, including their signaling pathways, are summarized in Fig. 15.2.

References

Angst BD, Marcozzi C, Magee AI (2001) The cadherin superfamily: diversity in form and function. J Cell Sci 114:629–641

Bhattacharjee G, Ahamed J, Pedersen B et al (2005) Regulation of tissue factor-mediated initiation of the coagulation cascade by cell surface grp78. Arterioscler. Thromb Vasc Biol 25:1737–1743

Bianco C, Strizzi L, Mancino M et al (2008) Regulation of Cripto-1 signaling and biological activity by caveolin-1 in mammary epithelial cells. Am J Pathol 172:345–357

Bickel PE (2002) Lipid raft and insulin signaling. Am J Physiol Endocrinol Metab 282:E1–E10

Bodman-Smith MD, Corrigal VM, Berglin E et al (2004) Antibody response to the human stress protein BiP in rheumatoid arthritis. Rheumatology 43:1283–1287

Burikhanov R, Zhao Y, Goswami A, Qiu S, Schwarze SR, Rangnekar VM (2009) The tumor suppressor Par-4 activates an extrinsic pathway for apoptosis. Cell 138:377–388

Chevalier M, Rhee H, Elguindi EC, Blond S (2000) Interaction of murine BiP/GRP78 with the DnaJ homologue MTJ1. J Biol Chem 275:19620–19627

Chinni SR, Falchetto R, Gercel-Taylor C, Shabanowitz J, Hunt DF, Taylor DD (1997) Humoral immune responses to cathepsin D and glucose-regulated protein 78 in ovarian cancer patients. Clin Cancer Res 1557–1564

Ciupitu AMT, Petersson M, O'Donnell CL, Williams K, Jindal S, Kiessling R, Welsh RM (1998) Immunization with a limphocytic choriomeningitis virus peptide mixed with heat shock protein 70 results in protective antiviral immunity and specific cytotoxic T lymphocytes. J Exp Med 187:685–691

Davidson DJ, Haskell C, Majest S et al (2005) Kringle 5 of human plasminogen induces apoptosis of endothelial and tumor cells through surface-expressed glucose-regulated protein 78. Cancer Res 65:4663–4672

de Ridder GG, Gonzalez-Gronow M, Ray R, Pizzo SV (2011) Autoantibodies against cell surface GRP78 promote tumor growth in a murine model of melanoma. Melanoma Res. 21:35–43

De Santis ML, Kannan S, Smith GH et al (1997) Cripto-1 inhibits beta-casein expressioin in mammary epithelial cells through a p21ras-and phosphatidylinositol 3'-kinase-dependent pathway. Cell Growth Differ 8:1257–1266

Di Guglielmo GM, Le Roy C, Goodfellow AF, Wrana JL (2003) Distinct endocytic pathgways regulate TGF-beta receptor signalling and turnover. Nat Cell Biol 5:419–421

Dudek J, Benedix J, Cappel S et al (2009) Functions and pathologies of Bip and its interaction partners. Cell Mol Life Sci 66:1556–1569

Ebert AD, Wechselberger C, Frank S et al (1999) "Cripto-1 induces phosphatidylinositol 3"-kinase-dependent phosphorylation of AKT- and glycogen synthase kinase 3beta in human cervical carcinoma cells. Cancer Res 59:4502–4505

Fu Y, Li J, Lee AS (2007) GRP78/BiP inhibits endoplasmic reticulum BIK and protects human breast cancer cells against estrogen starvation-induced apoptosis. Cancer Res 67:3734–3740

Gonzalez-Gronow M, Cuchacovich M, Llanos C, Urzua C, Gawdi G, Pizzo SV (2006) Prostate cancer cell proliferation in vitro is modulated by antibodies against glucose-regulated protein 78 isolated from patient serum. Cancer Res 66:11424–11431

Gonzalez-Gronow M, Kaczowka SJ, Payne S, Wang F, Gawdi G, Pizzo SV (2007) Plasminogen structural domains exhibit different functions when associated with cell surface GRP78 or the voltage-dependent anion channel. J Biol Chem 282:32811–32820

Gonzalez-Gronow M, Selim MA, Papalas J, Pizzo SV (2009) GRP78: A multifunctional receptor on the cell surface. Antiox Red Sig 11:2299–2306

Gray PC, Shani G, Aung K, Kelber J, Vale W (2006) Cripto binds transforming growth factor beta (TGF-beta) and inhibits TGF-beta signaling. Mol Cell Biol 26:9268–9278

Guderman T, Grosse R, Schultz G (2000) Contribution of receptor/G protein signaling to cell growth and transformation. Naunyn Schmiedebergs Arch Pharmacol 361:345–362

Hardy B, Battler A, Weiss C, Kudasi O, Raiter A (2008) Therapeutic angiogenesis of mouse hind limb ischemia by novel peptide activating GRP78 receptor on endothelial cells. Biochem Pharmacol 75:891–899

Honda T, Horie M, Daito T, Ikuta K, Tomonaga K (2009) Molecular chaperone BiP interacts with Borna disease virus glycoprotein at the cell surface. J Virol 83:12622–12625

Jindadamrongwech S, Theparit C, Smith DR (2004) Identification of GRP78 (BiP) as a liver celll expressed receptor element for dengue virus serotype 2. Arch Virol 149:915–927

Joshi MB, Philippova M, Ivavov D, Allenspach R, Erne P, Rensik TJ (2005) T-cadherin protects endothelial cells from oxidative stress-induced apoptosis. FASEB J 19:1737–1739

Kannan S, De Santis M, Lohmeyer M et al (1997) Cripto enhances the tyrosyne phosphorylation of Shc and activates mitogen-activated kinase (MAPK) in mammary epithelial cells. J Biol Chem 272:3330–3335

Kelber JA, Panopoulos AD, Shani G et al (2009) Blockade of Cripto binding to cell curface GRP78 inhibits oncogenic Cripto signaling via MAPK/PI3K and Smad2/3 pathways. Oncogene 28:2324–2336

Kim Y, Lillo AM, Steiniger SC et al (2006) Targeting heat shock proteins on cancer cells: selection, characterization, and cell-penetrating properties of a peptidic GRP78 ligand. Biochemistry 45:9434–9444

Kirkham M, Parton RG (2005) Clathrin-independent endocytosis: New insights into caveolae and non-caveolar lipid raft carriers. Biochim Biophys Acta 1746:350–363

Lee AS, Bell J, Ting J (1984) Biochemical characterization of the 94- and 78-kilodalton glucose-regulated proteins in hamster fibroblasts. J Biol Chem 259:4616–4621

Lee AS (1987) Coordinated regulation of a set of genes by glucose and calcium ionophores in mammalian cells. Trends Biochem Sci 12:20–23

Li Z, Menoret A, and Sristava P (2002) Roles of heat-shock proteins in antigen presentation and cross-presentation. Curr Opin Immunol 14:45–51

Limjindaporn T, Wongwiwat W, Noisakran S et al (2009) Interaction of dengue virus envelope protein with endoplasmic reticulum-resident chaperones facilitates dengue virus production. Biochem. Biophy Res Commun 379:196–200

Ma Y, Hendershot LM (2004) The role of the unfolded protein response in tumor development: Friend or foe? Nat Rev Cancer 4:966–977

Macario AJL, Conway de Macario E (2007) Molecular chaperones: multiple functions, pathologies, and potential applications. Front Biosci 12:2588–2600

Marincola FM, Jafee EM, Hicklin DJ, Ferrone S (2000) Escape of human solid tumors from T-cell recognition: Molecular mechanisms and functional significance. Adv Immunol 74:181–273

Markoff LJ, Innis BL, Houghten R, Henchal LSD (1991) Development of cross-reactive antibodies to plasminogen during the immune response to dengue virus infection. J Infect Dis 164:294–307

Mintz PJ, Kim J, Do KA et al (2003) Fingerprinting the circulating repertoire of antibodies from cancer patients. Nat Biotech 21:57–63

Misra UK, Pizzo SV (1998a) Binding of receptor-recognized forms of α_2-macroglobulin signaling receptor activates phosphatidylinositol 3-kinase. J Biol Chem 273:13399–13402

Misra UK, Pizzo SV (1998b) Ligation of the α_2M, signaling receptor elevates the levels of p21Ras-GTP in macrophages. Cell Signal 10:441–445
Misra UK, Pizzo SV (1999) Upregulation of macrophage plasma membrane and nuclear phospholipase D activity on ligation of the α_2-macroglobulin signaling receptor: Involvement of heterotrimeric and monomeric G proteins. Arch Biochem Biophys 36:68–80
Misra UK, Pizzo SV (2004a) Potentiation of signal transduction mitogenesis and cellular proliferation upon binding of receptor-recognized forms of α_2-macroglobulin to 1-LN prostate cancer cells. Cell 16:487–496
Misra UK, Pizzo SV (2004b) Regulation of phospholipase A2 activity in macrophages stimulated with receptor-recognized forms of α_2-macroglobulin. Role in mitogenesis and cell proliferation. J Biol Chem 277:4069–4078
Misra UK, Pizzo SV (2004c) Activation of Akt/PDK signaling in macrophages upon binding of receptor-recognized forms of α_2-macroglobulin to its cellular receptor. Effect of silencing the CREB gene. J Cell Biochem 93:1020–1032
Misra UK, Pizzo SV (2008) Heterotrimeric Gαq11 co-immunoprecipitates with surface anchored GRP78 from plasma membranes of α_2M^*-stimulated macrophages. J Cell Biochem 104:96–104
Misra UK, Pizzo SV (2010a) Ligation of cell surface GRP78 with antibody directed against the COOH-terminal domain of GRP78 suppresses Ras/MAPK and PI3-kinase/Akt signaling while promoting caspase activation in human prostate cancer cells. Cancer Biol Ther 9:1–11
Misra UK, Pizzo SV (2010b) Modulation of the unfolded protein response in prostate cancer cells by antibody-directed against the carboxyl-terminal domain of GRP78. Apoptosis 15:173–182
Misra UK, Pizzo SV (2010c) PTF-α inhibits antibody-induced activation of p53 and pro-apoptotic signaling in 1-LN prostate cancer cells. Biochim Biophys Res Commun 391:272–276
Misra UK, Chu CT, Rubenstein DS, Gawdi G, Pizzo SV (1993) Receptor-recognized α_2-macroglobulin-methyl amine elevates intracellular calcium, inositolphosphates and cyclic AMP in murine peritoneal macrophages. Biochem J 290:885–891
Misra UK, Gawdi G, Pizzo SV (1995) Ligation of the α_2-macroglobulin signaling receptor on macreophages induces protein phosphorylation and an increase in cytosolic pH. Biochem J 309:151–158
Misra UK, Gonzalez-Gronow M, Gawdi G, Hart JP, Johnson CE, Pizzo SV (2002) The role of GRP78 in alpha-2-macroglobulin induced signal transduction. Evidence from RNA interference that the low density lipoprotein receptor-related protein is associated with but not necessary for GRP78-mediated signal transduction. J Biol Chem 277:42082–42087
Misra UK, Gonzalez-Gronow M, Gawdi G, Wang F, Pizzo SV (2004) A novel receptor function for the heat shock protein GRP78: Silencing of GRP78 gene expression attenuates alpha-2M*-induced signaling. Cell Signal 16:929–938
Misra UK, Deedwania R, Pizzo SV (2005) Binding of activated α_2-macroglobulin to its cell surface receptor GRP78 in 1-LN prostate cancer cells regulates PAK-2-dependent activation of LIMK. J Biol Chem 280:26278–26286
Misra UK, Deedwania R, Pizzo SV (2006) Activation and cross-talk between Akt, NF-κB, and unfolded protein response signaling in 1-LN prostate cancer cells consequent to ligation of cell surface-associated GRP78. J Biol Chem 281:13694–13707
Misra UK, Mowery Y, Kaczowka S, Pizzo SV (2009) Ligation of cancer cell surface GRP78 with antibodies directed against its COOH- terminal domain up-regulates p53 activity and promotes apoptosis. Mol Cancer Ther 8:1350–1362
Misra UK, Kaczowka S, Pizzo SV (2010) Inhibition of NF-κB1 and NF-κB2 activation in prostate cancer cells treated with antibody against the carboxy terminal domain of GRP78: Effect of p53 upregulation. Biochim. Biophys Res Commun 392:538–542
Misra UK, Payne S, Pizzo SV (2011a) Ligation of prostate cancer cell surface GRP78 activates a proproliferative and antiapoptotic feedback loop:A role for secreted prostate-specific antigen. J Biol Chem 286:1248–1259
Misra UK, Mowery YM, Gawdi G, Pizzo SV (2011b) Loss of cell surface TFII-I promotes apoptosis in prostate cancer cells stimulated with activated α_2-macroglobulin. J Cell Biochem 112: 1685–1695

Miyake H, Hara I, Arakawa S, Kamidoro S (2000) Stress protein GRP78 prevents apoptosis induced by calcium ionophore, ionomycin, but not by glycosylation inhibitor, tunicamycin, in human prostate cancer cell, J Cell Biochem 77:396–408

Munro S, Pelham HRB (1986) An Hsp70-like protein in the ER: identity with the 78 kd glucose-regulated protein and immunoglobulin heavy chain binding protein. Cell 46:291–293

Neves SR, Ram PT, Iyengar R (2002) G protein pathways. Science 296:13694–13707

Olden K, Pratt RM, Jaworski C, Yamada KM (1979) Evidence for role of glycoprotein carbohydrates in membrane transport: specific inhibition by tunicamycin. Proc Natl Acad Sci U S A 76:791–795

Patel TB (2004) Single transmembrane spanning heterotrimeric G-protein-coupled receptors and their signaling cascades. Pharmacol Rev 56:371–385

Pelkmans L (2005) Secrets of caveolae- and lipid raft-mediated endocytosis revealed by mammalian viruses. Biochim Biophys Acta 1746:295–304

Philippova MP, Bochkov VN, Stambolsky DV, Tkachuk VA, Resink TJ (1998) T-cadherin and signal-transducing molecules co-localize in caveolin-rich membrane domains of vascular smooth muscle cells. FEBS Lett 429:207–210

Philippova M, Ivanov D, Joshi MB et al (2008) Identification of proteins associating with glycosylphosphatidylinositol-anchored T-cadherin on the surface of vascular endothelial cells: role for Grp78/Bip in T-cadherin-dependent cell survival. Mol Cell Biol 28:4004–4017

Pohle T, Brandlein S, Ruoff N, Müller-Hermelink HK, Vollmers HP (2004) Lipoptosis tumor-specific cell death by antibody-induced intracellular lipid accumulation. Cancer Res 64:3900–3906

Poppleton H, Shu H, Fulgham D, Bertics P, Patel TB (1996) Activation of Gsalpha by the epidermal growth factor receptor involves phosphorylation. J Biol Chem 271:6947–6951

Pouyssegur J, Shiu RPC, Pastan I (1977) Induction of two transformation-sensitive membrane polypeptides in normal fibroblasts by a block in glycoprotein synthesis or glucose deprivation. Cell 11:941–947

Raiter A, Weiss CD, Bechor Z et al (2010) Activation of GRP78 on endothelial cell membranes by an ADAM15-derived peptide induces angiogenesis. J Vasc Res 47:399–411

Rauschert N, Brändlein S, Holzinger E, Hensel F, Müller-Hermelink HK, Vollmers HP (2008) A new tumor-specific variant of GRP78 as target for antibody-based therapy. Lab Invest 88:375–386

Reddy R, Mao C, Baumeister P, Austin R, Kaufman RJ, Lee AS (2003) Endoplasmic reticulum chaperone protein GRP78 protects cells from apoptosis induced by topoisomerase inhibitors. J Biol Chem 278:20915–20924

Rutkowski DT, Kang SW, Goodman AG, Garrison JL, Taunton J, Katze MG et al (2007) The role of p58IPK in protecting the stressed endoplasmic reticulum. Mol Biol Cell 18:3681–3691

Sabharanjak S, Sharma P, Parton RG, Mayor S (2002) GPI-anchored proteins are delivered to recycling endosomes via a distinct cdc42-regulated, clathrin-independent pinocytic pathway. Dev Cell 2:411–423

Saloman DS, Bianco C, Ebert AD et al (2000) The EGF-CFC family: novel epidermal growth factor-related proteins in development and cancer. Endocr Relat Cancer 7:199–226

Scandra JJ, Subjeck JR, Hughes CS (1984) Induction of glucose-regulated proteins during anaerobic exposure and of heat-shock proteins after reoxygenation. Proc Natl Acad Sci U S A 81:4843–4947

Seliger B, Atkins D, Bock M, Ritz U, Ferrone S, Huber C, Storkel S (2003) Characterization of human lymphocyte antigen class I antigen-processing machinery defects in renal cell carcinoma lesions with special emphasis on transporter-associated with antigen-processing down-regulation. Clin Cancer Res 9:1721–1727

Selim MA, Burchette JL, Bowers EV et al (2011) Changes in oligosaccharide chains of autoantibodies to GRP78 expressed during progression of malignant melanoma stimulate melanoma cell growth and survival. Mel Res (in press)

Shani G, Fischer WH, Justice NJ, Kelber JA, Vale W, Gray PC (2008) GRP78 and Cripto form a complex at the cell surface and collaborate to inhibit transforming growth factor β signaling and enhance cell growth. Mol Cell Biol 28:666–677

Sharma P, Varma R, Sarasij RC et al (2004) Nanoscale organization of multiple GPI-anchored proteins in living cell membranes. Cell 116:577–589

Simons K, Toomre D (2000) Lipids rafts and signal transduction. Nat Rev Mol Cell Biol 1:31–29

Strizzi L, Bianco C, Normanno N, Salomon D (2005) Cripto-1: A multifunctional modulator during embryogenesis and oncogenesis. Oncogene 24:5731–5741

Stone KR, Smith RE, Joklik WK (1974) Changes in membrane polypeptides that occur when chick embryo fibroblasts and NRK cells are transformed with avian sarcoma viruses. Virology 58:86–100

Triantafilou M, Fradelizi D, Triantafilou K (2001) Major histocompatibility class one molecule associates with glucose-regulated protein (GRP78) on the cell surface. Hum Immunol 62:764–770

Triantafilou M, Fradelizi D, Wilson K, Triantafilou K (2002) GRP78, a co-receptor for Coxsackie virus A9, interacts with major histocompatibility complex class I molecules which mediate virus internalization. J Virol 76:633–643

Wang J, Yin Y, Hua H et al (2009) Blockade of GRP78 sensitizes breast cancer cells to microtubules-interfering agents that induce the unfolded protein response. J Cell Mol Med 13:3888–3897

Watanabe K, Bianco C, Strizzi L et al (2007) Growth factor induction of Cripto-1 shedding by glycosylphosphatidylinositol-phospholipase D and enhancement of endothelial cell migration. J Biol Chem 282:31643–31655

Watson LM, Chan AK, Berry LR et al (2003) Overexpression of the 78-kDa glucose-regulated protein/immunoglobulin-binding protein (GRP78/BiP) inhibits tissue factor procoagulant activity. J Biol Chem 278:17438–17447

Weinreb I, Goldstein D, Irish J, Perez-Ordoez B (2009) Expression patterns of Trk-A, Trk-B, GRP78 and p75NRT in olfactory neuroblastoma. Human Pathol 40:1330–1335

Whelan SA, Hightower IE (1985) Differential induction of glucose-regulated and heat-shock proteins: effects of pH and sulfhydryl-reducing agents on chicken embryo cells. J Cell Physiol 125:251–258

Xie P, Browning DD, Hay N, Mackman N, Ye RD (2000) Activation of NF-κB by bradykinin through a Galpha(q)- and Gceta gamma-dependent pathway that involves phosphoinositide 3-kinase and Akt. J Biol Chem 275:24907–24914

Ying M, Sannerud R, Flatmark T, Saraste J (2002) Colocalization of Ca2+-ATPase and GRP94 with p58 and the effects of thapsigargin on protein recycling suggests the participation of the pre-Golgi intermediate compartment in intracellular Ca2+ storage. Eur J Cell Biol 81:469–483

Yoshida I, Monji A, Tashiro K, Nakamura K, Inoue R, Kanba S (2006) Depletion of intracellular Ca2+ store itself may be a major factor in thapsigargin-induced ER stress and apoptosis in PC12 cells. Neurochem Int 48:696–702

Zhang LH, Zhang X (2010) Roles of GRP78 in physiology and cancer. J Cell Biochem 110:1299–1305

Zhao Y, Rangnekar VM (2008) Apoptosis and tumor resistance conferred by Par-4. Cancer Biol Ther 7:1867–1874

Zheng HC, Takahashi H, Li XH, Hara T, Masuda S, Guan YF, Takano Y (2008) Overexpression of GRP78 and GRP94 are markers for aggressive behavior and poor prognosis in gastric carcinoma. Hum Pathol 39:1042–1049

Zhuang L, Scolyer RA, Lee CS et al (2009) Expression of glucose-regulated to progression of melanoma. Histopathology 54:462–470

Chapter 16
Mycobacterium tuberculosis Hsp60 as a Key Virulence Factor in Tuberculosis

Richard W. Stokes

Abstract *Mycobacterium tuberculosis*, the etiological agent of tuberculosis, is a major pathogen of man with about one-third of the world's population being infected. *M. tuberculosis* resides within macrophages which are members of the host's cell-mediated immune response that supposedly protect against bacterial invasion. Obviously *M. tuberculosis* has strategies that enable it to survive in this hostile environment. Amongst the many virulence factors that *M. tuberculosis* possesses are two paralogues of the ubiquitous stress protein, chaperonin (Cpn) or Hsp60 that have been named Cpn60.1 and Cpn60.2. While *cpn60.2* is an essential gene involved in the maintenance of cell viability through "normal" chaperoning activities, *cpn60.1* is non-essential and appears to have little or no involvement in protein folding activities. Both Cpn60.1 and Cpn60.2 have varied "moonlighting" functions and can act as secreted signaling molecules, modulators of host immunity, surface located bacterial ligands and bacterial cell wall components. How these proteins leave the cytosol to function extracellularly or at the surface of the bacterial cell wall is still not clear. That they can act as bacterial virulence factors is becoming clear although recognition of Cpn60.2 by the host may instead mediate a host defence mechanism.

16.1 Introduction

All the other chapters in this book have concentrated on the cell stress proteins of mammals, largely *Homo sapiens*. However, bacteria express many homologs of the eukaryotic cell stress proteins and there is emerging evidence that these proteins can play a role in bacterial virulence. The vast majority of the genus, *Mycobacterium,* are saprophytic species that, like other Actinomycetes, are found in soil and water. Some species, however, are the causative agents of disease in man. While some of these (e.g. the *M. avium-intracellulare* complex, *M. kansasii* and *M. fortuitum*) are opportunistic pathogens of man, that predominantly cause disease in immunocompromised individuals, some species are major pathogens of man. These include *Mycobacterium tuberculosis*, *M. leprae* and *M. ulcerans* of which the best known

R. W. Stokes (✉)
Department of Microbiology & Immunology, University of British Columbia,
Life Sciences Centre, 2504–2350 Health Sciences Mall, Vancouver,
British Columbia, V6T 1Z3, Canada
e-mail: rstokes@mail.ubc.ca

B. Henderson, A. G. Pockley (eds.), *Cellular Trafficking of Cell Stress Proteins in Health and Disease*, Heat Shock Proteins 6, DOI 10.1007/978-94-007-4740-1_16,

and most important is *M. tuberculosis* (*M. tb*), the causative agent of tuberculosis (TB). TB causes more death in adults than does any other single bacterial species and it is estimated that in the year 2010 there was approximately ten million new cases of TB (Dye and Williams 2010) and that TB is the cause of around two million deaths worldwide (Anonymous 2009). The relatively recent emergence of multidrug resistant and extensively drug resistant TB has become a serious problem and threatens to make the disease incurable (Jain and Mondal 2008; Mitnick et al. 2008). Unfortunately, the recent AIDS pandemic has compounded the infection rates and morbidity of TB, due to the diminished CD4 T-cell mediated immunity of the AIDS patient (Getahun et al. 2010). Despite the recent successes in the global treatment of TB, it remains a devastating disease of mankind that continues to infect increasingly higher numbers of people (Dye and Williams 2010).

Mycobacterium tuberculosis is an intracellular pathogen that generally resides within macrophages (MΦs) and is predominantly found in the lung. MΦs are phagocytic cells that act as part of the effector arm of the host's cell-mediated immunity. They can kill invading microbes (though not as efficiently as do neutrophils) using a variety of mechanisms such as reactive oxygen intermediates (ROI), cationic peptides, LRG-47 and by undergoing apoptosis. Obviously, in some cases, *M. tb* survives the initial move to the lung and a subsequent intracellular environment. There it begins to replicate which results in bacterial antigens being processed by the host leading to the development of adaptive immunity resulting in the production of interferon-γ (IFNγ) by T-cells (Orme 2004; Cooper 2009; Torrado et al. 2011). The IFNγ acts on the infected MΦs to increase their ability to control bacterial replication through what is called classical activation. MΦ activation leads to an enhancement of the killing mechanisms seen in resting MΦs (such as ROI and apoptosis) and to the induction of new killing mechanisms mediated by reactive nitrogen intermediates (RNI) or phagosomal maturation to lysosomes. This exposure to IFNγ activated MΦs following the development of adaptive immunity marks a change in the pathogenesis of TB. The bacterial growth slows down drastically to the point where there is little bacterial replication and what little there is is countered by host immunity (Gill et al. 2009; Ehlers 2009). This coincides with a change to the physiology of *M. tb* towards a so called dormant (also called latent or chronic) phase. In this state, *M. tb* metabolism switches to the use of lipids instead of carbohydrates as a carbon source and to the glyoxalate shunt for energy production. However, although the host manages to control the replication of the *M. tb* bacilli, the bacteria still survive and can remain viable for the remaining lifespan of the infected TB patient. The host response to these persisting bacteria is to form an organised collection of cells that surround and wall off the microbes in a hypoxic environment where they are commonly found within lipid rich foamy MΦs, surrounded by lymphocytes, giant cells and fibroblasts. This is called a granuloma (or a tubercle in clinical terminology) and contains the TB bacilli, preventing their spread and replication. However, *M. tb* within a granuloma are not necessarily eradicated and in many cases can remain in a dormant physiological state for decades until some extrinsic factor results in diminished immune regulation of the granuloma that then facilitates reactivation of bacterial replication, a breakdown of the granuloma's integrity and breakthrough into the airways of the lung. At this point the patient is said to be productive and will cough up these bacteria which, if inhaled by a new host, complete the infectious cycle.

Mycobacterium tuberculosis has obviously evolved strategies for evading and subverting the antimicrobial effector mechanisms of the host immune response. The prevalence of TB demonstrates how successful these strategies are. However, only 10 % of immunocompetent individuals infected with *M. tb* develop clinical disease over their lifetime. Within the other 90 % of infected individuals, the bacterium survives in a dormant, yet viable state for decades. These data indicate that there is an ongoing conflict between pathogen and host with the balance being tipped in the favour of fulminating bacterial disease in some cases, whereas usually the host can keep the pathogen in check. Which way the balance tips in any given individual is greatly affected by environmental factors (Lienhardt 2001; van der Eijk et al. 2007), but also by genetic variation in host susceptibility (Newport and Levin 1999; Doffinger et al. 2006; Stein 2011) and the virulence of the infecting *M. tb* isolate (Collins and Smith 1969; North and Izzo 1993; Orme 1999; Sassetti and Rubin 2003). It has long been known that the virulence (defined as the ability to produce a progressive infection) of separate strains or isolates of *M. tb* can vary in animal models of tuberculosis (Steenken et al. 1934; Alsaadi and Smith 1973; Orme 1999). Recently, members of the so called Beijing family of *M. tb* strains have been shown to have greater infectivity and virulence in man and have rapidly spread around the world (Bifani et al. 2002; Lasunskaia et al. 2010). It is therefore logical to propose that specific bacterial genes are critical for the survival and virulence of *M. tb* within the host and that identification of these virulence genes and their products will facilitate the design of novel vaccines and therapies to treat tuberculosis by providing novel targets for pharmacological research. This has resulted in a concerted effort within the research community to identify *M. tb* virulence factors using a variety of methodologies (Braunstein et al. 2002; Smith 2003; Sharma and Tyagi 2007). Surveying a transposon mutant library for survival in MΦ (Rengarajan et al. 2005) or *in vivo* (Sassetti and Rubin 2003) has identified gene sets that appear to be required for survival under these experimental conditions which could therefore be considered the essential genes for virulence. However, alternate methodologies have identified other gene sets that appear to be virulence factors for the survival of *M. tb* in MΦ (Schnappinger et al. 2003; Li et al. 2008; Li et al. 2010) and *in vivo* (Talaat et al. 2004; Lamichhane et al. 2005). A direct comparison of the gene sets identified to be virulence factors in three separate MΦ infection studies showed only limited overlap (Li et al. 2010). Undoubtedly, differences in methodologies account for some of this variation but different interpretations of what defines a virulence factor should also be taken into account. Whether it is defined as a genomic difference between a virulent and an avirulent strain of *M. tb* or as an expression difference between broth grown and intracellular bacteria or between intracellular strains of varying virulence, the identification of a gene product as a virulence factor is open to interpretation.

Although an understanding of the molecular details controlling the pathogenesis of TB is by no means complete, recent research has begun to determine the microbiology and immunology of this phenomenon (Glickman and Jacobs 2001; Smith 2003;

de Chastellier 2009; Barry et al. 2009; Stokes and Waddell 2009). *Mycobacterium tuberculosis* bacilli within their host encounter numerous stresses including residence within an intracellular environment, exposure to MΦ killing mechanisms, exposure to the effector mechanisms of the host's adaptive immune response and changes to oxygen and nutrient availability (Ehrt and Schnappinger 2009). The pathogen responds to these stresses in several ways (Ehrt and Schnappinger 2009; Stokes and Waddell 2009) including the induction of stress proteins of which the Hsp60 homologue is perhaps the best known.

16.2 The Chaperonin 60 (Hsp60/Hsp65/Cpn60) Proteins of *Mycobacterium tuberculosis*

The mycobacterial Hsp60 homologue (often called Hsp65 in mycobacteria) was first identified as a member of the "common antigen" family (Thole et al. 1988a) and an immunodominant antigen in TB patients and in experimental animal infections (Young et al. 1988; Young 1990). That the Hsp60 of *M. tb* is a strong immunogen is not surprising as members of the Hsp60 family are major antigens in several pathogens and strong antibody responses to Hsp60 are to be found following bacterial, protozoan and helminth infections (Young 1990). Early studies on the mycobacterial Hsp60 included the identification of B cell and T cell epitopes, identifying numerous epitopes of varying degrees of species specificity and species cross-reactivity (Anderson et al. 1988; Thole et al. 1988b). It was shown that *M. tb* Hsp60 could stimulate T-cell responses in human subjects, irrespective of whether they were infected with *M. tb* (Lamb et al. 1986; Thole et al. 1988a). This raised the possibility that recognition of conserved epitopes in the Hsp60 family could lead to autoimmunity (Lamb et al. 1989; Dubaniewicz 2010).

Studies have shown that experimental adjuvant arthritis in rats and mice (a model of rheumatoid arthritis (RA) in humans) can be induced by the injection of intact mycobacteria or complete Freund's adjuvant (a mixture of *M. tb* components within a mineral oil vehicle) (McLean et al. 1990; Cohen 1991). In contrast, pre-immunization with recombinant Hsp60 or virally expressed Hsp60 leads to suppression of and/or protection from adjuvant arthritis (Billingham et al. 1990; Yang et al. 1990; Lopez-Guerrero et al. 1994; Haque et al. 1996). An explanation for how *M. tb* Hsp60 induces autoimmunity has not been determined but it appears that repeated exposure to bacteria, especially pathogens containing proteins with a high similarity to host mammalian antigens, affects the host's ability to discriminate between self and non-self antigens (Moudgil and Sercarz 1994). As mycobacterial and mammalian Hsp60 homologues share 60 % homology (Jindal et al. 1989) and mycobacterial infections are commonly chronic and may remain with the patient all their lives, it can be seen why mycobacterial Hsps are commonly implicated in autoimmune diseases. The autoimmunity induced by mycobacterial Hsp60 could merely be an unavoidable consequence of the homology seen between Hsps of all species and is of no advantage to host or pathogen. However, the possibility that the autoimmunity is an occasional "by product" of an *M. tb* virulence strategy should not be discounted. As will be shown below, Hsp60 of *M. tb* appears to be actively transported to the outer layers of the bacterial cell wall and beyond where it interacts with the host. The data showing that the extracellular Hsp60 is acting as an immunomodulator, a bacterial ligand mediating attachment to MΦs and a cell signaling molecule indicates that the Hsp60 of *M. tb* has a role in the pathogenesis of TB.

16.2.1 Mycobacterial Paralogues of Hsp60

Most bacteria contain a single Hsp60 gene but it is becoming clear that in approximately 30 % of the bacteria that have been currently sequenced, multiple copies of Hsp60 exist (Lund 2009). This is true for *M. leprae* (Rinke de Wit et al. 1992), *M. tb* (Kong et al. 1993), *M. bovis* (Wang et al. 2011) and *M. avium paratuberculosis* (Goyal et al. 2006) which all have two copies of Hsp60, whereas *M. smegmatis* has three copies (Rao and Lund 2010). Phylogenetic analysis suggests that the two paralogues resulted from a single gene duplication event followed by varied rates of evolutionary change (Hughes 1993) with the third homologue in *M. smegmatis* appearing to have been acquired by horizontal gene transfer (Rao and Lund 2010). The two paralogues of Hsp60 in pathogenic mycobacteria are designated Cpn60.1 (GroEL1, *M. tb* genome accession number Rv3417c) and Cpn60.2 (GroEL2, Hsp65, *M. tb* genome accession number Rv0440). The Cpn60.1 and Cpn60.2 proteins from *M. tuberculosis* only share 61 % sequence identity (Kong et al. 1993) while there is 95 % identity between Cpn60.2 of *M. tb* and *M. leprae* (Shinnick et al. 1987). This implies that Cpn60.1 and Cpn60.2 would have divergent functions (Qamra et al. 2005). Comparable to the GroEL function in *E.coli*, Cpn60.2 shows hydrophobicity-based protein folding activity and is acting as a "normal" chaperonin. However, this function seems to result from the formation of a Cpn60.2 homodimer that is less ATP-dependent than is the GroEL of *E. coli* (Qamra et al. 2004; Shahar et al. 2011). Both Cpn60.1 and Cpn60.2 behave as dimers *in vivo* and *in vitro* which is unlike other bacterial Hsp60s that exist as tetradecamers (Qamra et al. 2004; Shahar et al. 2011). As it appears that *M.tb* Cpn60.2 is acting as a GroEL equivalent, it was surprising to find that *cpn60.1* appears to be arranged in a putative operon with *cpn10* (GroES, *M. tb* genome accession number Rv3418c) (Kong et al. 1993), while *cpn60.2* is found elsewhere on the chromosome. However, recent studies show that the apical domains of *M. tb* Cpn60.1 and Cpn60.2 are conserved in their 3-D structure and appear to be like the *E. coli* GroEL. Thus, it seems that while Cpn60.2 functions as the general housekeeping chaperonin, Cpn60.1, like Cpn60.2, can also act as a chaperonin (Sielaff et al. 2010), although this is only based on structural homology.

Further support for the divergent functions of Cpn60.1 and Cpn60.2 came from the attempts to delete these genes in mycobacteria. It was found that a knockout mutant can be obtained for *cpn60.1* in both *M. tb* (Hu et al. 2008) and *M. bovis* BCG (Wang et al. 2011). In contrast, *cpn60.2* can not be deleted and has been shown to be an essential gene required for the survival of *M. tb* (Hu et al. 2008). The fact that *cpn60.2* is essential lends support to the idea that it acts as the main housekeeping chaperone for *M.tb* in much the same way as GroEL does in *E. coli*. The role of Cpn60.1 is less clear. While it can possibly act as a chaperonin (Sielaff et al. 2010), deletion of the gene in *M. tb* did not result in a dramatic phenotype (Hu et al. 2008) suggesting that any chaperonin activity is not essential to bacterial survival. Growth of the mutant in broth and in MΦs was found to be equal to that of the wild-type parent (Hu et al. 2008). However, the mutant failed to grow in mice as rapidly as did the wild type, attaining comparable bacterial load in the lung and spleen only at later time points. This was associated with differences in the granulomatous inflammation in both mice and guinea pigs, with the mutant infected lungs showing only minimal inflammation in mice at 15 weeks post-infection, even though bacterial numbers were similar to that of the wild type (Hu et al. 2008). This suggested that Cpn60.1 is essential for the induction of normal granuloma formation during *M.tb* infection. The finding that levels of TNFα, IFNγ, IL-6 and IL-12 in the lungs of mice infected with the mutant were significantly lower than that seen in mice infected with wild-type bacteria up to 15 weeks post-infection (Hu et al. 2008) suggests that inflammation is affected throughout the course of the infection and indicates that Cpn60.1 is important in the induction of this inflammation. It may seem counter-intuitive that a putative *M.tb* virulence factor would induce inflammation, a host response usually associated with defence against bacterial infection. However, it is important to note that *M.tb* has a level of resistance to the effector arm of cell-mediated immunity and, in fact, resides within the very cells that are part of this response.

The growth of a *M. bovis* BCG Cpn60.1 mutant in broth was equal to that of the wild type although more protein was secreted into the supernatant by the mutant (Wang et al. 2011). Cell wall lipids were altered in the mutant and it was more susceptible to hydrogen peroxide (Wang et al. 2011). When growth in mice was investigated, the mutant was slightly less persistent in the lungs and spleen but retained its ability to protect vaccinated mice against a challenge with *M. tb* (Wang et al. 2011). Thus, like *M.tb*, the growth of a *M. bovis* BCG Cpn60.1 mutant is not greatly affected. However, it was shown for BCG that Cpn60.1 was necessary for bacterial cell wall integrity and resistance to hydrogen peroxide, but is not essential for the vaccine potential of BCG.

16.3 Secretion of Hsp60

It can thus be seen that the two Hsp60 paralogues in *M.tb* differ in their essentiality for bacterial survival and also in their function. That these stress proteins have other roles besides that of acting as a chaperone is becoming clear, with increasing evidence that they are secreted signaling molecules, modulators of host immunity, surface located bacterial ligands and bacterial cell wall components. A useful term for these additional roles of *M.tb* Hsp60 has been suggested by Henderson and his colleagues (Cehovin et al. 2010; Henderson et al. 2010) who call them "moonlighting" functions—a term initially introduced by Connie Jeffery (Jeffery 1999). Some resistance to this idea that stress proteins may have other functions besides acting as chaperonins has been forthcoming and seems, at least in part, to be connected to the dogma that *M.tb* Hsp60 acts only as a chaperonin and is therefore located intracellularly where they can function to mediate protein folding and does not transfer across the plasma membrane. In fact, it is commonly believed that detection of Hsp60 in a culture supernatant is indicative of cell lysis (Sonnenberg and Belisle 1997). It is therefore worthwhile examining the evidence that both Cpn60.1 and Cpn60.2 are normally to be found both within the cytosol and on the outer layers of the cell wall where they can be shed or actively secreted into the extracellular environment.

With the demonstration that mycobacteria have multiple copies of Cpn60 and the increasing demonstrations that Cpn60 has "moonlighting" functions, it becomes easier to accept that Cpn60 may have functional roles that involve its location other than in the cytosol. Indeed, Cpn60 has been shown to be secreted and to be located within the outer layers of the cell wall of *M.tb*. The identification of MΦ receptors that mediate binding of intact mycobacteria via Cpn60.2 (Hickey et al. 2009, 2010) necessitates that the Cpn60.2 must be located at the cell surface of the mycobacteria. In addition to this evidence, it has been demonstrated that mycobacteria do contain several Hsps, including Cpn60, within their outer cell wall by using various methodologies such as electron microscopy (Esaguy and Aguas 1997), antibody binding (Gillis et al. 1985; Esaguy and Aguas 1997; Hickey et al. 2009) and proteomics (Stokes, unpublished observations and (Rosenkrands et al. 2000; Wolfe et al. 2010)). In fact, using isobaric tags for relative and absolute quantitation (iTRAQ), Cpn60.1 and Cpn60.2, along with Hsp70, Hsp10 and Hsp16 have been shown to be among the most prevalent of proteins within the outer cell wall capsular layer of *M.tb* (Stokes unpublished observations). This is further supported by studies analyzing proteins in the cell wall of *M.tb* (Wolfe et al. 2010), by demonstrating the presence of Hsp16 in the cell wall (Cunningham and Spreadbury 1998) and by protein gel analysis of *M.tb* capsule (Hickey et al. 2009). Furthermore, Cpn60.1 has been shown to be secreted by *M.tb* into the supernatant of broth cultures (Cehovin et al. 2010). Interestingly, at the same time point that Cpn60.1 first appears in culture filtrates (6 days), no Cpn60.2 can be found (Hickey et al. 2009; Cehovin et al. 2010) even though it is on the surface of the bacteria (Hickey et al. 2009). This would imply that Cpn60.1 is actively secreted, perhaps to facilitate its actions on host cells. However, it is worth noting that Cpn60.2 secretion (or release) can be induced by the removal of zinc from the culture medium (De Bruyn et al. 1989).

To date no mechanism for the active secretion of Cpn60.1 has been identified, nor has a mechanism for how Cpn60.2 and the other Hsps access their outer cell wall location been discovered. The means by which these, and, for that matter, the many other cell wall-located and secreted proteins exit the mycobacterial cytosol are poorly understood. Although significant progress has been made in identifying the protein secretion systems of mycobacteria (Abdallah et al. 2007; Digiuseppe Champion and Cox 2007), none of the systems identified appear (at least, as yet) to be involved in the transport of Cpn60 across the plasma membrane. However, possible mechanisms for the egress of Cpn60 and other Hsps can be postulated. For instance, secretion may be due to their hydrophobic surfaces allowing them to interact with membrane phospholipids and other lipidic molecules within the largely hydrophobic milieu of the lipid rich mycobacterial cell wall, as suggested for other bacteria (Hennequin et al. 2001). Indeed, one report has shown that GroEL, human Hsp70, Cpn60.2 and DnaK all have the capacity to induce the formation of pores in lipid bilayers (Alder et al. 1990). Additionally, GroEL can promote lipid bilayer stability during protein folding activity (Torok et al. 1997), indicating its ability to traverse the plasma membrane. Alternatively, Cpn60 may engage more specific export mechanisms such as 'hitch-hiker'-based export via the recently described mycobacterial Twin-Arginine Translocation (Tat) pathway (McDonough et al. 2005; Lee et al. 2006). Proof-of-principal for the secretion of Hsps exists, even though they have not been specifically applied to mycobacterial Hsp60. For example, *M. tb* Cpn10 protein appears to be secreted from the bacterium, and shares some structural elements common to the N-terminal region of Hsp60 (Hughes 1993). In addition, the active secretion of mammalian Hsp60 (Merendino et al. 2010) and Hsp70 (Mambula et al. 2007, see Chap. 6) demonstrate that Hsp60 could be secreted to the mycobacterial cell surface and beyond.

16.4 Moonlighting Functions of Bacterial Hsp60 Proteins

Once Hsp60 has traversed the plasma membrane and lipidic layers of the mycobacterial cell wall, what functions does it have? The growing literature on this topic would suggest several functions. The immunomodulatory function of Cpn60 in autoimmunity has already been covered above, but other cell-cell signaling mechanisms for Cpn60 have also been discovered. Hsp60 of the oral bacterium *Aggregatibacter actinomycetemcomitans* stimulates osteoclast function resulting in the breakdown of murine calvarial bone (Kirby et al. 1995; Henderson et al. 2003). Interestingly, the Hsp60 proteins from both humans and some other bacteria (e.g. *E.coli*) also have this function (Reddi et al. 1998; Meghji et al. 2003). However, this shared function in Hsp60 homologues is not repeated with *M.tb*. While *M. tb* Cpn10 (Meghji et al. 1997) has this activity, *M. tb* Cpn60.2 does not and *M. tb* Cpn60.1 actually inhibits osteoclast function (Winrow et al. 2008). Whether these observations are related in any way to the virulence of intracellular *M.tb* is not yet clear.

More obviously connected to the virulence of *M.tb* is the effect of Hsp60 on the induction of host cell MΦ production of cytokines, reactive oxygen intermediates (ROI) and reactive nitrogen intermediates (RNI). Early studies did not differentiate the two Hsp60 paralogues but still showed that Hsp60 of *M. tb* (actually the Hsp65 or Cpn60.2 protein) induced the production of TNFα, IL-6 and IL-8 by the human macrophage-like cell line, THP-1 (Friedland et al. 1993) and TNFα and IL-6 by murine peritoneal MΦs (Peetermans et al. 1995). Interestingly, murine MΦs also produced RNI in response to Hsp60 which was TNFα-dependent and inhibited intracellular replication of the protozoan pathogen, *Toxoplasma gondii* (Peetermans et al. 1995). However, whether RNI play any role in human MΦs is still a topic of some controversy (Fang 2004). Hsp60 treatment of human monocyte-derived MΦs induced the pro-inflammatory cytokines TNFα and IL-1β and increased the expression of the surface complement receptor 3, but did not result in increased induction of reactive oxygen intermediates or MHCII expression, indicating a lack of classical MΦ activation (Peetermans et al. 1994). As it is classical IFNγ-mediated activation that is able to conrol *M.tb* intracellular replication (Cooper and Flynn 1995; Doffinger et al. 2006), it would not be advantageous to *M.tb* to induce this response, while an increase in complement receptor 3 expression may aid the uptake of the bacteria in an advantageous manner (Stokes et al. 1993; Velasco-Velazquez et al. 2003).

Following the discovery that *M.tb* has two paralogues of Hsp60, it was possible to compare the ability of *M. tb* Cpn60.1 and Cpn60.2 to induce cytokine production by MΦs. While both Cpn60.1 and Cpn60.2 stimulate human MΦ to produce IL-1, IL-6, IL-8, IL-10, IL-12, TNFα and GM-CSF but not IL-4 or IFNγ, 100 fold less Cpn60.1 was required to stimulate comparable amounts of these cytokines (Lewthwaite et al. 2001). Furthermore, Cpn60.1 but not Cpn60.2, signalling was shown to involve CD14 (Lewthwaite et al. 2001). Both Cpn60.1 and Cpn60.2 have only a partial requirement for MyD88 to induce MΦ cytokine production. Additionally, both have a requirement for Toll-like Receptor (TLR)-4, with Cpn60.2 having an additional requirement for TLR2 (Cehovin et al. 2010). Additional studies showed that both Cpn60.1 and Cpn60.2 utilize the ERK/1 and MAPK signaling pathways to induce cytokine production by MΦ (Lewthwaite et al. 2007). When whole blood leucocyte populations are stimulated with Cpn60.1 and Cpn60.2, only IL-1β and IL-6 and not IL-8, IL-10, IL-12 or IFNγ were produced by the mixed cell population. In this model, Cpn60.2 was a more potent stimulator than was Cpn60.1 and was the only one that induced TNFα production (Cehovin et al. 2010). The contrasting results with those previously reported for MΦ (Lewthwaite et al. 2001) indicated that the interaction of Cpn60.1 and Cpn60.2 with whole blood was very different from that seen with pure MΦ.

16.5 Binding of Hsp60 to Immune Effector Cells

The demonstration that TLRs, CD14 and MyD88 are necessary for appropriate signaling to take place in response to Cpn60 does not mean that they are necessarily the receptors for Cpn60. In another model studying the interaction of lipopolysaccharide (LPS) with MΦ, it was shown that CD14, TLR4, MD2 and other cell surface moieties form an intricate complex that mediates binding and cell signaling in response to LPS (Triantafilou and Triantafilou 2005, see also Chap. 10). Perhaps a similar complex of MΦ surface receptors is needed to interact with mycobacterial cell wall glycolipids and proteins. Nevertheless, the search for MΦ receptors that bind Cpn60 has indicated a number of cell surface proteins that may bind to Cpn60.1 (Henderson and Mesher 2007). Binding of Hsp70 and Cpn60.1 from *M. bovis* BCG to DC-SIGN has also been reported (Carroll et al. 2010). In contrast, it appears that, in the absence of serum (a situation that would be found within the alveolar space where *M.tb* first encounters MΦs), Cpn60.2 binds strongly to the MΦ surface receptor, CD43 (sialophorin, leukosialin) and that this receptor/ligand interaction accounts for 30–40 % of all binding of *M. tb* bacilli to MΦs (Hickey et al. 2009, 2010). Whether this binding can be considered a true receptor/ligand interaction and not just an interaction of "sticky" chaperonins with a host glycoprotein is not unequivocally determined. However, two observations strongly suggest that this interaction is a specific binding of the two moieties: (i) although both Cpn60.1 and Cpn60.2 are present in large amounts in the outer cell wall capsule of *M.tb* (Stokes, unpublished data), only Cpn60.2 binds to isolated CD43 (Hickey et al. 2009) and (ii) Hsp70 was also shown to bind to isolated CD43 but does not mediate binding of whole bacteria to CD43 on MΦs (Hickey et al. 2009, 2010). It is interesting to note that Cpn60.1 was found to bind to approximately 90 % of circulating human monocytes compared to <50 % binding with Cpn60.2 (Cehovin et al. 2010). This may reflect the very different surface receptors found on monocytes and MΦ, although CD43 is expressed on both.

The finding that Cpn60.2 can interact with purified CD43 (Hickey et al. 2009, 2010), does not necessarily mean that they interact with MΦ surface CD43 in isolation. It is possible that CD43 interacts with mycobacteria within the context of a group of MΦ surface molecules, as was described above for LPS (Triantafilou and Triantafilou 2005, see Chap. 10). In this model, CD43 would co-operate with other surface MΦ receptors to facilitate efficient bacterial binding and/or signal transduction via interaction with one or more bacterial surface molecules. The demonstration that soluble CD43 can overcome the deficiency of mycobacterial binding to MΦ from CD43 knockout mice (Fratazzi et al. 2000) suggests that, although *M. tb* can bind CD43 directly, it may also interact with other MΦ receptors. In addition to Cpn60.2, numerous other mycobacterial cell wall constituents have been identified as ligands that mediate binding to MΦ and several MΦ receptors have been shown to be involved in this binding (El-Etr and Cirillo 2001; Schafer et al. 2009; Mishra et al. 2011). Alternatively, it may be that Cpn60.2 and CD43 do interact in isolation and that this interaction anchors the *M.tb*, thereby facilitating subsequent ligand-receptor interactions to effectively take place such as binding by the phagocytic CR3 receptor (Melo et al. 2000; Rooyakkers and Stokes 2005), or signaling via TLRs (Means et al. 1999; Thoma-Uszynski et al. 2001; Reiling et al. 2008). It is noteworthy here that CD43 often plays the role of an intercellular binding modulator, allowing some receptor-ligand interactions to take place more readily, while limiting other interactions (Ostberg et al. 1998).

Whether this interaction of CD43 with *M.tb* Cpn60.2 is to the advantage of the bacteria or the host is not clear yet. It is known that absence of CD43 results in more rapid bacterial growth in MΦs and a more severe pathology resulting from *M.tb* infection *in vivo* (Randhawa et al. 2005). Increased growth of *M.tb* in CD43 null MΦs is due to a reduction in TNFα -mediated apoptosis of these MΦs that then allows for greater bacterial replication (Randhawa et al. 2008). This would suggest that recognition of Cpn60.2 by CD43 is a host defence mechanism and not a bacterial virulence strategy. Recognition of an essential *M.tb* protein that results in induction of a mechanism to control the intracellular replication of the pathogen would be a good defence strategy for the host. As *M.tb* can not survive without Cpn60.2, it has little opportunity to avoid this immune defence mechanism. However, whether the induction of TNFα-mediated apoptosis via CD43 is facilitated by Cpn60.2 or some other *M.tb* surface moiety binding to the CD43 is not yet unequivocally determined. Another intriguing possibility is that the role of secreted Cpn60.1 may be to counter the host defence mechanisms initiated by recognition of Cpn60.2, thus providing one possible explanation for the evolution of two Cpn60 paralogues in *M.tb*.

Genetic evolution analyses provide evidence that an ancient mycobacterial ancestor gained an additional Cpn60 copy at some point and since that time Cpn60.1 has undergone a more rapid level of nonsynonomous mutation, apparently leading to a form that no longer functions in protein folding, while Cpn60.2 has evolved to facilitate protein folding without the need of Cpn10 (Hughes 1993; Qamra et al. 2004). That GroE (GroEL + GroES) is necessary for the formation and maintenance of the *E. coli* cell wall suggests that these chaperonins may have originally located to the cell wall to facilitate cell maintenance (McLennan and Masters 1998). In addition, Cpn60.1 from *M. smegmatis* has been implicated in the formation of mycolic acids, again suggesting a functional role within the mycobacterial cell wall (Ojha et al. 2005). These observations suggest that at least one reason that bacterial molecular chaperones leave the cytosolic space is to facilitate their role in maintenance of the cell wall. Thus, the additional "moonlighting" roles that molecular chaperones demonstrate may have evolved as a byproduct of this extracellular localization. An additional means by which the Cpn60 proteins may have attained additional functions relates to the fact that the mycobacteria contain multiple copies of these proteins. The finding that only Cpn60.2 is necessary for viability suggests that Cpn60.1 and Cpn60.2 have unique roles within the bacterium (Hu et al. 2008) and that Hsps can evolve to have additional functions if another functional copy is retained for housekeeping functions related to cellular viability (Hu et al. 2008).

References

Abdallah AM, Gey van Pittius NC et al (2007) Type VII secretion–mycobacteria show the way. Nat Rev Microbiol 5:883–891

Alder GM, Austen BM et al (1990) Heat shock proteins induce pores in membranes. Biosci Rep 10:509–518

Alsaadi AI, Smith D (1973) The fate of virulent and attenuated mycobacteria in Guinea Pigs infected by the respiratory route. Amer Rev Respir Dis 107:1041–1046

Anderson DC, Barry ME et al (1988) Exact definition of species-specific and cross-reactive epitopes of the 65-kilodalton protein of Mycobacterium leprae using synthetic peptides. J Immunol 141:607–613

Anonymous (2009) World Health Organization Report. http://www.who.int/tb/publications/global_report/2009/en/index.html

Barry CER, Boshoff HI et al (2009) The spectrum of latent tuberculosis: rethinking the biology and intervention strategies. Nat Rev Microbiol 7:845–855

Bifani PJ, Mathema JB et al (2002) Global dissemination of the *Mycobacterium tuberculosis* W-Beijing family strains. Trends Microbiol 10:45–52

Billingham ME, Carney S et al (1990) A mycobacterial 65-kD heat shock protein induces antigen-specific suppression of adjuvant arthritis, but is not itself arthritogenic. J Exp Med 171:339–344

Braunstein M, Bardarov S et al (2002) Genetic methods for deciphering virulence determinants of *Mycobacterium tuberculosis*. Methods Enzymol 358:67–99

Carroll MV, Sim RB et al (2010) Identification of four novel DC-SIGN ligands on *Mycobacterium bovis* BCG. Protein Cell 1:859–870

Cehovin A, Coates AR et al (2010) Comparison of the Moonlighting Actions of the Two Highly Homologous Chaperonin 60 Proteins of *Mycobacterium tuberculosis*. Infect Immun 78:3196–3206

Cohen IR (1991) Autoimmunity to chaperonins in the pathogenesis of arthritis and diabetes. Annu Rev Immunol 9:567–589

Collins FM, Smith MM (1969) A comparative study of the virulence of *Mycobacterium tuberculosis* measured in mice and guinea pigs. Amer Rev Respir Dis 100:631–639

Cooper AM (2009) Cell-mediated immune responses in tuberculosis. Annu Rev Immunol 27:393–422

Cooper AM, Flynn JL (1995) The protective immune response to *Mycobacterium tuberculosis*. Curr Opin Immunol 7:512–516

Cunningham AF, Spreadbury CL (1998) Mycobacterial stationary phase induced by low oxygen tension – cell wall thickening and localization of the 16-kilodalton a-crystallin homolog. J Bacteriol 180:801–808

De Bruyn J, Bosmans R et al (1989) Effect of zinc deficiency of the appearance of two immunodominant protein antigens (32 kDa and 65 kDa) in culture filtrates of mycobacteria. J Gen Microbiol 135:79–84

de Chastellier C (2009) The many niches and strategies used by pathogenic mycobacteria for survival within host macrophages. Immunobiol 214:526–542

Digiuseppe Champion PA, Cox JS (2007) Protein secretion systems in Mycobacteria. Cell Microbiol 9:1376–1384

Doffinger R, Patel SY et al (2006) Host genetic factors and mycobacterial infections: lessons from single gene disorders affecting innate and adaptive immunity. Microbes and Infection 8:1141–1150

Dubaniewicz A (2010) Mycobacterium tuberclosis heat shock proteins and autoimmunity in sarcoidosis. Autoimmun Rev 9:419–424

Dye C, Williams BG (2010) The population dynamics and control of tuberculosis. Science 328:856–861

Ehlers S (2009) Lazy, Dynamic or Minimally Recrudescent? On the Elusive Nature and Location of the Mycobacterium Responsible for Latent Tuberculosis. Infection 37:87–95

Ehrt S, Schnappinger D (2009) Mycobacterial survival strategies in the phagosome: Defense against host stresses. Cell Microbiol 11:1170–1178

El-Etr SH, Cirillo JD (2001) Entry mechanisms of mycobacteria. Frontiers in Bioscience 6:D737–D747

Esaguy N, Aguas AP (1997) Subcellular localization of the 65-kDa heat shock protein in mycobacteria by immunoblotting and immunogold ultracytochemistry. J Submicrosc Cytol Pathol 29:85–90

Fang FC (2004) Antimicrobial reactive oxygen and nitrogen species: concepts and controversies. Nat Rev Microbiol 2:820–832

Fratazzi C, Manjunath N et al (2000) A macrophage invasion mechanism for mycobacteria implicating the extracellular domain of CD43. J Exp Med 192:183–192

Friedland JS, Shattock R et al (1993) Mycobacterial 65-kd heat shock protein induces release of proinflammatory cytokines from human monocytic cells. Clin Exp Immunol 91:58–62

Getahun H, Gunneberg C et al (2010) HIV infection-associated tuberculosis: the epidemiology and the response. Clin Infect Dis 50(Suppl 3):S201–207

Gill WP, Harik NS et al (2009) A replication clock for *Mycobacterium tuberculosis*. Nat Med 15:211–214

Gillis TP, Miller RA et al (1985) Immunochemical characterization of a protein associated with Mycobacterium leprae cell wall. Infect Immun 49:371–377

Glickman MS, Jacobs WR (2001) Microbial pathogenesis of *Mycobacterium tuberculosis*: Dawn of a discipline. Cell 104:477–485

Goyal K, Qamra R et al (2006) Multiple gene duplication and rapid evolution in the groEL gene: functional implications. J Mol Evol 63:781–787

Haque MA, Yoshino S et al (1996) Suppression of adjuvant arthritis in rats by induction of oral tolerance to mycobacterial 65-kDa heat shock protein. Eur J Immunol 26:2650–2656

Henderson B, Lund PA et al (2010) Multiple moonlighting functions of mycobacterial molecular chaperones. Tuberculosis 90:119–124

Henderson B, Mesher J (2007) The search for the chaperonin 60 receptors. Methods 43:223–228

Henderson B, Nair SP et al (2003) Molecular pathogenicity of the oral opportunistic pathogen Actinobacillus actinomycetemcomitans. Annu Rev Microbiol 57:29–55

Hennequin C, Porcheray F et al (2001) GroEL (Hsp60) of *Clostridium difficile* is involved in cell adherence. Microbiology 147:87–96

Hickey TB, Thorson LM et al (2009) *Mycobacterium tuberculosis* Cpn60.2 and DnaK are located on the bacterial surface, where Cpn60.2 facilitates efficient bacterial association with macrophages. Infect Immun 77:3389–3401

Hickey TB, Ziltener HJ et al (2010) *Mycobacterium tuberculosis* employs Cpn60.2 as an adhesin that binds CD43 on the macrophage surface. Cell Microbiol 12:1634–1647

Hu Y, Henderson B et al (2008) A *Mycobacterium tuberculosis* mutant lacking the groEL homologue Cpn60.1 is viable but fails to induce an inflammatory response in animal models of infection. Infect Immun 76:1535–1546

Hughes AL (1993) Contrasting evolutionary rates in the duplicate chaperonin genes of *Mycobacterium tuberculosis* and *M. leprae*. Mol Biol Evol 10:1343–1359

Jain A, Mondal R (2008) Extensively drug-resistant tuberculosis: current challenges and threats. FEMS Immunol Med Microbiol 53:145–150

Jeffery CJ (1999) Moonlighting proteins. Trends Biochem Sci 24:8–11

Jindal S, Dudani AK et al (1989) Primary structure of a human mitochondrial protein homologous to the bacterial and plant chaperonins and to the 65-kilodalton mycobacterial antigen. Mol Cell Biol 9:2279–2283

Kirby AC, Meghji S et al (1995) The potent bone-resorbing mediator of *Actinobacillus actinomycetemcomitans* is homologous to the molecular chaperone GroEL. J Clin Invest 96:1185–1194

Kong TH, Coates ARM et al (1993) *Mycobacterium tuberculosis* expresses 2 chaperonin-60 homologs. Proc Natl Acad Sci USA 90:2608–2612

Lamb JR, Bal V et al (1986) The identification of T cell epitopes in Mycobacterium tuberculosis using human T lymphocyte clones. Leprosy Rev 57:131–137

Lamb JR, Ivanyi J et al (1989) The mycobacterial groel stress protein: a common target of T cell recognition in infection and autoimmunity. J Autoimmun 2:93–100

Lamichhane G, Tyagi S et al (2005) Designer Arrays for Defined Mutant Analysis To Detect Genes Essential for Survival of *Mycobacterium tuberculosis* in Mouse Lungs. Infect Immun 73:2533–2540

Lasunskaia E, Ribeiro SC et al (2010) Emerging multidrug resistant Mycobacterium tuberculosis strains of the Beijing genotype circulating in Russia express a pattern of biological properties associated with enhanced virulence. Microbes Infect 12:467–475

Lee PA, Tullman-Ercek D et al (2006) The bacterial twin-arginine translocation pathway. Annu Rev Microbiol 60:373–395

Lewthwaite JC, Clarkin CE et al (2001) *Mycobacterium tuberculosis* Chaperonin 60.1 Is a More Potent Cytokine Stimulator than Chaperonin 60.2 (Hsp 65) and Contains a CD14-Binding Domain. Infect Immun 69:7349–7355

Lewthwaite JC, Coates ARM et al (2007) Highly homologous *Mycobacterium tuberculosis* chaperonin 60 proteins with differential CD14 dependencies stimulate cytokine production by human monocytes through cooperative activation of p38 and ERK1/2 mitogen-activated protein kinases. Int Immunopharmacol 7:230–240

Li AH, Lam WL et al (2008) Characterization of genes differentially expressed within macrophages by virulent and attenuated *Mycobacterium tuberculosis* identifies candidate genes involved in intracellular growth. Microbiology 154:2291–2303
Li AH, Waddell SJ et al (2010) Contrasting Transcriptional Responses of a Virulent and an Attenuated Strain of *Mycobacterium tuberculosis* Infecting Macrophages. PLoS ONE 5:e11066
Lienhardt C (2001) From exposure to disease: the role of environmental factors in susceptibility to and development of tuberculosis. Epidemiologic Revs 23:288–301
Lopez-Guerrero JA, Ortiz MA et al (1994) Therapeutic effect of recombinant vaccinia virus expressing the 60-kd heat-shock protein on adjuvant arthritis. Arthr Rheum 37:1462–1467
Lund PA (2009) Multiple chaperonins in bacteria – why so many? FEMS Microbiol Rev 33:785–800
Mambula SS, Stevenson MA et al (2007) Mechanisms for Hsp70 secretion: Crossing membranes without a leader: Heat Shock Proteins In Extracellular Signaling. Methods 43:168–175
McDonough JA, Hacker KE et al (2005) The twin-arginine translocation pathway of *Mycobacterium smegmatis* is functional and required for the export of mycobacterial beta-lactamases. J Bacteriol 187:7667–7679
McLean IL, Archer JR et al (1990) Specific antibody response to the mycobacterial 65 kDa stress protein in ankylosing spondylitis and rheumatoid arthritis. Br J Rheum 29:426–429
McLennan N, Masters M (1998) GroE is vital for cell-wall synthesis. Nature 392:139
Means TK, Wang SY et al (1999) Human toll-like receptors mediate cellular activation by *Mycobacterium tuberculosis*. J Immunol 163:3920–3927
Meghji S, Lillicrap M et al (2003) Human chaperonin 60 (Hsp60) stimulates bone resorption: structure/function relationships. Bone 33:419–425
Meghji S, White PA et al (1997) *Mycobacterium tuberculosis* chaperonin 10 stimulates bone resorption: a potential contributory factor in Pott's disease. J Exp Med 186:1241–1246
Melo MD, Catchpole IR et al (2000) Utilization of CD11b knockout mice to characterize the role of complement receptor 3 (CR3, CD11b/CD18) in the growth of *Mycobacterium tuberculosis* in macrophages. Cell Immunol 205:13–23
Merendino AM, Bucchieri F et al (2010) Hsp60 is actively secreted by human tumor cells. PLoS ONE 5:e9247
Mishra AK, Driessen NN et al (2011) Lipoarabinomannan and related glycoconjugates: structure, biogenesis and role in *Mycobacterium tuberculosis* physiology and host-pathogen interaction. FEMS Microbiol Revs 35:1126–1157
Mitnick CD, Appleton SC et al (2008) Epidemiology and treatment of multidrug resistant tuberculosis. Semin Respir Crit Care Med 29:499–524
Moudgil KD, Sercarz EE (1994) The T cell repertoire against cryptic self determinants and its involvement in autoimmunity and cancer. Clin Immunol Immunopathol 73:283–289
Newport M, Levin M (1999) Genetic susceptibility to tuberculosis. J Infect 39:117–121
North RJ, Izzo AA (1993) Mycobacterial virulence – virulent strains of *Mycobacteria tuberculosis* have faster *in vivo* doubling times and are better equipped to resist growth-inhibiting functions of macrophages in the presence and absence of specific immunity. J Exp Med 177:1723–1733
Ojha A, Anand M et al (2005) GroEL1: a dedicated chaperone involved in mycolic acid biosynthesis during biofilm formation in mycobacteria. Cell 123:861–873
Orme I (2004) Adaptive immunity to mycobacteria. Curr Opin Microbiol 7:58–61
Orme IM (1999) Virulence of recent notorious *Mycobacterium tuberculosis* isolates. Tuber Lung Dis 79:379–381
Ostberg JR, Barth RK et al (1998) The Roman god Janus: a paradigm for the function of CD43. Immunol Today 19:546–550
Peetermans WE, Raats CJ et al (1994) Mycobacterial heat-shock protein 65 induces proinflammatory cytokines but does not activate human mononuclear phagocytes. Scand J Immunol 39:613–617
Peetermans WE, Raats CJ et al (1995) Mycobacterial 65-kilodalton heat shock protein induces tumor necrosis factor alpha and interleukin 6, reactive nitrogen intermediates, and toxoplasmastatic activity in murine peritoneal macrophages. Infect Immun 63:3454–3458

Qamra R, Mande SC et al (2004) *Mycobacterium tuberculosis* GroEL homologues unusually exist as lower oligomers and retain the ability to suppress aggregation of substrate proteins. J Mol Biol 342(2):605–617

Qamra R, Srinivas V et al (2005) The unusual chaperonins of Mycobacterium tuberculosis. Tuberculosis 85:385–394

Randhawa AK, Ziltener HJ et al (2005) CD43 is required for optimal growth inhibition of Mycobacterium tuberculosis in macrophages and in mice. J Immunol 175:1805–1812

Randhawa AK, Ziltener HJ et al (2008) CD43 controls the intracellular growth of Mycobacterium tuberculosis through the induction of TNF-alpha-mediated apoptosis. Cell Microbiol 10:2105–2117

Rao T, Lund PA (2010) Differential expression of the multiple chaperonins of *Mycobacterium smegmatis*. FEMS Microbiol Lett 310:24–31

Reddi K, Meghji S et al (1998) The *Escherichia coli* chaperonin 60 (groEL) is a potent stimulator of osteoclast formation. J Bone Miner Res 13:1260–1266

Reiling N, Ehlers S et al (2008) MyDths and un-TOLLed truths: sensor, instructive and effector immunity to tuberculosis. Immunol Lett 116:15–23

Rengarajan J, Bloom BR et al (2005) Genome-wide requirements for *Mycobacterium tuberculosis* adaptation and survival in macrophages. Proc Natl Acad Sci U S A 102:8327–8332

Rinke de Wit TF, Bekelie S et al (1992) Mycobacteria contain two groEL genes: the second Mycobacterium leprae groEL gene is arranged in an operon with groES. Mol Microbiol 6:1995–2007

Rooyakkers AW, Stokes RW (2005) Absence of complement receptor 3 results in reduced binding and ingestion of *Mycobacterium tuberculosis* but has no significant effect on the induction of reactive oxygen and nitrogen intermediates or on the survival of the bacteria in resident and interferon-gamma activated macrophages. Microb Pathog 39:57–67

Rosenkrands I, King A et al (2000) Towards the proteome of *Mycobacterium tuberculosis*. Electrophoresis 21:3740–3756

Sassetti CM, Rubin EJ (2003) Genetic requirements for mycobacterial survival during infection. Proc Natl Acad Sci U S A 100:12989–12994

Schafer G, Jacobs M et al (2009) Non-opsonic recognition of *Mycobacterium tuberculosis* by phagocytes. J Innate Immun 1:231–243

Schnappinger D, Ehrt S et al (2003) Transcriptional Adaptation of *Mycobacterium tuberculosis* within Macrophages: Insights into the Phagosomal Environment. J Exp Med 198:693–704

Shahar A, Melamed-Frank M, Kashi Y, Shimon L, Adir N (2011) The dimeric structure of the Cpn60.2 chaperonin of *Mycobacterium tuberculosis* at 2.8 Å reveals possible modes of function. J Mol Biol 412:192–203

Sharma D, Tyagi JE (2007) The value of comparative genomics in understanding mycobacterial virulence. J Biosci 32:185–189

Shinnick TM, Sweetser D et al (1987) The etiologic agents of leprosy and tuberculosis share an immunoreactive protein antigen with the vaccine strain *Mycobacterium bovis* BCG. Infect Immun 55:1932–1935

Sielaff B, Lee KS et al (2010) Structural and Functional Conservation of *Mycobacterium tuberculosis* GroEL Paralogs Suggests that GroEL1 is a Chaperonin. J Mol Biol 405:831–839

Smith I (2003) *Mycobacterium tuberculosis* pathogenesis and molecular determinants of virulence. Clin Microbiol Rev 16:463–496

Sonnenberg MG, Belisle JT (1997) Definition of *Mycobacterium tuberculosis* culture filtrate proteins by two-dimensional polyacrylamide gel electrophoresis, N-terminal amino acid sequencing, and electrospray mass spectrometry. Infect Immun 65:4515–4524

Steenken W, Oatway WH et al (1934) Biological studies of the tubercle bacillus III. Dissociation and pathogenicity of the R and S variants of the human tubercle bacillus (H37). J Exp Med 60:515–540

Stein CM (2011) Genetic epidemiology of tuberculosis susceptibility: impact of study design. PLoS Pathog 7:e1001189

Stokes RW, Haidl ID et al (1993) Mycobacteria-macrophage interactions. Macrophage phenotype determines the nonopsonic binding of *Mycobacterium tuberculosis* to murine macrophages. J Immunol 151:7067–7076

Stokes RW, Waddell SJ (2009) Adjusting to a new home: *Mycobacterium tuberculosis* gene expression in response to an intracellular lifestyle. Future Microbiol 4:1317–1335

Talaat AM, Lyons R et al (2004) The temporal expression profile of *Mycobacterium tuberculosis* infection in mice. Proc Natl Acad Sci U S A 101:4602–4607

Thole JER, Hindersson P et al (1988a) Antigenic relatedness of a strongly immunogenic 65 kDA mycobacterial protein antigen with a similarly sized ubiquitous bacterial common antigen. Microb Pathog 4:71–83

Thole JER, Van Schooten WCA et al (1988b) Use of recombinant antigens expressed in Escherichia coli K-12 to map B-cell epitopes on the immunodominant 65-kilodalton protein of Mycobacterium bovis BCG. Infect Immun 56:1633–1640

Thoma-Uszynski S, Stenger S et al (2001) Induction of direct antimicrobial activity through mammalian toll-like receptors. Science 291:1544–1547

Torok Z, Horvath I et al (1997) Evidence for a lipochaperonin: association of active protein-folding GroESL oligomers with lipids can stabilize membranes under heat shock conditions. Proc Natl Acad Sci U S A 94:2192–2197

Torrado E, Robinson T et al (2011) Cellular response to mycobacteria: balancing protection and pathology. Trends Immunol 32:66–72

Triantafilou M, Triantafilou K (2005) The dynamics of LPS recognition: complex orchestration of multiple receptors. J Endotoxin Res 11:5–11

van der Eijk EA, van de Vosse E et al (2007) Heredity versus environment in tuberculosis in twins: the 1950s United Kingdom Prophit Survey Simonds and Comstock revisited. Am J Respir Care Med Crit 176:1281–1288

Velasco-Velazquez MA, Barrera D et al (2003) Macrophage–*Mycobacterium tuberculosis* interactions: role of complement receptor 3. Microb Pathog 35:125–131

Wang XM, Lu C et al (2011) Biochemical and immunological characterization of a cpn60.1 knockout mutant of *Mycobacterium bovis* BCG. Microbiology 157:1205–1219

Winrow VR, Mesher J et al (2008) The two homologous chaperonin 60 proteins of *Mycobacterium tuberculosis* have distinct effects on monocyte differentiation into osteoclasts. Cell Microbiol 10:2091–2104

Wolfe LM, Mahaffey SB et al (2010) Proteomic definition of the cell wall of *Mycobacterium tuberculosis*. J Proteome Res 9:5816–5826

Yang XD, Gasser J et al (1990) Prevention of adjuvant arthritis in rats by a nonapeptide from the 65-kD mycobacterial heat-shock protein. Clin Exp Immunol 81:189–194

Young D, Lathigra R et al (1988) Stress proteins are immune targets in leprosy and tuberculosis. Proc Natl Acad Sci U S A 85:4267–4270

Young DB (1990) The immune response to mycobacterial heat shock proteins. Autoimmunity 7:237–244

Chapter 17
Heat Shock Protein 90 Versus Conventional Growth Factors in Acute and Diabetic Wound Healing

Wei Li, Kathryn O'Brien, David T. Woodley and Mei Chen

Abstract Diabetic foot lesions are responsible for more hospitalizations than any other complication of diabetes in the United States of America. The number of diabetic foot ulcer-caused lower limb amputations is approaching 100,000/year, a rapid increase due to an aging population and the rising incidence of obesity. Cost-effect treatments are currently limited. For decades, the conventional wisdom is that growth factors constitute the driving forces of wound healing. Therefore, more than 30 growth factors have been extensively studied in animal models and a dozen of these growth factors have been subjected to clinical trials. Only PDGF-BB received the US FDA approval for treatment of diabetic ulcers in 1997. However, the modest efficacy, high cost and risks of causing cancer by PDGF-BB (becaplermin gel) have limited its use in clinical practice. This reality continues to be overlooked or ignored. An unconventional wound-healing molecule, extracellular heat shock protein-90alpha (eHsp90α), has recently been reported. Unlike restricted cell type specificity of PDGF-BB, eHsp90α is a common pro-motility factor of all skin cells, eHsp90α also overrides TGFβ inhibition and hyperglycemia. Topical application of eHsp90α accelerated both acute and diabetic wound closure far more effectively than PDGF-BB. We discuss what makes Hsp90α superior to conventional growth factors in wound healing.

17.1 The Clinical Problem

Wound healing is a significant health, economic and social issue in the United States and in the world generally (Boulton et al. 2005). The annual healthcare costs for treatment of chronic skin wounds (all kinds) exceed US$ 25 billion in the US (Sen et al. 2009). For instance, the average cost of healing a single diabetic ulcer is US$ 8,000, that of an infected ulcer is US$ 17,000, and that of a major amputation is US$ 45,000. More than 80,000 amputations were performed in 2008 on diabetic patients and approximately 50 % of the people with amputations will develop ulcerations and

W. Li (✉) · K. O'Brien · D. T. Woodley · M. Chen
The Department of Dermatology and the USC/Norris Comprehensive Cancer Center,
The University of Southern California Keck School of Medicine, 1441 Eastlake Avenue,
Room-6320, CA 90033, Los Angeles, USA
e-mail: wli@usc.edu

B. Henderson, A. G. Pockley (eds.), *Cellular Trafficking of Cell Stress Proteins in Health and Disease*, Heat Shock Proteins 6, DOI 10.1007/978-94-007-4740-1_17,

infections in the contralateral limb within 18 months. An alarming 58 % will have a contralateral amputation 3–5 years after the first amputation. In addition, the 3-year mortality after a first amputation was estimated to be as high as 20–50 %. These numbers have not changed much in the past 30 years. An elderly diabetic patient residing at home with a chronic heel pressure ulcer would require a certified wound specialist for debridement, dressing selection and patient education; an orthotist for pressure relief footwear; a physical therapist for gait training and home modification; a nutritionist to monitor diet and enhance overall nutritional status; a vascular surgeon to evaluate patency of blood vessels; a primary care physician to manage glycemic control and other co-morbidities; an infectious disease specialist for evaluation of osteomyelitis; a podiatrist or plastic surgeon for surgical intervention; and a home care nurse for periodic wound assessment and coordination of community services (Harrington et al. 2000). This so-called "the cost to achieve the desired outcome" for a single diabetic foot ulcer could run over US$ 10,000. These currently available treatments mostly address three basic issues: debridement, offloading, and infection control. These costs continue rising rapidly due to: (i) increasing health care costs; (ii) aging of the population and; (iii) the rise in the incidence of diabetes and obesity. More importantly, the quality of life of the patients due to morbidity of non-healing leg ulcers is significantly compromised, because of wound odor, infection, and pain. These issues lead to social isolation and diminished self-image in those suffering from diabetic and other chronic skin wounds (Alvarez et al. 2002). Thus, there is a pressing need to develop new and effective wound healing drugs.

17.2 Diabetic Wound Healing: Conventional Wisdom and Disappointing Facts

Since the discovery of the first "growth factor" in the late 1970s, it has become widely believed that the local growth factors in the wound bed constitute the main driving forces for wound healing (Werner and Grose 2003), i.e. the lateral migration and proliferation of epidermal keratinocytes to close the wound and the inward migration and expansion of dermal fibroblasts and dermal microvascular endothelial cells into the wounded space to remodel and to build a new vascularized neodermis (Martin 1997; Singer and Clark 1999). The reason is that these growth factors were thought to either appear only when skin is wounded or to rise significantly from their basal concentrations in response to the injury. Over the past 2 decades, more than 30 growth factors have been subjected to extensive pre-clinical studies alone or in combinations (Grose and Werner 2004). Since the first report of a clinical trial of EGF (epidermal growth factor) in human wound healing in 1989 (Brown et al. 1989), more than a dozen similar trials have been completed. These trials include: (1) EGF on partial thickness wounds of skin grafts (Brown et al. 1989), on traumatic corneal epithelial defects (Pastor and Calonge 1992), on tympanic membrane with chronic perforation (Ramsay et al. 1995) and on advanced diabetic foot ulcers (Fernandez-Mantequin et al. 2007; Mohan 2006); (2) basic FGF (bFGF) on partial-thickness burn wounds of children (Greenhalgh and Rieman 1994), on second degree burns (Fu et al. 1998) and on diabetic ulcers (Uchi et al. 2009); (3) acidic FGF (aFGF) on partial thickness burns and skin graft donor sites (Ma et al. 2007); (4) GM-CSF plus bFGF on pressure ulcers (Robson et al. 2000); and (5) PDGF-BB on chronic pressure and diabetic ulcers (Pierce et al. 1992; LeGrand 1998; Smiell et al. 1999; Wieman et al. 1998). It must be pointed out that, despite the fact that most of these double-blinded and randomized trials reported promising clinical efficacies in humans, only human recombinant PDGF-BB has ultimately received the US Food and Drug Administration (FDA) approval for treatment of limb diabetic ulcers (Regranex™/becaplermin gel 0.01 %, Ortho-McNeil Pharmaceutical, Raritan, NJ) (LeGrand 1998). A multicenter, randomized, parallel trials showed that becaplermin at $\sim$100 μg/g gel of PDGF-BB improved at the best 15 % in complete wound closure (50 % treated versus 36 % placebo) (Steed 1995; Smiell et al. 1999; Wieman et al. 1998). These results did not support a cost-effect beneficiary of becaplermin for clinical practice (Nagai and Embil 2002; Mandracchia et al. 2001). In 2008, the US FDA added a black box warning to increased risks for cancer mortality in patients who need extensive treatments ($\geq$3 tubes) of the becaplermin gel. This cancer-causing side effect may not be surprising in consideration of the fact that the usage of PDGF-BB in becaplermin gel is 10^{4-5} higher than the PDGF-BB levels in human serum (1–10 ng/ml) (R&D System, PDGF-BB Quantikin) and overexpression of PDGF-BB (c-sis) or its viral form, v-sis, is known to cause cell transformation (Bejcek et al. 1992). Unfortunately, these outcomes of growth factor-based therapies continue to be downplayed or simply overlooked. Li and colleagues speculated that there must be fundamental reasons underlining the ineffectiveness of conventional growth factor therapies in promoting wound healing (Cheng et al. 2011).

17.3 The Odds Against Effectiveness of Conventional Growth Factor Therapy

Using PDGF-BB as a representative of human growth factors, recent studies have begun to reveal some physiological hurdles that might block the effectiveness of growth factor therapy. First, ideally, a single growth factor-based wound-healing agent should be capable of recruiting multiple types of human skin cells into the wound bed. In contrast, if this factor selectively acts on some, but not all, of the skin cell types required for wound healing, it would be less effective in this multicell type healing process. PDGF-BB acts only on dermal fibroblasts, but not keratinocytes or dermal microvascular endothelial cells, due to a complete lack of both PDGFRα and PDGFRβ on the latter two types of cells. Only human dermal fibroblasts express the PDGFRs (Cheng et al. 2011). If one extrapolates these *in vitro* findings to equivalent wound healing events *in vivo*, it suggests that PDGF-BB cannot have a direct role in recruitments of keratinocytes for wound re-epithelialization and dermal microvascular endothelial cells for wound neovascularization. If this were true in reality, and applied also to other growth factors in general, the implications would need to be taken seriously.

Second, the tumor suppressing signals of the TGFβ family cytokines include anti-motility and anti-growth actions. Therefore, many tumors lose responsiveness to TGFβ through down-regulation or mutations of the TGFβ receptors or elimination of the downstream Smad4 protein, which transduces signals from the TGFβ receptors to gene expression in the nucleus. These alterations in TGFβ signaling pathways in cancer cells eliminated anti-proliferation and anti-migration signals, which otherwise prevent the cancer cells from expanding, invading and metastasizing (Hanahan and Weinberg 2000). Similarly in wound healing, Bandyopadhyay et al. (2006) showed that human serum, the recognized source for PDGF-BB, could not promote migration of human dermal cells due to the presence of TGFβ3. *In vitro,* TGFβ3 potently blocked PDGF-BB-stimulated dermal fibroblast and VEGF-A-induced dermal endothelial cells migration (Bandyopadhyay et al. 2006) This finding implies that the inhibitory signals of TGFβ must be overridden during wound healing, but growth factors are obviously incapable of doing so.

Third, it is known that all forms of diabetes are characterized by chronic hyperglycemia in the circulation (Brownlee 2001), which is blamed, at least partially, for the delayed healing of diabetic wounds (Peppa et al. 2009) Normal glycemia is around 5.5 mmol/l (100 mg/dl) of glucose after fasting. Hyperglycemia (or hyperglycaemia) is a condition in which an excessive amount of glucose (above 10 mmol/l) circulates in the blood plasma. Chronic glucose levels exceeding 7 mmol/l (125 mg/dl) may already cause organ damage. However, diabetic symptoms often may not start to become noticeable until even higher values, such as 15–20 mmol/l, of the plasma glucose are reached. Due to ethical, practical and technical concerns, humans are not an option for pre-clinical studies for potentially new wound healing treatments. One of the commonly used mouse models, the db/db mouse, has 20–30 mmol/l blood glucose levels by the age of 6–8 weeks. More importantly, a standard 1.5 cm × 1.5 cm full-thickness excision skin wound takes ~50 days to close in db/db mice, in comparison to ~18 days to close for the same wound in normal mice (Olerud 2008). The same processes that affect cells and molecules *in vivo* can also be studied *in vitro*. The amount of glucose in commercial cell culture formulations ranges from 5.5 mmol/l to 55 mmol/l. For example, Basal Medium Eagle (BME), Fischer's Medium and Eagle's Minimum Essential Medium (EMEM) contain 5.5 mmol/l D-glucose. In contrast, DMEM (Hi), GMEM and IMDM all contain 25 mmol/l of D-glucose. Mascardo had first shown that hyperglycemia blocks serum-induced polarization and directed migration of wounded endothelial cells (Masccardo et al. 1988). Hamed and co-workers, reported that hyperglycemia and oxidized-LDL strongly inhibited serum-induced migration of endothelial progenitor cells isolated from type 2 diabetic patients (Hamed et al. 2010). O'Brien et al. (unpublished) recently observed that hyperglycemia blocks human dermal fibroblast migration in response to major conventional wound healing signals, including serum, hypoxia and

PDGF-BB. These results suggest that conventional growth factors may not function well in diabetic wounds.

Having analyzed the potentially fundamental hurdles facing conventional growth factor therapies, we introduce and discuss a recently reported and unconventional wound-healing agent—the secreted form of heat shock protein 90-alpha (Hsp90α). In both wound healing and cancer, "extracellular Hsp90" (or eHsp90) proteins are distinct from the intracellular Hsp90 chaperones. Firstly, one is found inside, and the other, outside, the cell. Secondly, one acts as a chaperone and the other as a pro-motility factor (and possibly more). Thirdly, one depends upon its N-terminal ATPase and C-terminal dimerization/co-factor binding domains for proper folding the client proteins, whereas the other uses a short peptide epitope between the highly charged linker region (LR) and the middle domain (M) for promotion of cell motility. We will emphasize that eHsp90 carries unique properties, which are absent from conventional growth factors, to effectively heal both acute and diabetic wounds.

17.4 Discovery of a Secreted Form of Heat Shock Protein-90 (Hsp90) in Wound Healing

17.4.1 An Unexpected Finding

Inflammation, granulation tissue formation, re-epithelialization and tissue remodeling are the proposed sequential events of skin wound healing (Singer and Clark 1999). Defects in any of these events could result in delayed/unhealed wounds or in reverse hypertrophic scars (Tredget et al. 1997). Throughout these repair (not regeneration) processes, a critical rate-limiting step is initiation of the resident cell migration around the wound edge. Within hours of injury, human keratinocytes laterally migrate across the wound bed from the cut edge to close the wound, the process known as re-epithelialization. A few days later, dermal cells, including dermal fibroblasts and dermal microvascular endothelial cells, move into the wound to produce and deposit large amounts of matrix proteins, to contract and remodel the newly closed wound and to build new blood vessels. Initiation of skin cell migration is triggered by microenvironmental changes in the wound milieu. One of the most dramatic microenvironmental changes by its magnitude (in both volume and contents) is the transition of plasma released to the wounded bed from broken blood vessels to platelet-degranulated serum that has acquired many new factors absent from plasma. Why is this plasma-to-serum transition relevant for wound healing? In intact skin, the epidermal and dermal cells are bathed in interstitial fluid, largely a filtrate of plasma. Following acute skin wounding, however, the cells at the cut edge of the wound experience a transition from the initial components of plasma to a new environment of platelet-degranulated serum for the first time. As schematically depicted in Fig. 17.1, during this transition of plasma to serum (A), a number of vasoactive mediators, growth factors and cytokines appeared for the first time, such as TGFβ1,

a

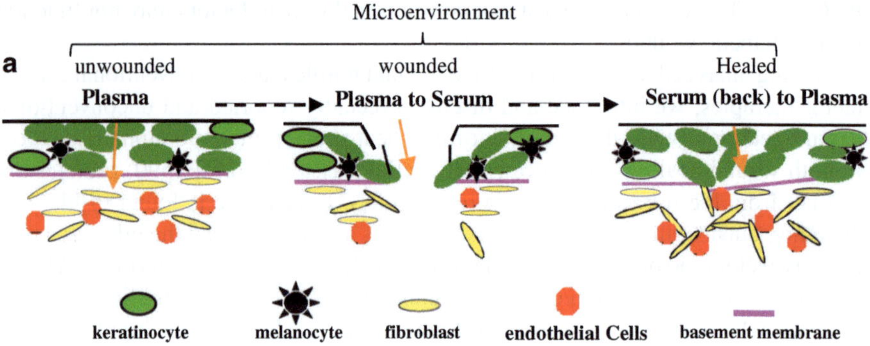

b **Reported Skin Cell Growth Factors detected in Human Plasma vs. Human Serum ***

growth factor	human plasma (ng/ml)	human serum (ng/ml)	target cell(s)	source
bFGF	UD	UD	fibroblast	bFGF Qt, R & D
IL-1α	UD	UD	fibroblast	IL-1α Qt, R & D
KGF	0.09	0.09	keratinocyte	KGF Qt, R & D
Insulin	0.2 ~ 1.08	0.2 ~1.08	all	ARUP Lab.
HGF	0.787	1.257	keratinocyte	HGF Qt, R & D
IGF-1	86	105	all	IGF-1 Qt, R & D
TGF β1	0~1.26	**40.6**	fibroblast	TGFβ1 Qt, R & D
TGF β2	0.3	0.4	fibro. & endo	TGFβ2 Qt, R & D
TGF β3	UD	**0.15-0.2**	fibro. & endo	Hering et al. 2001
TGF α	UD	UD to 0.032	keratinocyte	TGFα Qt, R & D
EGF	0.013	**0.336**	kera. & fibro.	EGF Qt, R & D
HB-EGF	ND	ND	keratinocyte	none
IL-8	0.004	0.012	keratinocyte	IL-8 Qt, R & D
PDGF-BB	0.05	**3.4**	fibroblast	PDGF-bb Qt, R & D
VEGF	0.05	**0.4**	Endothelial	VEGF Qt, R & D

*Numbers in red are significantly higher in serum than plasma. UD = undetectable; ND = no data; Qt = Quantikine kits; R&D = R&D Systems, Inc

Fig. 17.1 The plasma-serum-plasma cycle has an important role in wound healing. **a** The soluble environment of unwounded skin is mostly filtrated factors of plasma in blood vessels plus local secretion. Many critical factors, such as TGFβ and PDGF-BB, are either low or undetectable. In skin wounds, it is mostly platelet-degranulated serum that has acquired many new factors. Whence the wound is closed, it will gradually return to plasma environment. **b** A partial list of the growth factors reported for a role in wound healing show comparisons of their levels in human plasma vs. human serum. Only a small portion of them show changes during the plasma-serum transition

TGFβ3, EGF, VEGF and PDGF-BB. Whereas, many other growth factors remain unchanged. It is arguable whether these unchanged levels of growth factors are part of the initial driving forces of wound healing. Although technically too complicated to individually evaluate the importance of each of them *in vivo*, these factors have long been thought to formulate the driving force that cause a switch of non-motile to motile mode of the various types of skin cells.

In 2006, Bandyopadhyay et al. made a surprising observation that ultimately led to discovery of eHsp90α as a new generation of skin wound healing agent (see later sections). For years, fetal bovine serum (FBS) has been used to culture human cells and as a source of growth factors in experiments. These authors argued for the first time that human cells are never in contact with FBS in reality and, instead, it is the human serum that represents the main soluble microenvironment in the wound bed. Of course, the critical question was whether or not human serum and FBS are entirely exchangeable in terms of studying wound healing. Therefore, they compared the effects of human serum vs. FBS on migration of the three major human skin cell types, epidermal keratinocytes and dermal fibroblasts and microvascular endothelial cells. They found that FBS non-discriminatively stimulated migration of all the three types of human skin cells. However, human serum only promoted keratinocyte migration, whereas it halted the migration of the other two dermal cell types. They further demonstrated that the "halting" signal in human serum comes from TGFβ3 (not TGFβ1 or TGFβ2) and the sensitivity of the two types of human dermal cells to TGFβ3 is due to their 7–15 fold higher levels of the TβIIR than the epidermal keratinocytes (Bandyopadhyay et al. 2006). These findings explain the previous observations that keratinocytes migration (within hours) proceeds ahead of dermal fibroblasts and endothelial cell migration (after 3–4 days) during wound healing. After keratinocytes close the wound, the dermal cells then move into the wound to remodel the wound and to build new blood vessels. Another important message of this study was that TGFβ3 in human serum would block the action of conventional growth factors that are also present in human serum. They, therefore, concluded that the conventional growth factors from human serum may not be the driving force of wound healing.

17.4.2 Discovery of eHsp90α from Secretion of Migrating Human Keratinocytes

What is the source of the factor(s) that is responsible for recruiting dermal cells into the wound bed even in the presence of TGFβ? Li and colleagues speculated that this factor comes from the secretion of migrating keratinocytes at the wound edge. They reasoned that migrating keratinocytes secrete a factor into the wound bed. Whence this secreted factor reaches a threshold concentration, it acts as a chemotaxis factor to recruit the surrounding dermal cells into the wound. First, these authors demonstrated that a serum-free conditioned medium of migrating human keratinocytes has a robust pro-motility activity on all the human dermal cells, which could not be blocked by added human recombinant TGFβ3. Second, they generated 10 l of the conditioned medium and subjected it to multiple rounds of Fast Protein Liquid Chromatography (FPLC) to purify the pro-motility activity-containing protein. Their results showed that this activity came from secreted Hsp90α (Cheng et al. 2008). They further proved that depletion of Hsp90α from the conditioned medium alone with a specific neutralizing antibody against Hsp90α (but not Hsp90β) completely eliminated the pro-motility activity on human dermal cells from the medium and, in reverse, the addition of increasing human recombinant Hsp90α overrode the antibody and resumed the pro-motility activity in a dose-dependent manner.

17.4.3 Topical Application of eHsp90α Strongly Promoted Acute Wound Healing in Nude Mice

Is eHsp90α a naturally occurring wound healing factor and is it superior to conventional growth factor therapy? Li and colleagues topically treated full thickness wounds in nude mice with either full-length Hsp90α or US FDA approved becaplermin (PDGF-BB) gel and found Hsp90α shortened acute wound closure from 18 days (placebo control) to 10 days, whereas becaplermin showed little effect (Li et al. 2007; Cheng et al. 2011). This observation is consistent with previous report that PDGF-BB does not improve normal wound healing (Lynch et al. 1987). In contrast, inhibition of eHsp90α signaling by blocking its cell surface receptor, LRP-1, delayed normal wound healing. The primary reasons for the authors to choose the athymic hairless mice were: (1) to minimize the host innate immune response to human peptides and, therefore, to avoid immune response-caused wound contraction and; (2) to minimize the effect of inflammatory response following the injury on wound healing, so as to detect the specific effect of the topically applied peptides.

17.5 History of eHsp90

In fact, it has been more than two decades, since eHsp90 was first reported as a cell-surface bound tumor antigen (reviewed by Csermely et al. 1998). However, for years skepticism remained as to whether eHsp90 proteins are the result of pathophysiological processes in the cells or of release from a small portion of dead cells in culture (Multhoff and Hightower 1996). A primary reason for the skepticism is that Hsp90 does not fit into the conventional category of actively secreted proteins, such as growth factors, extracellular matrix (ECM) components and matrix metalloproteinases (MMPs). First, Hsp90 has nether a signal peptide (SP) for secretion via the endoplasmic reticulum (ER)/Golgi protein secretory pathway nor a recognizable transmembrane sequence for membrane anchoring (see Chap. 1 for further discussion of molecular chaperone secretion). Second, there have been reports that Hsp90 could indeed be released to the extracellular environment following cell necrosis (Basu and Srivastava 2000), which in turn binds and helps antigen recognition and triggers immune responses. Further information on the cellular trafficking of Hsp90 is to be found in Chaps. 7 and 9.

Studies over the past 10 years, however, have begun to provide stronger arguments that cells export Hsp90 for biological purposes. The take-home message is that cells secrete Hsp90 when they encounter environmental stresses. So far, the reported environmental cues that trigger cells to secrete Hsp90 include: reactive oxygen species (ROS), heat, hypoxia, gamma-irradiation and injury-released growth factors. First, Hightower and Guidon reported that heat-shocked rat embryonic cells secreted Hsp90 and Hsp70, which could not be blocked by monensin or colchicine, inhibitors of the conventional protein trafficking pathway (Hightower and Guidon 1989—detailed in Chap. 1). When Clayton and colleagues used proteomic methods to analyze the peptide contents of B cell-secreted exosomes under either a physiological temperature (37 °C) or after heat shock (42 °C for 3 h), they found that heat induces Hsp90 secretion among other heat shock proteins into the nano-vesicles called exosomes (Clayton et al. 2005). Similar observations were made for Hsp70 via the exosome-dependent trafficking pathway in different cell types (Lancaster and Febbraio 2005). Second, Liao et al. (2000) reported that treatment of rat vascular smooth muscle cells with LY83583, an oxidative stress generator, for two hours caused secretion of Hsp90 and a late phase activation of $ERK_{1/2}$ by the oxidative stress. Third, Yu and colleagues found that γ irradiation induces secretion of Hsp90β via a p53-dependent event into conditioned medium, suggesting a "DNA damage → p53 → secretion of Hsp90β" pathway (Yu et al. 2006). Fourth, Li and colleagues showed that hypoxia causes increased skin fibroblast migration and Hsp90α secretion via hypoxia-inducible factor-1α (HIF-1α). Blockade of the extracellular Hsp90α function by neutralizing antibodies completely inhibited hypoxia's pro-motility effect on the cells, suggesting that secretion of Hsp90α is part of the hypoxia's pro-motility signaling (Li et al. 2007). Cheng and co-workers showed that TGFα rapidly induces Hsp90α membrane translocation and secretion to an extracellular environment in primary human keratinocytes (Cheng et al. 2008). Since TGFα level are undetectable in human plasma (i.e. unwounded skin) and dramatically increased in serum (i.e. wounded skin), keratinocytes only come in contact with TGFα after the skin is wounded (Li et al. 2007). Therefore, the action of TGFα on keratinocytes is considered as a pathophysiological condition. Cheng et al. (2008) went further to identify LRP-1/CD91 as the cell surface receptor that mediates Hsp90α-stimulated migration of human keratinocytes and dermal fibroblasts. They demonstrated that cells without LRP-1 completely lost pro-motility responses to recombinant Hsp90α or hypoxia (Cheng et al. 2008; Woodley et al. 2009). Since overexpression of HIF-1α has been reported in many tumor cells, it was proposed that the overexpressed HIF-1α should cause constitutive surface expression and secretion of Hsp90α in these cells. This is indeed the case. Kuroita and colleagues reported purification of Hsp90α from conditioned media of human hybridoma SH-76 cells (Kuroita et al. 1992). Eustace et al. (2004) reported that Hsp90α, but not Hsp90β, was secreted into the conditioned media of HT-1080 tumor cells (Eustace et al. 2004). Wang et al. (2009) reported secretion of Hsp90α by MCF-7 human breast cells. Suzuki and Kulkarni (2010) found Hsp90β secreted

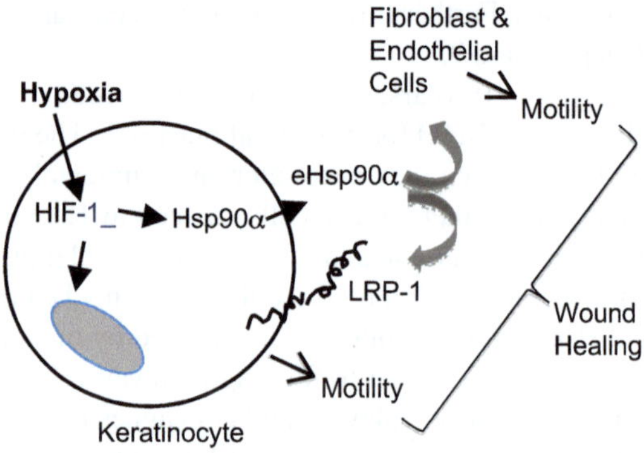

Fig. 17.2 The Hypoxia-eHsp90α-LRP-1autocrine mechanism to promote wound healing. Hypoxia drives Hsp90α secretion via HIF-1α. Secreted Hsp90α binds LRP-1 and promotes migration of keratinocytes and the surrounding LRP-1-positive dermal cells during wound healing. The mechanism, by which hypoxia causes secretion of the intracellular Hsp90α, remains to be further studied. (Modified with permission from Woodely et al. 2009)

by MG63 osteosarcoma cells. Chen et al. (2010) reported secretion of Hsp90α by colorectal cancer cell line, HCT-8. The results of a study by Tsutsumi and colleagues implied secretion of Hsp90α by a variety of tumor cell lines (Tsutsumi et al. 2008). Finally, recent studies from the author's laboratory demonstrated that breast cancer cells, MDA-MB-231 and MDA-MB-468, overexpress HIF-1α that causes constitutive secretion of Hsp90α in a HIF-1-depemndent fashion (Sahu et al. 2012). Taken together, it is clear that normal cells secrete Hsp90 for tissue repair, whereas tumor cells constitutively secrete Hsp90 for tissue invasion and metastasis (Reviewed by Li et al. 2012).

17.6 eHsp90, Hyperglycemia and Diabetic Wound Healing

The clinically important questions are whether skin cells in diabetic wounds fail to secrete Hsp90α and whether supplementation of eHsp90 would jumpstart diabetic wound healing. Studies over the past several years have begun to draw a linkage between HIF-1-regulated Hsp90 secretion and diabetic wound healing. One of the critical environmental stimuli for normal wound healing is relative hypoxia (Knighton et al. 1981; Elson et al. 2000; Tandara and Mustoe 2004). HIF-1α is a master transcription factor that regulates tissue adaptive responses to hypoxia (Semenza 2007a). Li et al. showed that HIF-1α mediates hypoxia-induced Hsp90α secretion in human dermal fibroblasts and keratinocytes. Depletion of HIF-1α completely eliminated the Hsp90α secretion by these cells. In reverse, exogenous expression of a constitutively activated HIF-1α, CA-HIF-1α, was sufficient to replace hypoxia to cause a constitutive secretion of HIF-1α in these cells. This mechanism of HIF-1α-driven Hsp90α secretion and action is summarized in Fig. 17.2. The secreted Hsp90α will act either as an autocrine or paracrine factor to promote cell motility and wound healing.

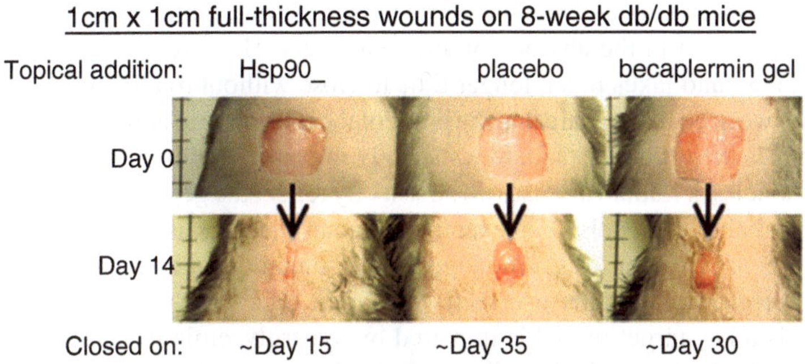

Fig. 17.3 eHsp90α shortens 2/3 of the time of diabetic wound closure. Full thickness excision wounds (1.2 cm × 1.2 cm) were created on the back of db/db mice and treated with placebo (10 % carboxymethycellulose gel) or the same gel containing an optimized concentration of human recombinant eHsp90 or becaplermin only once on day 0. The images of one of wounds on day 0 and day 14 are shown. A single treatment with eHsp90 on day 0 shortened the wound closure time from ~50 days to 14–18 days

The above finding is relevant to diabetic wound healing. Impaired reaction to hypoxia has been reported as a contributing factor to the impaired healing of diabetic ulcers (Botusan et al. 2008). In diabetic ulcers, hyperglycemia impairs HIF-1α stability and function (Catrina et al. 2004; Fadini et al. 2006; Gao et al. 2007). Lower levels of HIF-1α protein were found in foot ulcer biopsies in patients with diabetes (Catrina et al. 2004). These results suggest that a critical defect in diabetic wound healing is the result of destabilization of the HIF-1α protein. Indeed, Botusan et al. (2008) have recently demonstrated that, in dermal fibroblasts isolated from *db/db* mice, hyperglycemia causes destabilization of HIF-1α via the von Hippel-lidau (VHL) pathway and delays wound healing in *db/db* diabetic mice. More convincingly, they showed that forced stabilization of HIF-1α is necessary and sufficient for resuming diabetic wound healing (Botusan et al. 2008). Thus, it is concluded that the delayed diabetic wound healing is a result of destabilization of HIF-1α. Theoretically, one could manipulate HIF-1α stability in diabetic wounds in humans. However, strategies to directly target the intracellularly located HIF-1α in the past proved to be challenging (Poon et al. 2009; Semenza 2007b). To bypass this defect and rescue diabetic wound healing, one would need to identify downstream effectors of HIF-1α. We propose that the HIF-1α-triggered secretion of Hsp90α is impaired in diabetic wounds, resulting in delayed wound healing. Supplementations of eHsp90α should bypass the defect point and resume diabetic wound healing (Li et al. 2012). Cheng et al. (2011) have recently found that topical application of recombinant Hsp90α proteins shortened the time of 1 cm × 1 cm full thickness wounds in db/db mice from 35 days to ~18 days. In comparison, bercaplermin gel (FDA-approved for treatment of diabetic ulcers) treatment showed improvement of diabetic wound closure from 35 days to ~30 days. Moreover, Cheng's results show that eHsp90α-treated wound exhibited significantly more re-epithelialization than the placebo- or becaplermin-treated wounds. A typical example of this comparison is shown in Fig. 17.3. Topical application of eHsp90 caused the wound closure significantly faster than that of placebo or becaplermin treatment. Note that these wounds show significantly healing for the first 2 weeks even in the absence of any treatments. However, the rest of the open area of the wound takes much longer time to close without treatments.

Lastly, we need to be aware that the proof of the relevance of animal model research to humans is the ultimate standard, especially in consideration of the fact that many animal models for human diseases do not exactly reflect the genetic setting in humans, including diabetic models. For example, human diabetes is a polygenic disease, whereas the *db/db* mouse is a monogenic (i.e. mutation in a single gene) diabetic model. Therefore, it remains to be seen whether the "HIF-1α → secretion of Hsp90α" pathway is a key target of and is impaired by hyperglycemia in diabetic wounds in humans and whether topical application of eHsp90α is effective in overriding the damaging effects of hyperglycemia and resumes diabetic wound healing in humans.

17.7 eHsp90 has Three Unique Properties, Absent from Conventional Growth Factors, to More Effectively Heal Wounds

What has made eHsp90α superior to conventional growth factor therapy, specifically becaplermin/PDGF-BB? Cheng et al. (2011) provided three lines of evidence for the unique properties of eHsp90α. First, eHsp90 is a common pro-motility factor for all three types of human skin cells. Following skin injury, the lateral migration of keratinocytes closes wound and subsequent inward migration of dermal fibroblasts and dermal microvascular endothelial cells into the wound remodels the damaged tissue and builds a new blood vessels. In contrast, most growth factors, such as PDGF-BB and VEGF-A, have various degrees of cell type specificity. Second, eHsp90 remained equally effective on stimulation of migration of all three types of human skin cells even in the presence of TGFβ. No conventional growth factors can override the inhibition signals of TGFβ. Third, it is known that all forms of diabetes are characterized by chronic hyperglycemia in the circulation, which is blamed for the delayed healing of diabetic wounds. A reported damage by hyperglycemia was destabilization of the HIF-1α protein, the key regulator of Hsp90α secretion, in the wound (Catrina et al. 2004; Botusan et al. 2008). Cheng et al. (2011) recently found that hyperglycemia blocked hypoxia and serum-stimulated human dermal fibroblast migration. However, eHsp90 not only enhanced cell migration under normal glycemia, but also "rescued" the migration of the cells cultured under hyperglycemia (Cheng et al. 2011). We believe that the effectiveness of eHsp90 in diabetic wound healing is to bypass the hyperglycemia-caused HIF-1α down-regulation and jumpstart migration of the cells that otherwise cannot respond to the environmental hypoxia.

Based on the above findings, we propose a new paradigm, as schematically shown in Fig. 17.4, to explain how epidermal and dermal cells migrate into the wound. Within hours following skin injury, keratinocytes start to migrate laterally across the wound (possibly induced by serum) and at the same time secrete Hsp90α. At the same time, however, dermal fibroblasts and dermal macrovascular endothelial

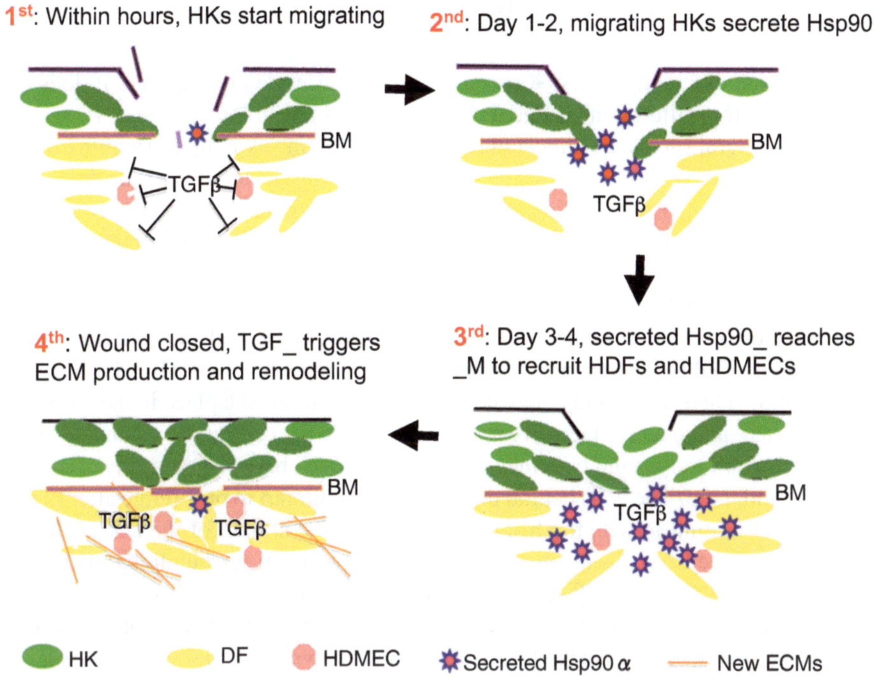

Fig. 17.4 A model of how released Hsp90, but not conventional growth factors, promotes re-epithelialization and recruits dermal cells into the wound during wound healing. (1st Step) Injury triggers release of TGFβ from several sources, the immotile to motile transition of keratinocytes and release of growth factors from serum and Hsp90α from migrating keratinocytes. The growth factors will not be able to recruit the dermal cells at the wound edge to the wound bed due to the presence of TGFβ. (2nd Step) Migrating keratinocytes continue secreting Hsp90α into the wound bed. (3rd Step) Whence the secreted Hsp90α reaches the threshold concentration of >0.1 μM, it will drive inward migration of dermal fibroblasts and dermal endothelial cells. (4th Step) The keratinocytes close the wound. The moved-in fibrobalsts will start remodeling the wound and endothelial cells to build new blood vessels. *HK* human keratinocyte, *HDF* human dermal fibroblast, *HDMECs* human dermal microvascular endothelial cells

cells at the wound edge cannot immediately move into the wound bed due to the presence of TGFβ3 (1st Step). Keratinocytes continue migrating and secreting Hsp90α (2nd Step). Whence eHsp90α reaches the threshold concentration of 100 nM (Cheng et al. 2008), the dermal cells migrate inwardly and chemotactically into the wound bed from the surrounding wound edge even in the presence of TGFβ3 (3rd Step). Finally, the migrating keratinocytes completely close the wound and the newly arrived dermal fibroblasts start to remodel the wounded tissue and dermal microvascular endothelial cells to re-build new blood vessels (Step 4). At this stage, TGFβ1 plays a critical role in stimulation of the fibroblasts to synthesize and deposit ECMs. Thus, migrating keratinocyte-secreted eHsp90α, instead of conventional growth factors from serum, is the recruiting factor of the dermal cells into the wound bed.

17.8 eHsp90, a Motogen But not a Mitogen

The fact that extracellular Hsp90α is a motogen, but not a mitogen (i.e. it does not stimulate cell proliferation) (Li et al. 2007) makes a physiological sense. First, keratinocyte migration occurs almost immediately following skin injury (within hours), whereas the inward migration of dermal cells is not detected until four days later (Singer and Clark 1999). Following the initial epidermal closure of the wound, the dermal neovascularization and remodeling processes would take many months to complete. Second, cell migration proceeds cell proliferation during wound healing. Third, it is known that, when a cell is migrating, it cannot proliferate at the same time. Fourth, growth factor-stimulated proliferation of both epidermal and dermal cells are inhibited by TGFβ that appears immediately following skin injury (Bandyopadhyay et al. 2006). Then, when and where does cell proliferation take place in the wounded skin? Based on the above facts, we reason that, while the cells at the wound edge are moving toward the wound bed, they left "empty space" between themselves and the cells behind them. The cells behind the migrating cells start to proliferate after losing the contact inhibition with the front moving cells. The stimuli of the cell proliferation likely come from plasma growth factors diffused from surrounding unwounded blood vessels, where TGFβ levels are low or undetectable. Thus, the role of cell proliferation in wound healing is to re-fill the space generated by the front-migrating cells. The specific role of eHsp90α is to help to achieve the initial wound closure as quickly as possible to prevent infection, water loss, and severe environmental stress. Many other factors, including conventional growth factors, will participate in the remaining long and tedious wound remodeling processes.

17.9 Mechanisms of Action by eHsp90

Among a handful of targets for eHsp90 reported since 2004, only studies of LRP-1 as a cell surface receptor for eHsp90α are directly related to wound healing. Therefore, we focus on LRP-1 here (see Li et al. 2012 for other targets of eHsp90). It was estimated that secreted Hsp90α from keratinocytes could readily reach an optimal working concentration (0.05–0.1 μM) that maximally stimulates cell migration *in vitro* (Cheng et al. 2008). In these migration assays, human recombinant Hsp90α exhibited saturating and subsequently declined in terms of its effects on human skin cell migration, when an excess amount of eHsp90α was added. These data suggested that eHsp90α acts by binding to a receptor-like molecule on the cell surface following Michaelis–Menten kinetics. To demonstrate whether the widely expressed cell surface receptor, LRP-1, mediates eHsp90α signaling to promote cell migration, Cheng and co-workers used four independent approaches. First, neutralizing antibodies against LRP-1, which block its ligand binding site, inhibited eHsp90-stimulated cell migration. Second, RAP (LRP-1-associated protein) inhibitor, which is known to block ligand binding to LRP-1, nullified eHsp90's effect. Third, down-regulation of endogenous LRP-1 by RNAi made the cells fail to migrate in response to eHsp90. Fourth, re-introduction of an RNAi-resistant mini-LRP-1 cDNA into the LRP1-downregulated cells was able to rescue the migration response of the cells to eHsp90α and hypoxia. In addition, GST-eHsp90 directly pulled down LRP-1 via its pro-motility fragment between the LR and the M domain of Hsp90α (Cheng et al. 2008; Woodley et al. 2009). These findings led the investigators to propose the following working model: "Hypoxia → HIF-1α → Hsp90α secretion → LRP-1 → cell motility" pathway for wound healing, as previously shown in Fig. 17.2.

However, the complexity of LRP-1 signal transduction comes from both its large ligand binding repertoire and its interactions with numerous other cell surface proteins. For instance, besides gp96 and eHsp90α, other extracellular heat shock proteins that also bind LRP-1 include calreticulin, eHsp60 and eHsp70 (Basu et al. 2001; Ogden et al. 2001; Vandivier et al. 2002; Habich et al. 2000). In addition, opposite effects for LRP-1 signaling have been reported. For instance, LRP-1 has been shown to play a critical role in both PDGF-BB-stimulated ERK1/2 activation and cell proliferation and also TGFβ-stimulated anti-proliferation (Lillis et al. 2005). In contrast, Li et al. (2007) showed that eHsp90α had little mitogenic effect. While eHsp90α dramatically increased cell migration, Hsp70, gp96 and calreticulin exhibited either a modest or no stimulation of cell migration (Li et al. 2007). There is a clear need to identify the specific binding site in LRP-1 for eHsp90, in order to further understand the mechanism of action by eHsp90.

17.10 Why eHsp90?

A fundamental question is why Hsp90α has evolved as an extracellular factor for wound healing? Two speculations/explanations are offered. First, Hsp90α accounts for 1–2 % of the total cellular protein, a luxury that evolution seldom tolerates otherwise. Therefore, it was speculated that function of Hsp90 is not restricted only to an intracellular chaperone, but rather it plays another unrecognized role that would require such a large amount of stored Hsp90 (Csermely et al. 1998). It is now becoming clear that the new role for Hsp90α is to prepare the cells to deal with environmental insults, such as tissue injury. This notion is supported by the observation that normal cells do not secrete Hsp90α under physiological conditions, instead they secrete Hsp90α in response to all environmental stress signals, including reactive oxygen species (ROS), heat, hypoxia, gamma-irradiation or tissue injury-released cytokines. In contrast, tumor cells constitutively secrete Hsp90 for motility and invasion (Li et al. 2012). Second, tissue repair and cancer progression must overcome the inhibitory signals of the TGFβ family cytokines, present constantly in surrounding inflammatory tissues. Conventional growth factors (such as PDGF-BB and VEGF) are unable to override the inhibition signals of TGFβ, the possible reason why they are less effective in wound healing in vivo. Hsp90α, however, is able to override the inhibition by TGFβ and to promote cell migration in the presence of TGFβ (Cheng et al. 2008). We believe that these unique properties of eHsp90α are the main reason for its superior effect over the FDA-approved becaplermin in skin wound healing.

Acknowledgments This study was supported by NIH grant GM/AR066193-01 (to W. L.) and AR46538 (to D. T. W.).

References

Alvarez OM, Meehan M, Ennis W, Thomas DR, Ferris FD, Kennedy KL, Rogers R, Bradley M, Baker JJ, Fernandez-Obregon A, Rodeheaver G (2002) Chronic wounds: palliative management for the frail population. Wounds 14(p Suppl):713–732

Bandyopadhyay B, Fan JF, Guan SX, Li Y, Fedesco M, Chen M, Woodley DT, Li W (2006) A "traffic control" role for TGFbeta3: orchestrating dermal and epidermal cell motility during wound healing. J Cell Biol 172:1093–1105

Basu, S, Srivastava PF (2000) Heat shock proteins: the fountainhead of innate and adaptive immune responses. Cell Stress Chaperones 5:443–451

Basu S, Binder RJ, Ramalingam T, Srivastava PK (2001) CD91 is a common receptor for heat shock proteins gp96, hsp90, hsp70, and calreticulin. Immunity 14:303–313

Bejcek BE et al (1992) The v-sis oncogene product but not platelet-derived growth factor (PDGF) A homodimers activate PDGF alpha and beta receptors intracellularly and initiate cellular transformation. J Biol Chem 267:3289–3293

Botusan IR, Sunkari VG, Savu O, Catrina AI, Grünler J, Lindberg S, Pereira T, Ylä-Herttuala S, Poellinger L, Brismar K, Catrina SB (2008) Stabilization of HIF-1alpha is critical to improve wound healing in diabetic mice. Proc Natl Acad Sci U S A 105:19426–19431

Boulton AJ, Vileikyte L, Ragnarson-Tennvall G, Apelqvist J (2005) The global burden of diabetic foot disease. Lancet 366:1719–1724

Brown GL et al (1989) Enhancement of wound healing by topical treatment with epidermal growth factor. N Engl J Med 321:76–79

Brownlee M (2001) Biochemistry and molecular cell biology of diabetic complications. Nature 414:813–820

Catrina SB, Okamoto K, Pereira T, Brismar K, Poellinger L (2004) Hyperglycemia regulates hypoxia-inducible factor-1alpha protein stability and function. Diabetes 53:3226–3232

Chen JS, Hsu YM, Chen CC, Chen LL, Lee CC, Huang TS (2010) Secreted heat shock protein 90alpha induces colorectal cancer cell invasion through CD91/LRP-1 and NF-kappaB-mediated integrin alphaV expression. J Biol Chem 285:25458–25466

Cheng CF et al (2008) Transforming growth factor alpha (TGFalpha)-stimulated secretion of HSP90alpha: using the receptor LRP-1/CD91 to promote human skin cell migration against a TGFbeta-rich environment during wound healing. Mol Cell Biol 28:3344–3358

Cheng CF et al (2011) A fragment of secreted from Hsp90α carries properties that enable it to accelerate effectively both acute and diabetic wound healing in mice. J Clin Invest 121:4348–4361

Clayton A, Turkes A, Navabi H, Mason MD, Tabi Z (2005) Induction of heat shock proteins in B-cell exosomes. J Cell Sci 118:3631–3638

Csermely P, Schnaider T, Soti C, Prohaszka Z, Nardai G (1998) The 90-kDa molecular chaperone family: structure, function, and clinical applications. A comprehensive review. Pharmacol Ther 79:129–68

Elson DA, Ryan HE, Snow JW, Johnson R, Arbeit JM (2000) Coordinate up-regulation of hypoxia inducible factor (HIF)-1alpha and HIF-1 target genes during multi-stage epidermal carcinogenesis and wound healing. Cancer Res 60:6189–6195

Eustace BK, Sakurai T, Stewart JK, Yimlamai D, Unger C, Zehetmeier C, Lain B, Torella C, Henning SW, Beste G, Scroggins BT, Neckers L, Ilag LL, Jay DG. 2004. Functional proteomic screens reveal an essential extracellular role for hsp90 alpha in cancer cell invasiveness. Nat Cell Biol 6:507–514

Fadini GP, Sartore S, Schiavon M, Albiero M, Baesso I, Cabrelle A, Agostini C, Avogaro A (2006) Diabetes impairs progenitor cell mobilisation after hindlimb ischaemia-reperfusion injury in rats. Diabetologia 49:3075–3084

Fernandez-Montequin JI et al (2007) Intralesional injections of Citoprot-P (recombinant human epidermal growth factor) in advanced diabetic foot ulcers with risk of amputation. Int Wound J 4:333–343

Fu X et al (1998) Randomised placebo-controlled trial of use of topical recombinant bovine basic fibroblast growth factor for second-degree burns. Lancet 352:1661–1664

Gao Z, Sasaoka T, Fujimori T, Oya T, Ishii Y, Sabit H, Kawaguchi M, Kurotaki Y, Naito M, Wada T, Ishizawa S, Kobayashi M, Nabeshima Y, Sasahara M (2005) Deletion of the PDGFR-beta gene affects key fibroblast functions important for wound healing. J Biol Chem 280:9375–9389

Gao W, Ferguson G, Connell P, Walshe T, Murphy R, Birney YA, O'Brien C, Cahill PA (2007) High glucose concentrations alter hypoxia-induced control of vascular smooth muscle cell growth via a HIF-1alpha-dependent pathway. J Mol Cell Cardiol 42:609–619

Greenhalgh DG, Rieman M (1994) Effects of basic fibroblast growth factor on the healing of partial-thickness donor sites. A prospective, randomized, double-blind trial. Wound Repair Regen 2:113–121

Grose R, Werner S (2004) Wound-healing studies in transgenic and knockout mice. Mol Biotechnol 28:147–166

Hanahan D, Weinberg RA (2000) The hallmarks of cancer. Cell 100:57–70

Habich C, Baumgart K, Kolb H, Burkart V (2000) The receptor for heat shock protein 60 on macrophages is saturable, specific, and distinct from receptors for other heat shock proteins. J Immunol 168:569–576

Hamed S, Brenner B, Abassi Z, Aharon A, Daoud D, Roguin A (2010) Hyperglycemia and oxidized-LDL exert a deleterious effect on endothelial progenitor cell migration in type 2 diabetes mellitus. Thromb Res 126:166–174

Harrington C, Zagari MJ, Corea J, Klitenic J (2000) A cost analysis of diabetic lower-extremity ulcers. Diabetes Care 23:1333–1338

Hightower LE, Guidon PT Jr (1989) Selective release from cultured mammalian cells of heat-shock (stress) proteins that resemble glia-axon transfer proteins. J Cell Physiol 138:257–266

Knighton DR, Silver IA, Hunt TK (1981) Regulation of wound-healing angiogenesis-effect of oxygen gradients and inspired oxygen concentration. Surgery 90:262–270

Kuroita TH, Tachibana H, Ohashi S, Shirahata H, Murakami (1992) Growth stimulating activity of heat shock protein 90 alpha to lymphoid cell lines in serum-free medium. Cytotechnology 8(1992):109–117

Lancaster GI, Febbraio MA (2005) Exosome-dependent trafficking of HSP70: a novel secretory pathway for cellular stress proteins. J Biol Chem 280:23349–23355

LeGrand EK (1998) Preclinical promise of becaplermin (rhPDGF-BB) in wound healing. Am J Surg 176:48S–54S

Li W, Li Y, Guan S, Fan J, Cheng C, Bright A, Chin C, Chen M, Woodley DT (2007) Extracellular heat shock protein-90alpha: linking hypoxia to skin cell motility and wound healing. EMBO J 26:1221–1233

Li W, Sahu D, Tsen F (2012) Secreted heat shock protein-90 (Hsp90) in wound healing and cancer. Biochim Biophys Acta 1823:730–741

Liao DF et al (2000) Purification and identification of secreted oxidative stress-induced factors from vascular smooth muscle cells. J Biol Chem 275:189–196

Lillis AP, Mikhailenko I, Strickland DK (2005) Beyond endocytosis: LRP function in cell migration, proliferation and vascular permeability. J Thromb Haemost 3:1884–1893

Lynch SE, Nixon JC, Colvin RB, Antoniades HN (1987) Role of platelet-derived growth factor in wound healing: synergistic effects with other growth factors. Proc Natl Acad Sci U S A 84:7696–7700

Ma B et al (2007) Randomized, multicenter, double-blind, and placebo-controlled trial using topical recombinant human acidic fibroblast growth factor for deep partial-thickness burns and skin graft donor site. Wound Repair Regen 15:795–799

Mandracchia VJ, Sanders SM, Frerichs JA (2001) The use of becaplermin (rhPDGF-BB) gel for chronic nonhealing ulcers. A retrospective analysis. Clin Podiatr Med Surg 18:189–209

Martin P (1997) Wound healing-aiming for perfect skin regeneration. Science 276:75–81

Mandracchia VJ, Sanders SM, Frerichs JA (2001) The use of becaplermin (rhPDGF-BB) gel for chronic nonhealing ulcers. A retrospective analysis. Clin Podiatr Med Surg 18:189–209 (viii)

Mascardo RN (1988) The effects of hyperglycemia on the directed migration of wounded endothelial cell monolayers. Metabolism 37:378–85

Mohan VK (2006) Recombinant human epidermal growth factor (REGEN-D 150): effect on healing of diabetic foot ulcers. Diabetes Res Clin Pract 78:405–411

Multhoff G, Hightower LE (1996) Cell surface expression of heat shock proteins and the immune response. Cell Stress Chaperones 1:167–176

Nagai MK, Embil JM (2002) Becaplermin: recombinant platelet derived growth factor, a new treatment for healing diabetic foot ulcers. Expert Opin Biol Ther 2:211–218

Ogden CA, deCathelineau A, Hoffmann PR, Bratton D, Ghebrehiwet B, Fadok VA, Henson PM (2001) C1q and mannose binding lectin engagement of cell surface calreticulin and CD91 initiates macropinocytosis and uptake of apoptotic cells. J Exp Med 194:81–795

Olerud JE (2008) Models for diabetic wound healing and healing into percutaneous devices. J Biomater Sci Polym Ed 19:1007–1020

Pastor JC, Calonge M (1992) Epidermal growth factor and corneal wound healing. A multicenter study. Cornea 11:311–314

Peppa M, Stavroulakis P, Raptis SA (2009) Advanced glycoxidation products and impaired diabetic wound healing. Wound Repair Regen 17:461–472

Pierce GF, Tarpley JE, Yanagihara D, Mustoe TA, Fox GM, Thomason A (1992) Platelet-derived growth factor (BB homodimer), transforming growth factor-beta 1, and basic fibroblast growth factor in dermal wound healing. Neovessel and matrix formation and cessation of repair. Am J Pathol 140:1375–1388

Poon E et al (2009) Targeting the hypoxia-inducible factor (HIF) pathway in cancer. Expert Rev Mol Med 11:e26

Ramsay HA, Heikkonen EJ, Laurila PK (1995) Effect of epidermal growth factor on tympanic membranes with chronic perforations: a clinical trial. Otolaryngol Head Neck Surg 113:375–379

Robson MC, Hill DP, Smith PD, Wang X, Meyer-Siegler K, Ko F, VandeBerg JS, Payne WG, Ochs D, Robson LE (2000) Sequential cytokine therapy for pressure ulcers: clinical and mechanistic response. Ann Surg 231:600–611

Smiell JM, Wieman TJ, Steed DL, Perry BH, Sampson AR, Schwab BH (1999) Efficacy and safety of becaplermin (recombinant human platelet-derived growth factor-BB) in patients with nonhealing, lower extremity diabetic ulcers: a combined analysis of four randomized studies. Wound Repair Regen 7:335–346

Semenza GL (2007a) Life with oxygen. Science 318:62–64

Semenza GL (2007b) Evaluation of HIF-1 inhibitors as anticancer agents. Drug Discov Today 12:853–869

Sen CK et al (2009) Human skin wounds: a major and snowballing threat to public health and the economy. Wound Repair Regen 17:763–771

Singer AJ, Clark RA (1999) Cutaneous wound healing. N Engl J Med 341:738–746

Steed DL (1995) Clinical evaluation of recombinant human platelet-derived growth factor for the treatment of lower extremity diabetic ulcers. Diabetic Ulcer Study Group. J Vasc Surg 21:71–78 (discussion 79–81)

Suzuki S, Kulkarni AB (2010) Extracellular heat shock protein HSP90beta secreted by MG63 osteosarcoma cells inhibits activation of latent TGF-beta1. Biochem Biophys Res Commun 398:525–5331

Tandara AA, Mustoe TA (2004) Oxygen in wound healing-more than a nutrient. World J Surg 28:294–300

Tredget EE, Nedelec B, Scott PG, Ghahary A (1997) Hypertrophic scars, keloids, and contractures. The cellular and molecular basis for therapy. Surg Clin North Am 77:701–730

Tsutsumi S, Scroggins B, Koga F, Lee MJ, Trepel J, Felts S, Carreras C, Neckers L (2008) A small molecule cell-impermeant Hsp90 antagonist inhibits tumor cell motility and invasion. Oncogene 27:2478–2487

Uchi H et al (2009) Clinical efficacy of basic fibroblast growth factor (bFGF) for diabetic ulcer. Eur J Dermatol 19:461–468

Vandivier RW, Ogden CA, Fadok VA, Hoffmann PR, Brown KK, Botto M, Walport MJ, Fisher JH, Henson PM, Greene KE (2002) Role of surfactant proteins A, D, and C1q in the clearance of apoptotic cells in vivo and in vitro: calreticulin and CD91 as a common collectin receptor complex. J Immunol 169:3978–3986

Wang X, Song X, Zhuo W, Fu Y, Shi H, Liang Y, Tong M, Chang G, Luo Y (2009) The regulatory mechanism of Hsp90alpha secretion and its function in tumor malignancy. Proc Natl Acad Sci U S A 106:21288–21293

Werner S, Grose R (2003) Regulation of wound healing by growth factors and cytokines. Physiol Rev 83:835–870

Wieman TJ, Smiell JM, Su Y (1998) Efficacy and safety of a topical gel formulation of recombinant human platelet-derived growth factor-BB (becaplermin) in patients with chronic neuropathic diabetic ulcers. A phase III randomized placebo-controlled double-blind study. Diabetes Care 21:822–827

Woodley DT, Fan J, Cheng CF, Li Y, Chen M, Bu G, Li W (2009) Participation of the lipoprotein receptor LRP1 in hypoxia-HSP90alpha autocrine signaling to promote keratinocyte migration. J Cell Sci 122:1495–1498

Yu X, Harris SL, Levine AJ (2006) The regulation of exosome secretion: a novel function of the p53 protein. Cancer Res 66:4795–4801

Chapter 18
Circulating Chaperones in Health and Disease

Zoltán Prohászka

Abstract Heat-shock proteins (HSPs), released from living or dying cells by active and passive mechanisms, are present in the circulation of healthy individuals and patients with different diseases. Recent clinical studies provided evidence for circulating HSPs as promising biomarkers of inflammation, infection, cellular damage and in different forms of cardiovascular disease. The group of malignant human diseases represents probably the most important research area today to find the clinical utility of measuring circulating HSP70 levels. This chapter summarizes observational human clinical studies of this HSP-biomarker field with major emphasis on circulating stress-inducible HSP70 (HSPA1A).

18.1 Introduction

In this chapter, observations on plasma/serum chaperones as biomarkers of different human diseases or ageing will be summarized. These studies are typically small or medium sized, cross-sectional case series or case-control studies allowing only the description of the ongoing pathological processes. Only a few prospective studies were carried out in the past on the subject of extracellular heat-shock proteins. In the majority of the studies the stress-inducible form of the 70 kD HSP (HSPA1A) was measured, and in other, less common studies, members of the other HSP families were investigated. The major emphasis of this summary will be laid on extracellular HSP70, and data on small heat-shock proteins and HSP60 will only be summarized at the end.

Biomarkers are indicators of a given biological state or process, and are used, therefore, in clinical practice for multiple purposes. Their presence may indicate past or ongoing pathological processes, and the changes are useful to follow the effects of pharmacological or other interventions. Beyond their theoretical applicability, new or emerging biomarkers have to be well traceable and measurable substances with significant additive power to the existing markers of the field. As for extracellular

Z. Prohászka (✉)
Research Laboratory, 3rd Department of Medicine, Semmelweis University,
Budapest, Kútvölgyi st. 4, 1125 Budapest, Hungary
e-mail: prohoz@kut.sote.hu

B. Henderson, A. G. Pockley (eds.), *Cellular Trafficking of Cell Stress Proteins in Health and Disease*, Heat Shock Proteins 6, DOI 10.1007/978-94-007-4740-1_18,

HSPs, validated immunoassays with appropriate analytical properties are available, and accumulating data indicate that these markers may find their indications in the future to meet well-defined, novel clinical needs.

18.2 Extracellular HSP70 and Longevity

According to original observations and research data, heat-shock proteins (also called stress proteins) were believed to be exclusively intracellular proteins playing important roles in the maintenance of cellular homeostasis. In 1989, Hightower and Guidon (1989) observed that stimulated, cultured rat embryo cells rapidly released a small group of proteins that included several heat-shock proteins. The release of these proteins was not blocked by either monensin or colchicine, inhibitors of the common secretory pathway. The history of the secretion of Hsp70 is given in Chap. 1. Stimulated by these and other observations, and with the development of assay tools to measure extracellular HSPs, a substantial number of clinical studies were published on various aspects of extracellular HSPs.

Among the first of these studies Pockley et al. (1998) published important data on the presence of HSP70 and HSP60 (Pockley et al. 1999) in the peripheral circulation of *normal individuals*, and these results were extended later by reporting negative correlation between HSP70 levels and age (Rea et al. 2001). This study has identified an apparent decrease in serum HSP70 and HSP60 levels with *increasing age*, and authors concluded that increased age is associated with reduced capacity to maintain homeostasis in all physiological systems, including a progressive decline in the ability to produce heat shock proteins. These findings were corroborated recently by Njemini and colleagues, showing association of low or undetectable levels of HSP70 with increasing age (Njemini et al. 2011a). Furthermore, in this study it was also shown that in healthy subjects, increased HSP70 concentrations are associated with inflammatory markers such as CRP and IL-6. Moreover, it has also been described that in the absence of serious inflammatory conditions, low levels of serum HSP70 are associated with *exceptional longevity* and biological ageing (Terry et al. 2006). It will be intriguing to determine whether this decline in HSP70 plays an active role in ageing or is only a marker of the process.

18.3 Association of Circulating HSP70 with Tissue Injury, Infection and Inflammation

There are several reports on the association of circulating HSP70 with various clinical states characterized by tissue injury, infection and inflammation. These clinical states include divergent diseases such as *sepsis* (Gelain et al. 2011), *multiorgan-failure* (Kimura et al. 2004), *myocardial infarction* (Dybdahl et al. 2005), *pregnancy complications* (Madach et al. 2008; Molvarec et al. 2006), *oxidant status* (Gelain et al. 2011; Molvarec et al. 2009), *infections* (Njemini et al. 2003, 2011b), *chronic heart failure* (Genth-Zotz et al. 2004; Gombos et al. 2008), *open-heart surgery* (Pizon et al. 2006) or *sudden sensorineural hearing loss* (Park et al. 2006). Furthermore, data from the Febbraio laboratory demonstrated that *physical exercise* results in the appearance of HSP70 in the circulation prior to any increase in gene or protein expression in contracting skeletal muscle (Walsh et al. 2001). Recent studies have confirmed these findings demonstrating that both intensity and duration of exercise influence the concentration of released HSP70 in the plasma (Fehrenbach et al. 2005). In addition, Suzuki and co-workers elegantly demonstrated the concomitant release of HSP70 and other markers of hemolysis and tissue damage after an *Ironman triathlon race* (Suzuki et al. 2006b). Finally, the group of Alexzander Asea demonstrated the elevation of HSP70 levels in the circulation of patients with *sickle cell disease*, a disorder characterized also by hemolysis and tissue damage (Adewoye et al. 2005).

It is clear from the above studies, that tissue necrosis and apoptosis is strongly associated with the release of HSP70, and possibly with other intracellular stress proteins. Besides the known markers of tissue injury, HSP70 is also a good biomarker of this damage. Moreover, HSP70 is not only a simple marker, but may also be considered as a mediator of injury and inflammation, since it has well-documented initiating effects on different arms of innate immunity, including complement activation (Prohaszka et al. 2002), stimulation of antigen presenting cells by cell surface receptors (Calderwood et al. 2007), and NK cells (Elsner et al. 2007). Collectively, the phrase 'danger signal' activity of HSP70 was introduced (reviewed by Henderson and Pockley 2010) to indicate, that the release of heat-shock proteins to the extracellular space is an alarm sign and may initiate the protective stress and inflammatory responses of the host. More recently the term Stress Observation System (SOS) for the mechanism for sensing extracellular HSP was also suggested, and proposed as a form of cellular communication during stress conditions (De Maio 2011).

18.4 Circulating HSP70 and Cardiovascular Disease

In 2003 two independent, observations indicated that serum HSP70 levels may predict the development cardiovascular disease. Pockley and co-workers reported that the progression of *carotid atherosclerosis*, as measured by intima-media thickness, was less prevalent in patients with high HSP70 levels (Pockley et al. 2003). In addition, Zhu et al. (2003) presented observational data on the association of high HSP70 levels and lower risk of *coronary artery disease*. These above and other data were carefully analyzed and reviewed recently (Pockley et al. 2009), ending-up in a conclusion that HSP70 appear to protect against cardiovascular disease, by its inherent anti-inflammatory activity and by its direct effects on endothelial cells. However, recent literature data indicate that increased HSP70 levels are linked to elevated risk and severity of *acute coronary syndrome* (Zhang et al. 2010) and to *vascular calcification* score in atherosclerotic patients (Krepuska et al. 2011). Nevertheless, there are two important points that have to be taken into account while interpreting the results of studies on cardiovascular disease and HSP70. First, proteolytic enzymes released during neutrophil activation may influence plasma HSP70 levels (Martin-Ventura et al. 2007). Second, since most of the above studies on circulating HSP70 were done using serum samples where full in-tube activation of neutrophils occurred, neutrophil counts and sample processing could influence the results. Therefore, these potential biasing factors have to be considered while planning of future studies on extracellular HSP70.

18.5 HSP70 as a Cancer Biomarker

Probably the most intriguing and, only recently recognized, group of diseases, where HSP70 plays an important and prominent role as a biomarker, are human malignancies. Although expression of HSP70 is relatively low in normal cells under physiological circumstances, it is abundantly expressed in most cancer cells (reviewed in Ciocca and Calderwood 2005 and Multhoff et al. 1995). While induction of cytosolic HSP70, in concert with other members of the stress response, confers protection to further stress, its membrane expression on tumor cells elicits anti-tumor immune responses in human (Schmitt et al. 2007). Membrane associated HSP70 detected specifically on tumor cells provides a recognition target for the cytolytic attack mediated by natural killer (NK) cells (Gehrmann et al. 2003; Multhoff et al. 1995). Accordingly, the amount of membrane-bound HSP70 correlated positively with susceptibility to cell lysis by NK cells (Gehrmann et al. 2004). Furthermore, the active secretion of HSC70 (Barreto et al. 2003) and HSP70 (Mambula and Calderwood 2006) was documented in the case of viable tumor cells. Therefore, it was reasonable to hypothesize that extracellular HSP70 levels could provide valuable clinical information in different malignant diseases. Accordingly, there is a growing interest in studying serum HSP70 as a biomarker in malignant human diseases, and the number of well-designed clinical studies is increasing in this field.

Abe and associates were the first to describe data on the diagnostic application of plasma HSP70 in *prostate cancer* (Abe et al. 2004). Plasma HSP70 levels in the patients with localized untreated disease were significantly higher than those in the control group. Although the primary cutoff point (1.15 ng/ml) significantly distinguished the localized untreated patients from the control group, plasma HSP70 levels did not prove more effective than PSA as a predictor for diagnosis or stratification of patients with prostate cancer in the context of group comparisons. The authors concluded that HSP70 is a good marker of prostate cancer, although its clinical utility is uncertain. It is possible that when used in conjunction with PSA it might prove useful in identifying patients with early-stage prostate cancer who might otherwise be missed by PSA screening alone. Recently, Hurwitz et al. (2010) showed in a clinical setting that exposure of patients with prostate cancer to radiation resulted in the release of Hsp72, which is believed to come primarily from the tumor. No such elevations followed hormone therapy. Thus blood levels of Hsp72 may be a useful biomarker of the effectiveness of radiation therapy (Hurwitz et al. 2010).

Suzuki and co-workers observed an increasing risk of *lung cancer* across increasing quartiles of serum Hsp70 levels in Japanese males (Suzuki et al. 2006a). Serum HSP70 levels were measured in 189 cases and 377 controls in the Japan Collaborative Cohort Study for Evaluation of Cancer Risk. However, this association was not significant in females. Moreover, no significant association was found between serum hsCRP levels and risk of lung cancer in either sex. The authors concluded that serum Hsp70 may offer an attractive biomarker for predicting the risk of lung cancer. In addition, the group led by Maher Albitar published two important papers on their studies on plasma HSP70 levels in patients with different forms of leukemia. It was demonstrated in their first study with large number of Philadelphia chromosome positive *chronic myeloid leukemia* patients, that circulating HSP70 levels are higher in the patients than in the controls (Yeh et al. 2009). The elevation of HSP70 was significantly correlating to disease progression and/or treatment resistance with imatinib (an ATP-binding site antagonist of Bcr-Abl). These findings support the involvement of HSP70 expression in the development of imatinib resistance and HSP70 could therefore be a novel clinical biomarker in CML. In the second, more recent, study of the group the important clinical relevance of plasma HSP70 in acute leukemia was shown (Yeh et al. 2010). High levels of HSP70 in the plasma of patients with *acute myeloid leukemia* or *acute lymphoid leukemia* was demonstrated. Furthermore, Hsp70 correlated positively with risk factors such as white-blood cell count, lactate-dehydrogenase activity or beta-2 microglobulin, all of which have prognostic values and are associated with adverse outcomes in AML and ALL. Based on the data it was suggested that circulating HSP70 reflects the severity of the disease as well as the condition of the patient.

In 2010 our group reported on the predictive value of measuring baseline (before initiation of surgical or other treatment) levels of soluble HSP70 in the sera of patients with *colorectal cancer* (Kocsis et al. 2010). High concentration of HSP70 was associated with poor survival—patients with a serum concentration above the median (1.65 ng/mL) had more than twice higher age-, gender-, and localization-adjusted risk not to survive the almost 3-years-long follow-up period. Importantly, this association between high soluble HSP70 level and poor survival was found to be independent of disease stage defined by TNM staging systems, even after adjustment for individual stages. The adjusted chance for survival of the patients with high HSP70 serum concentration was approximately half of those with lower HSP70 levels. Moreover, we found an interaction between high HSP70 serum level and stage of the disease—that is, the association between high HSP70 level and survival is pronounced in the group of patients with less advanced disease at baseline (i.e., with tumors not invading other organs or structures and/or propagating beyond the visceral peritoneum, as well as with four or more tumor-free regional lymph nodes). Interestingly, we found that the association between serum HSP70 and mortality was dominant in women under the age of 70. It was observed in many large studies including a meta-analysis that the risk of colorectal cancer in current users of postmenopausal hormones is significantly 30–40 % lower compared with non-users (Grodstein et al. 1999; Newcomb and Storer 1995; Newcomb et al. 2007). It is tempting to speculate that high level of serum HSP70, as a potential marker of anti-tumor immune response, may in some way be modulating the protective function of estrogen in pre-menopausal women.

Extending our above observations are novel data showing that measuring some acute-phase reactants, in addition to HSP70, can markedly increase the predictive value of HSP70 on the survival of patients with colorectal cancer (Kocsis et al. 2011). The risk of early mortality of patients who had high C-reactive protein and complement C1-INH levels, in addition to a high HSP70 level, was significantly higher [HR 2.83 (1.13–6.87)] than the effect of high HSP70 alone. These data indicate, that HSP70 may be used as partners in biomarker combinations to stratify different groups of patients with increased efficacy and accuracy (see also below supporting results obtained in proteomic studies).

The identification of early, molecular markers, which can predict cancer-related mortality, is an urgent and important goal in the development of cancer diagnostics and therapy. Ideally, identifying patients with high-risk carcinomas and a low chance of survival at initial diagnosis could afford early and individualized therapy that would improve clinical outcome. Our above findings indicate that the prognostic value (adjusted odds ratios of about 2) for the overall survival of the soluble HSP70 measurement is comparable to that of the known tumor markers, such as CEA and CA 19-9 for disease-free survival in early stage colorectal cancer (Ogata et al. 2009; Wang et al. 2002a, b).

In the first study that systematically searched for a diagnostic serum profile in *glioblastoma multiforme* (GBM) patients (Elstner et al. 2011), serum concentrations of 14 proteins, which were pre-selected by screening gene and protein expression profiles of astrocytomas for proteins potentially released by the tumor cells, was analyzed. Included were also two classes of cytoplasmic proteins associated with cell stress (HSP70) and neural stem cells (FABP7). Using the complete dataset to identify patterns of proteins by applying data mining, it was possible to associate combinations of serum proteins with the clinical parameters 'presence of a GBM'. The profile with a relative small number of proteins (BMP2, CXCL10, HSP70) was sufficient to correctly assign 96 % of the GBM and 89 % of the control subjects, and HSP70 was included in a pattern capable to assign a survival prognosis.

The research to find plasma markers by proteomic approach, published by Liao and colleagues demonstrated, that HSP70, sICAM-1 and SAA—confirmed with ELISA on sera and immunohistochemistry—are potential *nasopharyngeal carcinoma* (NPC) metastasis-specific serum biomarkers which may be of great underlying significance in clinical detection and management of NPC (Liao et al. 2008).

An interesting observation obtained in animal model, was recently reported by Dakappagari on the applicability of serum HSP70 as potential biomarker for *HSP90 inhibitors* (Dakappagari et al. 2010). HSP90 inhibitors are under investigation in multiple human clinical trials for the treatment of cancers, including myeloma, breast cancer, prostate, lung, melanoma, gastrointestinal stromal tumor and acute myeloid leukaemia. Using a novel, highly sensitive and specific electrochemiluminescent ELISA for HSP70, maximal secretion of HSP70 by tumor cells was observed between 48 and 72 h after exposure to HSP90 inhibitors. In *in vivo* studies a 3–4-fold increase in serum HSP70 was observed following treatment with BIIB021 in tumor-bearing mice. Together these data suggest that serum HSP70 could be a future biomarker in human studies to assess reliably the pharmacological effects of HSP90 and/or proteasome inhibitors in clinical trials, especially under conditions where collection of tumor biopsies is not feasible. Supporting the validity of this approach are the observations of Noonan et al. (2008), indicating that Hsp70B' (having 100 % amino acid identity to HSP72) was expressed on the surface of HT-29 and CRL-1809 but not SW-480 human colon cell lines in response to proteasome inhibition as detected using flow cytometry.

HSP70 is currently in the focus of ongoing research to find its role as a potential anti-cancer immunotherapeutic target and biomarker of ongoing disease. Elevated expression of HSP70 has been reported in several malignant tumors such as endometrial, breast, gastric and colorectal cancers, osteosarcoma, and renal cell tumors as well as Bcr-Abl-positive leukemia (Ciocca and Calderwood 2005; Milicevic et al. 2007; Yeh et al. 2009). Increased HSP70 expression was associated with high-grade tumors, and also with metastasis, poor prognosis, and resistance to chemotherapy or radiation therapy. On the one hand, tumor-specific cytoplasmic membrane-associated HSP70 expression has been shown to function as a target of immune reaction and induce NK and CTL responses (Gehrmann et al. 2003; Schmitt et al. 2007), while on the other hand, high intracellular HSP70 levels contribute to cell protection mechanisms by regulating protein homeostasis and apoptosis. In this context, either cytosolic or membrane-bound HSP70 released or secreted into the blood stream by tumor cells could serve as a prognostic factor for cancer progression. Furthermore, HSP70 released from tumor cells may allow the transfer of antigenic peptides to APCs or may enhance the ability of APCs ability to process and present tumor antigens to specific T cells, thereby augmenting the immune responses (Srivastava 2005). Thus, extracellular HSP70 in plasma could also represent both an indicator of immune response and an indicator of tumor load or effect of tumor treatment.

18.6 HSP60 and HSP27 in Cardiovascular Diseases

Several concordant observations indicate the serum levels of HSP60 are associated with cardiovascular risk factors (Lewthwaite et al. 2002; Shamaei-Tousi et al. 2007), coronary artery disease (Zhang et al. 2008) and ischemic cardiomyopathy (Niizeki et al. 2008). HSP60 might be a marker of myocardial damage in patients with chronic heart failure and ischemic heart disease.

Small heat-shock proteins, including HSP27, as biomarkers of cardiovascular disease and myocardial ischemia were suggested by several studies, recently reviewed in (Ghayour-Mobarhan et al. 2012). Furthermore, in patients with glucose intolerance (Burut et al. 2010) and diabetes mellitus (Gruden et al. 2008) the presence of macrovascular complications and diabetic neuropathy was linked to higher levels of HSP27.

Whether heat-shock proteins released during myocardial damage, or associated with cardiovascular risk factors, bear any additive power to the currently existing and used cardiac biomarkers, or are independent markers of the affected clinical states, is not clear today, and future studies are required to delineate their potential clinical utility.

18.7 Conclusion

Accumulating evidence indicate that heat shock proteins are released into the circulation of healthy individuals, as well as of patients with different diseases. Increasing number of clinical studies shed more light onto the recent research area of the HSP-biomarker field. It seems apparent, that the level of circulating HSPs decline with increasing age, and the levels of HSP70 are strongly related to inflammatory mediators and markers of cellular damage. Furthermore, extracellular HSPs are implicated in the pathogenesis of cardiovascular disease, but it is not clear, whether this association is independent, or dependent on the association of circulating HSP with inflammation and cellular injury.

The group of malignant human diseases represents probably the most important research area today to find the clinical utility of measuring circulating HSP70 levels. Both, hypothesis- and data-driven studies highlighted the significance of HSP70 as a marker of presence of disease, its severity or prognosis. HSP70 as a good biomarker will most probably help clinicians in the future to stratify cancer patients into high-risk and low-risk groups and will therefore certainly help offer more effective treatment for the affected individuals.

Acknowledgments The author of this chapter expresses his gratitude thanks to Professor George Füst for his stimulating discussions and continuous support on studies with extracellular heat-shock proteins, and to Szigeti Antalné and Márta Kókai for expert technical support in the last years.

References

Abe M, Manola JB, Oh WK, Parslow DL, George DJ, Austin CL, Kantoff PW (2004) Plasma levels of heat shock protein 70 in patients with prostate cancer: a potential biomarker for prostate cancer. Clin Prostate Cancer 3:49–53

Adewoye AH, Klings ES, Farber HW, Palaima E, Bausero MA, McMahon L, Odhiambo A, Surinder S, Yoder M, Steinberg MH, Asea A (2005) Sickle cell vaso-occlusive crisis induces the release of circulating serum heat shock protein-70. Am J Hematol 78:240–242

Barreto A, Gonzalez JM, Kabingu E, Asea A, Fiorentino S (2003) Stress-induced release of HSC70 from human tumors. Cell Immunol 222:97–104

Burut DF, Borai A, Livingstone C, Ferns G (2010) Serum heat shock protein 27 antigen and antibody levels appear to be related to the macrovascular complications associated with insulin resistance: a pilot study. Cell Stress Chaperones 15:379–386

Calderwood SK, Mambula SS, Gray PJ Jr, Theriault JR (2007) Extracellular heat shock proteins in cell signaling. FEBS Lett 581:3689–3694

Ciocca DR, Calderwood SK (2005) Heat shock proteins in cancer: diagnostic, prognostic, predictive, and treatment implications. Cell Stress Chaperones 10:86–103

Dakappagari N, Neely L, Tangri S, Lundgren K, Hipolito L, Estrellado A, Burrows F, Zhang H (2010) An investigation into the potential use of serum Hsp70 as a novel tumour biomarker for Hsp90 inhibitors. Biomarkers 15:31–38

De Maio A (2011) Extracellular heat shock proteins, cellular export vesicles, and the stress observation system: a form of communication during injury, infection, and cell damage. It is never known how far a controversial finding will go! Dedicated to Ferruccio Ritossa. Cell Stress Chaperones 16:235–249

Dybdahl B, Slordahl SA, Waage A, Kierulf P, Espevik T, Sundan A (2005) Myocardial ischaemia and the inflammatory response: release of heat shock protein 70 after myocardial infarction. Heart 91:299–304

Elsner L, Muppala V, Gehrmann M, Lozano J, Malzahn D, Bickeboller H, Brunner E, Zientkowska M, Herrmann T, Walter L, Alves F, Multhoff G, Dressel R (2007) The heat shock protein HSP70 promotes mouse NK cell activity against tumors that express inducible NKG2D ligands. J Immunol 179:5523–5533

Elstner A, Stockhammer F, Nguyen-Dobinsky TN, Nguyen QL, Pilgermann I, Gill A, Guhr A, Zhang T, von Eckardstein K, Picht T, Veelken J, Martuza RL, von Deimling A, Kurtz A (2011) Identification of diagnostic serum protein profiles of glioblastoma patients. J Neurooncol 102:71–80

Fehrenbach E, Niess AM, Voelker K, Northoff H, Mooren FC (2005) Exercise intensity and duration affect blood soluble HSP72. Int J Sports Med 26:552–557

Gehrmann M, Schmetzer H, Eissner G, Haferlach T, Hiddemann W, Multhoff G (2003) Membrane-bound heat shock protein 70 (Hsp70) in acute myeloid leukemia: a tumor specific recognition structure for the cytolytic activity of autologous NK cells. Haematologica 88:474–476

Gehrmann M, Brunner M, Pfister K, Reichle A, Kremmer E, Multhoff G (2004) Differential up-regulation of cytosolic and membrane-bound heat shock protein 70 in tumor cells by anti-inflammatory drugs. Clin Cancer Res 10:3354–3364

Gelain DP, de Bittencourt Pasquali MA, Comim MC, Grunwald MS, Ritter C, Tomasi CD, Alves SC, Quevedo J, Dal-Pizzol F, Moreira JC (2011) Serum heat shock protein 70 levels, oxidant status, and mortality in sepsis. Shock 35:466–470

Genth-Zotz S, Bolger AP, Kalra PR, von Haehling S, Doehner W, Coats AJ, Volk HD, Anker SD (2004) Heat shock protein 70 in patients with chronic heart failure: relation to disease severity and survival. Int J Cardiol 96:397–401

Ghayour-Mobarhan M, Saber H, Ferns GA (2012) The potential role of heat shock protein 27 in cardiovascular disease. Clin Chim Acta 413:15–24

Gombos T, Forhecz Z, Pozsonyi Z, Janoskuti L, Prohaszka Z (2008) Interaction of serum 70-kDa heat shock protein levels and HspA1B (+1267) gene polymorphism with disease severity in patients with chronic heart failure. Cell Stress Chaperones 13:199–206

Grodstein F, Newcomb PA, Stampfer MJ (1999) Postmenopausal hormone therapy and the risk of colorectal cancer: a review and meta-analysis. Am J Med 106:574–582

Gruden G, Bruno G, Chaturvedi N, Burt D, Schalkwijk C, Pinach S, Stehouwer CD, Witte DR, Fuller JH, Perin PC (2008) Serum heat shock protein 27 and diabetes complications in the EURODIAB prospective complications study: a novel circulating marker for diabetic neuropathy. Diabetes 57:1966–1970

Henderson B, Pockley AG (2010) Molecular chaperones and protein-folding catalysts as intercellular signaling regulators in immunity and inflammation. J Leukoc Biol 88:445–462

Hightower LE, Guidon PT Jr (1989) Selective release from cultured mammalian cells of heat-shock (stress) proteins that resemble glia-axon transfer proteins. J Cell Physiol 138:257–266

Hurwitz MD, Kaur P, Nagaraja GM, Bausero MA, Manola J, Asea A (2010) Radiation therapy induces circulating serum Hsp72 in patients with prostate cancer. Radiother Oncol 95:350–358

Kimura F, Itoh H, Ambiru S, Shimizu H, Togawa A, Yoshidome H, Ohtsuka M, Shimamura F, Kato A, Nukui Y, Miyazaki M (2004) Circulating heat-shock protein 70 is associated with postoperative infection and organ dysfunction after liver resection. Am J Surg 187:777–784

Kocsis J, Madaras B, Toth EK, Fust G, Prohaszka Z (2010) Serum level of soluble 70-kD heat shock protein is associated with high mortality in patients with colorectal cancer without distant metastasis. Cell Stress Chaperones 15:143–151

Kocsis J, Meszaros T, Madaras B, Toth EK, Kamondi S, Gal P, Varga L, Prohaszka Z, Fust G (2011) High levels of acute phase proteins and soluble 70 kDa heat shock proteins are independent and additive risk factors for mortality in colorectal cancer. Cell Stress Chaperones 16:49–55

Krepuska M, Szeberin Z, Sotonyi P, Sarkadi H, Fehervari M, Apor A, Rimely E, Prohaszka Z, Acsady G (2011) Serum level of soluble Hsp70 is associated with vascular calcification. Cell Stress Chaperones 16:257–265

Lewthwaite J, Owen N, Coates A, Henderson B, Steptoe A (2002) Circulating human heat shock protein 60 in the plasma of British civil servants: relationship to physiological and psychosocial stress. Circulation 106:196–201

Liao Q, Zhao L, Chen X, Deng Y, Ding Y (2008) Serum proteome analysis for profiling protein markers associated with carcinogenesis and lymph node metastasis in nasopharyngeal carcinoma. Clin Exp Metastasis 25:465–476

Madach K, Molvarec A, Rigo J Jr, Nagy B, Penzes I, Karadi I, Prohaszka Z (2008) Elevated serum 70 kDa heat shock protein level reflects tissue damage and disease severity in the syndrome of hemolysis, elevated liver enzymes, and low platelet count. Eur J Obstet Gynecol Reprod Biol 139:133–138

Mambula SS, Calderwood SK (2006) Heat shock protein 70 is secreted from tumor cells by a nonclassical pathway involving lysosomal endosomes. J Immunol 177:7849–7857

Martin-Ventura JL, Leclercq A, Blanco-Colio LM, Egido J, Rossignol P, Meilhac O, Michel JB (2007) Low plasma levels of HSP70 in patients with carotid atherosclerosis are associated with increased levels of proteolytic markers of neutrophil activation. Atherosclerosis 194:334–341

Milicevic ZT, Petkovic MZ, Drndarevic NC, Pavlovic MD, Todorovic VN (2007) Expression of heat shock protein 70 (HSP70) in patients with colorectal adenocarcinoma–immunohistochemistry and Western blot analysis. Neoplasma 54:37–45

Molvarec A, Prohaszka Z, Nagy B, Szalay J, Fust G, Karadi I, Rigo J Jr (2006) Association of elevated serum heat-shock protein 70 concentration with transient hypertension of pregnancy, preeclampsia and superimposed preeclampsia: a case-control study. J Hum Hypertens 20:780–786

Molvarec A, Rigo J Jr, Lazar L, Balogh K, Mako V, Cervenak L, Mezes M, Prohaszka Z (2009) Increased serum heat-shock protein 70 levels reflect systemic inflammation, oxidative stress and hepatocellular injury in preeclampsia. Cell Stress Chaperones 14:151–159

Multhoff G, Botzler C, Wiesnet M, Muller E, Meier T, Wilmanns W, Issels RD (1995) A stress-inducible 72-kDa heat-shock protein (HSP72) is expressed on the surface of human tumor cells, but not on normal cells. Int J Cancer 61:272–279

Newcomb PA, Storer BE (1995) Postmenopausal hormone use and risk of large-bowel cancer. J Natl Cancer Inst 87:1067–1071

Newcomb PA, Zheng Y, Chia VM, Morimoto LM, Doria-Rose VP, Templeton A, Thibodeau SN, Potter JD (2007) Estrogen plus progestin use, microsatellite instability, and the risk of colorectal cancer in women. Cancer Res 67:7534–7539

Niizeki T, Takeishi Y, Watanabe T, Nitobe J, Miyashita T, Miyamoto T, Kitahara T, Suzuki S, Sasaki T, Bilim O, Ishino M, Kubota I (2008) Relation of serum heat shock protein 60 level to severity and prognosis in chronic heart failure secondary to ischemic or idiopathic dilated cardiomyopathy. Am J Cardiol 102:606–610

Njemini R, Lambert M, Demanet C, Mets T (2003) Elevated serum heat-shock protein 70 levels in patients with acute infection: use of an optimized enzyme-linked immunosorbent assay. Scand J Immunol 58:664–669

Njemini R, Bautmans I, Onyema OO, Van Puyvelde K, Demanet C, Mets T (2011a) Circulating heat shock protein 70 in health, aging and disease. BMC Immunol 12:24

Njemini R, Smitz J, Demanet C, Sosso M, Mets T (2011b) Circulating heat shock protein 70 (Hsp70) in elderly members of a rural population from Cameroon: association with infection and nutrition. Arch Gerontol Geriatr 53:359–363

Noonan EJ, Fournier G, Hightower LE (2008) Surface expression of Hsp70B' in response to proteasome inhibition in human colon cells. Cell Stress Chaperones 13:105–110

Ogata Y, Murakami H, Sasatomi T, Ishibashi N, Mori S, Ushijima M, Akagi Y, Shirouzu K (2009) Elevated preoperative serum carcinoembrionic antigen level may be an effective indicator for needing adjuvant chemotherapy after potentially curative resection of stage II colon cancer. J Surg Oncol 99:65–70

Park SN, Yeo SW, Park KH (2006) Serum heat shock protein 70 and its correlation with clinical characteristics in patients with sudden sensorineural hearing loss. Laryngoscope 116:121–125

Pizon MT, Gburek T, Pizon M, Sztefko K (2006) Kinetics of plasma heat shock protein HSP-70 release in coronary artery surgery: on-pump versus off-pump. Minerva Chir 61:483–491

Pockley AG, Shepherd J, Corton JM (1998) Detection of heat shock protein 70 (Hsp70) and anti-Hsp70 antibodies in the serum of normal individuals. Immunol Invest 27:367–377

Pockley AG, Bulmer J, Hanks BM, Wright BH (1999) Identification of human heat shock protein 60 (Hsp60) and anti-Hsp60 antibodies in the peripheral circulation of normal individuals. Cell Stress Chaperones 4:29–35

Pockley AG, Georgiades A, Thulin T, de Faire U, Frostegard J (2003) Serum heat shock protein 70 levels predict the development of atherosclerosis in subjects with established hypertension. Hypertension 42:235–238

Pockley AG, Calderwood SK, Multhoff G (2009) The atheroprotective properties of Hsp70: a role for Hsp70-endothelial interactions? Cell Stress Chaperones 14:545–553

Prohaszka Z, Singh M, Nagy K, Kiss E, Lakos G, Duba J, Fust G (2002) Heat shock protein 70 is a potent activator of the human complement system. Cell Stress Chaperones 7:17–22

Rea IM, McNerlan S, Pockley AG (2001) Serum heat shock protein and anti-heat shock protein antibody levels in aging. Exp Gerontol 36:341–352

Schmitt E, Gehrmann M, Brunet M, Multhoff G, Garrido C (2007) Intracellular and extracellular functions of heat shock proteins: repercussions in cancer therapy. J Leukoc Biol 81:15–27

Shamaei-Tousi A, Steptoe A, O'Donnell K, Palmen J, Stephens JW, Hurel SJ, Marmot M, Homer K, D'Aiuto F, Coates AR, Humphries SE, Henderson B (2007) Plasma heat shock protein 60 and cardiovascular disease risk: the role of psychosocial, genetic, and biological factors. Cell Stress Chaperones 12:384–392

Srivastava PK (2005) Immunotherapy for human cancer using heat shock protein-peptide complexes. Curr Oncol Rep 7:104–108

Suzuki K, Ito Y, Wakai K, Kawado M, Hashimoto S, Seki N, Ando M, Nishino Y, Kondo T, Watanabe Y, Ozasa K, Inoue T, Tamakoshi A (2006a) Serum heat shock protein 70 levels and lung cancer risk: a case-control study nested in a large cohort study. Cancer Epidemiol Biomarkers Prev 15:1733–1737

Suzuki K, Peake J, Nosaka K, Okutsu M, Abbiss CR, Surriano R, Bishop D, Quod MJ, Lee H, Martin DT, Laursen PB (2006b) Changes in markers of muscle damage, inflammation and HSP70 after an Ironman Triathlon race. Eur J Appl Physiol 98:525–534

Terry DF, Wyszynski DF, Nolan VG, Atzmon G, Schoenhofen EA, Pennington JY, Andersen SL, Wilcox MA, Farrer LA, Barzilai N, Baldwin CT, Asea A (2006) Serum heat shock protein 70 level as a biomarker of exceptional longevity. Mech Ageing Dev 127:862–868

Walsh RC, Koukoulas I, Garnham A, Moseley PL, Hargreaves M, Febbraio MA (2001) Exercise increases serum Hsp72 in humans. Cell Stress Chaperones 6:386–393

Wang WS, Lin JK, Chiou TJ, Liu JH, Fan FS, Yen CC, Lin TC, Jiang JK, Yang SH, Wang HS, Chen PM (2002a) CA19-9 as the most significant prognostic indicator of metastatic colorectal cancer. Hepatogastroenterology 49:160–164

Wang WS, Lin JK, Lin TC, Chiou TJ, Liu JH, Yen CC, Chen WS, Jiang JK, Yang SH, Wang HS, Chen PM (2002b) Tumor marker CEA in monitoring of response to tegafur-uracil and folinic acid in patients with metastatic colorectal cancer. Hepatogastroenterology 49:388–392

Yeh CH, Tseng R, Zhang Z, Cortes J, O'Brien S, Giles F, Hannah A, Estrov Z, Keating M, Kantarjian H, Albitar M (2009) Circulating heat shock protein 70 and progression in patients with chronic myeloid leukemia. Leuk Res 33:212–217

Yeh CH, Tseng R, Hannah A, Estrov Z, Estey E, Kantarjian H, Albitar M (2010) Clinical correlation of circulating heat shock protein 70 in acute leukemia. Leuk Res 34:605–609

Zhang X, He M, Cheng L, Chen Y, Zhou L, Zeng H, Pockley AG, Hu FB, Wu T (2008) Elevated heat shock protein 60 levels are associated with higher risk of coronary heart disease in Chinese. Circulation 118:2687–2693

Zhang X, Xu Z, Zhou L, Chen Y, He M, Cheng L, Hu FB, Tanguay RM, Wu T (2010) Plasma levels of Hsp70 and anti-Hsp70 antibody predict risk of acute coronary syndrome. Cell Stress Chaperones 15:675–686

Zhu J, Quyyumi AA, Wu H, Csako G, Rott D, Zalles-Ganley A, Ogunmakinwa J, Halcox J, Epstein SE (2003) Increased serum levels of heat shock protein 70 are associated with low risk of coronary artery disease. Arterioscler Thromb Vasc Biol 23:1055–1059

Index

B. Henderson, A. G. Pockley (eds.), *Cellular Trafficking of Cell Stress Proteins in Health and Disease*, Heat Shock Proteins 6, DOI 10.1007/978-94-007-4740-1,

MIX
Papier aus verantwortungsvollen Quellen
Paper from responsible sources
FSC® C105338

If you have any concerns about our products,
you can contact us on
ProductSafety@springernature.com
In case Publisher is established outside the EU,
the EU authorized representative is:
Springer Nature Customer Service Center GmbH
Europaplatz 3, 69115 Heidelberg, Germany
Printed by Libri Plureos GmbH
in Hamburg, Germany